Advances in Experimental Medicine and Biology

Volume 1431

Advances in Experimental Medicine and Biology provides a platform for scientific contributions in the main disciplines of the biomedicine and the life sciences. This series publishes thematic volumes on contemporary research in the areas of microbiology, immunology, neurosciences, biochemistry, biomedical engineering, genetics, physiology, and cancer research. Covering emerging topics and techniques in basic and clinical science, it brings together clinicians and researchers from various fields.

Advances in Experimental Medicine and Biology has been publishing exceptional works in the field for over 40 years, and is indexed in SCOPUS, Medline (PubMed), EMBASE, BIOSIS, Reaxys, EMBiology, the Chemical Abstracts Service (CAS), and Pathway Studio.

2022 CiteScore: 6.2

Dongmei Cui • Edgar R. Meyer •
Paul M. Rea
Editors

Biomedical Visualisation

Volume 17 – Advancements in Technologies and Methodologies for Anatomical and Medical Education

Springer

Editors
Dongmei Cui
Department of Advanced Biomedical Education
University of Mississippi Medical Center
Jackson, MS, USA

Edgar R. Meyer
Department of Advanced Biomedical Education
University of Mississippi Medical Center
Jackson, MS, USA

Paul M. Rea
Anatomy Facility, School of Medicine,
Dentistry and Nursing, College of Medical, Veterinary and Life Sciences
University of Glasgow
Glasgow, UK

ISSN 0065-2598 ISSN 2214-8019 (electronic)
Advances in Experimental Medicine and Biology
ISBN 978-3-031-36726-7 ISBN 978-3-031-36727-4 (eBook)
https://doi.org/10.1007/978-3-031-36727-4

This Springer imprint is published by the registered company Springer Nature Switzerland AG
The registered company address is: Gewerbestrasse 11, 6330 Cham, Switzerland

Preface

The past decade of the twenty-first century has arguably seen the most rapid and drastic changes and developments in technology in our planet's history. In fact, the recent global pandemic necessitated the use of virtual platforms and interfaces that still remain commonplace post-pandemic, thus serving as testaments to human beings' capacity for innovations and advancements: adaptations to change, enhanced interactions with one another, and transformed perspectives and "visualizations" of the world. These advancements include digital technologies which have revolutionized the healthcare industry, including health professions education. Virtual, augmented, and mixed reality have especially influenced the delivery and evaluation of content in anatomical sciences education, but so have digital communication, assessment, and learning methods.

The editors of this book felt compelled to provide within these pages a collection of educational research and review articles, which focus on new and emerging technologies, educational methodologies and assessments, and digital learning environments and options used in anatomical sciences education. The editors hope this book will also provide useful information for anatomy educators seeking to learn more about computer-based technologies that they might wish to implement in or modify for their classrooms and laboratories. In addition, anatomy educators can read insights from other anatomy instructors who have created and curated their own virtual resource database, who have adopted evidence-based instructional practices, or modified novel teaching and assessment techniques to improve the quality of their teaching and the effectiveness of their students' learning. Readers are also encouraged to reflect on the current and future benefits, challenges, and implications of digital enhancements to the anatomy learning environment, explorations of virtual tools and resources, and educational research trends in biomedical visualizations—all pertaining to the anatomical sciences.

As many anatomy educators in the health sciences begin to teach new generations of students, such as members of Generation Z, or Gen Z for short, they must embrace their own innovative, problem-solving, and critical thinking skills to imagine and "visualize" novel ways of reaching students who might not engage with traditional teaching methods or educational technologies the same as previous student generations. After all, in the next two decades, anatomy educators will be teaching the upcoming generations Alpha and Beta, and only time will tell what new demands these generational differences will pose for the future of anatomy education. Therefore, this publication marks an opportune time to bring together scholars and educators from different countries to showcase their knowledge, wisdom, and contributions to anatomical sciences education for the benefit of current and future educators who will be tasked with teaching the next generation of learners in the health sciences. But readers can expect many more useful volumes of Biomedical Visualizations to follow this one to reflect the ever-changing, ever-evolving nature of how humans learn from, interact with, and view each other and their world.

Jackson, MS Dongmei Cui
Jackson, MS Edgar R. Meyer
Glasgow, UK Paul M. Rea

Contents

About the Authors

Dongmei Cui M.D (hon), Ph.D. is an Associate Professor at the University of Mississippi Medical Center. She led the development of 3D Virtual Anatomy Research Laboratory and mentors undergraduate students, graduate students (MS and PhD), and medical students conducting 3D educational research at the UMMC. Her educational research philosophy is to develop innovative 3D teaching tools to facilitate the acquisition of anatomical knowledge. The educational research papers from her laboratory have been published in several journals, one of them was selected as the cover page for Anatomical Science Education, and another was featured in the News Letter of the American Association of Anatomists (AAA). She was the lead author of the textbook titled "Atlas of Histology with Functional and Clinical Correlations." This textbook was ranked as one of the few leading histology textbooks by Doody's Core Titles library referral service in 2012 and 2016. The textbook has been translated into five languages (Spanish, Italian, Indonesian, complex Chinese and Turkish editions) and is currently used for teaching histology worldwide. Her "Histology Flash Cards with Clinical Correlations" was very popular among students. She is also the lead author of the recent textbook "Histology from a Clinical Perspective." Those works were published by Wolters Kluwer/Lippincott Williams and Wilkins. She has participated in national and international meetings as a presenter and invited speaker. She also helped to organize symposium for the international meetings and was chair/co-chair and speaker at the 18th and 19th Congress of the International Federation of Associations of Anatomists (IFAA) and the 25th International Symposia on Morphological Sciences (ISMS). Dr. Cui is a member of the American Association of Anatomists and served on the editorial board of Journal of Anatomical Sciences Education (2018–2022). She is a leader for International Federation of Associations of Anatomists Task Force to develop Core Syllabus for the Histology in Medical Curriculum. She teaches Histology and Cell Biology to medical students and graduate students. She teaches Microscopic Anatomy to dental students for many years. She has developed an integrated curriculum and has served as Course Director for pre-matriculation course and currently a Course Director for Medical Review of Histology/with Clinical Correlation course and Course Director for Dental Histology (Human Microscopic and Development Anatomy).

Edgar R. Meyer, M.A.T., Ph.D., is an Assistant Professor with an appointment in the School of Medicine, Department of Advanced Biomedical Education at the University of Mississippi Medical Center (UMMC). He also serves as the Director of the Master of Science in Biomedical Sciences (M.S.-BMS) Program in the School of Graduate Studies in the Health Sciences at UMMC. In this role, he engages in one-on-one advising with graduate students intending to enroll in health professional programs to pursue careers in the health sciences. His research interests revolve around virtual anatomy; diversity, equity, and inclusion; and outreach endeavors that serve K-12 student and teacher populations. He has experience teaching embryology, histology, gross anatomy, and/or neuroanatomy to medical, dental, physician associate, nurse anesthesia, and graduate students.

Paul M. Rea Paul is Professor of Digital and Anatomical Education at the University of Glasgow. He is Director of Innovation, Engagement and Enterprise within the School of Medicine, Dentistry and Nursing. He is also a Senate Assessor for Student Conduct, Council Member on Senate and coordinates the day-to-day running of the Body Donor Program and is a Licensed Teacher of Anatomy, licensed by the Scottish Parliament.He is qualified with a medical degree (MBChB), an MSc (by research) in craniofacial anatomy/surgery, a PhD in neuroscience, the Diploma in Forensic Medical Science (DipFMS), and an MEd with Merit (Learning and Teaching in Higher Education). He is a Senior Fellow of the Higher Education Academy, Fellow of the Institute of Medical Illustrators (MIMI) and a registered medical illustrator with the Academy for Healthcare Science.Paul has published widely and presented at many national and international meetings, including invited talks. He has been the lead Editor for Biomedical Visualiz(s)ation over 15 published volumes and is the founding editor for this book series. This has resulted in almost over 110,000 downloads across these volumes, with contributions from over 450 different authors, across approximately 100 institutions from 23 countries across the globe. It has over 500 citations from these volumes. He is Associate Editor for the European Journal of Anatomy and has reviewed for 25 different journals/publishers. He is the Public Engagement and Outreach lead for anatomy coordinating collaborative projects with the Glasgow Science Centre, NHS and Royal College of Physicians and Surgeons of Glasgow. Paul is also a STEM ambassador and has visited numerous schools to undertake outreach work.His research involves a long-standing strategic partnership with the School of Simulation and Visualization—The Glasgow School of Art. This has led to multi-million-pound investment in creating world leading 3D digital datasets to be used in undergraduate and postgraduate teaching to enhance learning and assessment. This successful collaboration resulted in the creation of the world's first taught MSc Medical Visualization and Human Anatomy combining anatomy and digital technologies, for which Paul was the Founding Director having managed this for 12 years. The Institute of Medical Illustrators also accredits this postgraduate degree. Paul has led college-wide, industry, multi-institutional and NHS research linked projects for students.

Past and Current Learning and Teaching Resources and Platforms

1

Dongmei Cui, Gongchao Yang, Edgar R. Meyer, and Norma Ojeda

Abstract

For over two centuries, the educational landscape both nationally and globally has changed tremendously. The more traditional teaching and learning resources and platforms, such as traditional textbooks, chalkboards and whiteboards, overhead transparency and carousel projectors, and traditional classroom settings, have been either replaced or supplemented in the anatomical sciences by integrated and virtual eBooks, online learning management (OLM) platforms, and virtual learning and meeting apps. Virtual teaching and learning, especially proliferated with the advent and aftermath of the COVID-19 pandemic, and institutions worldwide that had already been utilizing virtual class and lab sessions in their anatomy curricula expanded virtual course offerings. Many institutions have retained virtual course offerings even after the pandemic, given the distance learning benefits. The future of anatomy education holds many promising possibilities given the voracious speed with which technology is advancing. One such promising advancement is the full, seamless incorporation of virtual three-dimensional (3D) immersive and semi-immersive learning into anatomy laboratories and classroom settings globally as well as into students' laptops and handheld devices for easy use at home or anywhere.

Keywords

Teaching and learning resources · Teaching and learning platforms · Traditional teaching · Virtual teaching and learning · Anatomy curricula · Histology curricula

1.1 Introduction

Medical education has changed rapidly in the last 10 years, from traditional course-based curricula to integrated systems-based curricula. In order to teach more information in less time, teaching hours including lecture hours and laboratory hours have been largely reduced (Drake et al. 2009; Drake 2014). For gross anatomy and microscopic anatomy (histology) teaching, the laboratory component is an important part of these courses. However, many medical schools have reduced laboratory hours, particularly in histology and gross anatomy classes (Cottam 1999; Drake et al. 2002; Verhoeven et al. 2002;

D. Cui (✉) · E. R. Meyer · N. Ojeda
Department of Advanced Biomedical Education, University of Mississippi Medical Center, Jackson, MS, USA
e-mail: dcui@umc.edu; emeyer@umc.edu; nojeda@umc.edu

G. Yang
Department of Advanced Biomedical Education and Department of Academic Information Services, University of Mississippi Medical Center, Jackson, MS, USA
e-mail: gyang@umc.edu

D. Cui et al. (eds.), *Biomedical Visualisation*, Advances in Experimental Medicine and Biology 1431,
https://doi.org/10.1007/978-3-031-36727-4_1

Moxham and Plaisant 2007; Yeung et al. 2012; Drake et al. 2014). A vast amount of anatomical research has relied upon microscopic technologies, and histology has been regarded as a prerequisite subject for the understanding of pathology (Cui and Moxham 2021). Many medical schools use virtual microscopy to replace physical conventional microscopy, and some schools have moved their histology laboratory online or to student self-study outside of the classroom (Cotter 2001; Boutonnat et al. 2006; Campbell et al. 2010; Bloodgood 2012; Barbeau et al. 2013; Caruso 2021). Traditional anatomy education is centered around cadaveric dissection, augmented by two-dimensional images in textbooks and on projection screens, with the goal of enabling the student to construct an internal, mental, visuospatial representation of the skeleton, internal organs, the circulatory system, and the nervous system of the human body (Cui et al. 2017). These advancements have highlighted the need to develop teaching techniques that will increase the efficiency of the education process (Drake 1998; Drake et al. 2009; Alpern et al. 2011; Hopkins et al. 2011). In recent years, many teaching techniques, programs, and software have been developed to support lectures in and out of the classroom and laboratory. Virtual learning has become most popular for a new generation of students who have grown up with and adapted well to new technology. Teaching platforms have changed from the traditional textbook, blackboard/whiteboard, overhead projector, or carousel slide projector to the use of PowerPoint slide presentation, meeting apps for virtual teaching, and recorded lecture videos that can be accessed synchronously during real-time virtual and/or in-person teaching and/or asynchronously outside of normal class hours. These virtual synchronous and asynchronous options greatly increased during the COVID-19 pandemic (Attardi et al. 2021). Although traditional teaching was mainly taught in the classroom, face to face, current teaching can now provide both a face-to-face attendance option inside of the classroom and/or a virtual attendance option for students, allowing for an increase in potential possibilities for reaching underserved students in more remote areas of the world. In such virtual or hybrid learning formats, lectures are typically given via online platforms, such as BigBlueButton, Microsoft Teams, Webex Conferencing, and Zoom. Many educators also post recorded lectures and class sessions on online learning management platforms, such as Canvas or Blackboard, for students to view in their time of need or choosing.

1.2 Traditional Teaching Resources and Platforms

1.2.1 Traditional Textbooks

Most traditional textbooks for teaching gross anatomy have been written from a regional perspective of anatomy. These textbooks mainly focus on the back, thorax, abdomen, pelvis and perineum, lower limb and upper limb, and head and neck regions (O'Rahilly 1986; Williams et al. 1989; Netter 2006). Most of these books contain blackand-white or color illustrations, and some contain radiographic images (radiograms) and black-and-white and/or colored photos taken from patients. Traditional anatomy textbooks also emphasize the basic needs for teaching and learning anatomical structures and their associations and orientations in three-dimensional space (e.g., muscles and their actions, attachments, innervations, and vascularization; important spaces, their borders, and their contents; and important landmarks and the critical adjacent structures). Histology textbooks, on the other hand, were commonly written based on a systemic approach, including in-depth discussion of the different organ systems and the structure and functions of their component organs, tissues, cells, and features as well as content on cell biology and basic tissues, such as epithelial tissue, connective tissue, muscle tissue, and nervous tissue. These textbooks typically include the following organ systems: circulatory, lymphoid (or lymphatic), respiratory, urinary, integumentary, digestive, endocrine, and female and male reproductive. In addition, these textbooks often feature specialized connective tissues, such as cartilage and bone and blood and hemopoietic tissue, as well as special regions of anatomy (i.e., eye, ear, and oral cavity) that contain small and intricate structures that are important to appreciate from a microscopic perspective just as much as a macroscopic perspective. Textbooks written 50 years ago usually include black-and-white illustrations or black-and-white photographs of tissue specimens and relatively few color images; these images also usually lack labels or include labels as abbreviations for important structures. Many early histology textbooks include a considerable amount of detailed textual information, and some contain fewer images and illustrations compared with more current textbooks (Le Gros Clark 1958; Bloom and Fawcett 1968; Grant 1972; Ham and Cormack 1979; Hollinshead and Rosse 1985). For example, Le Gros Clark's textbook, *The Tissue of the Body: An Introduction to the Study of Anatomy*, which introduced histology (microscopic anatomy) in the context of anatomy contains 132 figures, most of which are black and white images and fewer than 6 of which are color images. The textbook is composed of basic tissue chapters on epithelium/mucous membranes and glands, connective tissue, cartilage, bone, muscle, blood vessels, blood, lymphatic tissue, skin, and nervous tissue and the nervous system. This book also focused intently on the details of the basic tissues of histology and their associations with anatomical structures, but not as intently on the details of organ systems. For instance, the book includes a section on the tissue of joints, such as synarthroses, amphiarthroses, diarthroses, and of ligaments. In addition, the skin chapter discusses the structure of skin, regeneration of skin, flexure lines, pigmentation of skin, hair, cutaneous glands, nails as well as more gross anatomical aspects such as the innervation, blood supply, and lymphatic drainage of the skin. In fact, the textbook includes very detailed textual descriptions of nervous innervation to the skin (Le Gros Clark 1958).

As far as assignments corresponding to textbooks are concerned, for traditional textbooks, instructors have historically given students a reading assignment before in-class lectures. Before the advent of carousel, overhead, and digital projectors, instructors typically drew images or sketches of tissues on the blackboard to allow their students to copy the images and labels during class or lab time, and instructors would often request drawings as assignments or on assessments. Although some instructors might still incorporate drawing into their histology teaching, drawing is either conducted on whiteboards or glass boards or even as annotations on PowerPoint slides or virtual histology lab platforms either in real time or for recorded sessions. In gross anatomy, instructors still use similar drawing methods during instruction. Instructors using virtual platforms for labs might incorporate annotations into their virtual presentation formats.

1.2.2 Blackboards and Whiteboards

The blackboard was popularly used for traditional teaching for over a century, having become popular since it reduced the number of materials for teachers and students during class instruction (Krause 2000). And then a little after the mid-twentieth century, green boards were invented after which point in time blackboards and green boards were commonly referred to as "chalk boards" (Laliberte 2017). For the past 20–30 years, many instructors have written and drawn on whiteboards which also are now being gradually replaced by glass boards due to the fact that they leave minimal residue that requires cleaning after habitual erasing. Through each iteration of writing/drawing board usage during instruction, students have been generally busy taking notes while following their instructors. In some classes, class presidents assign students to prepare lecture notes or transcripts from lecture recordings and then disseminate them to the rest of the class. In this regard, students take turns to prepare these lecture notes or transcripts. A number of classrooms in our university still have the green boards, namely in old lecture rooms and amphitheaters, while newer classrooms and lab spaces include whiteboards or glass boards (Fig. 1.1). Instructors do not usually use these

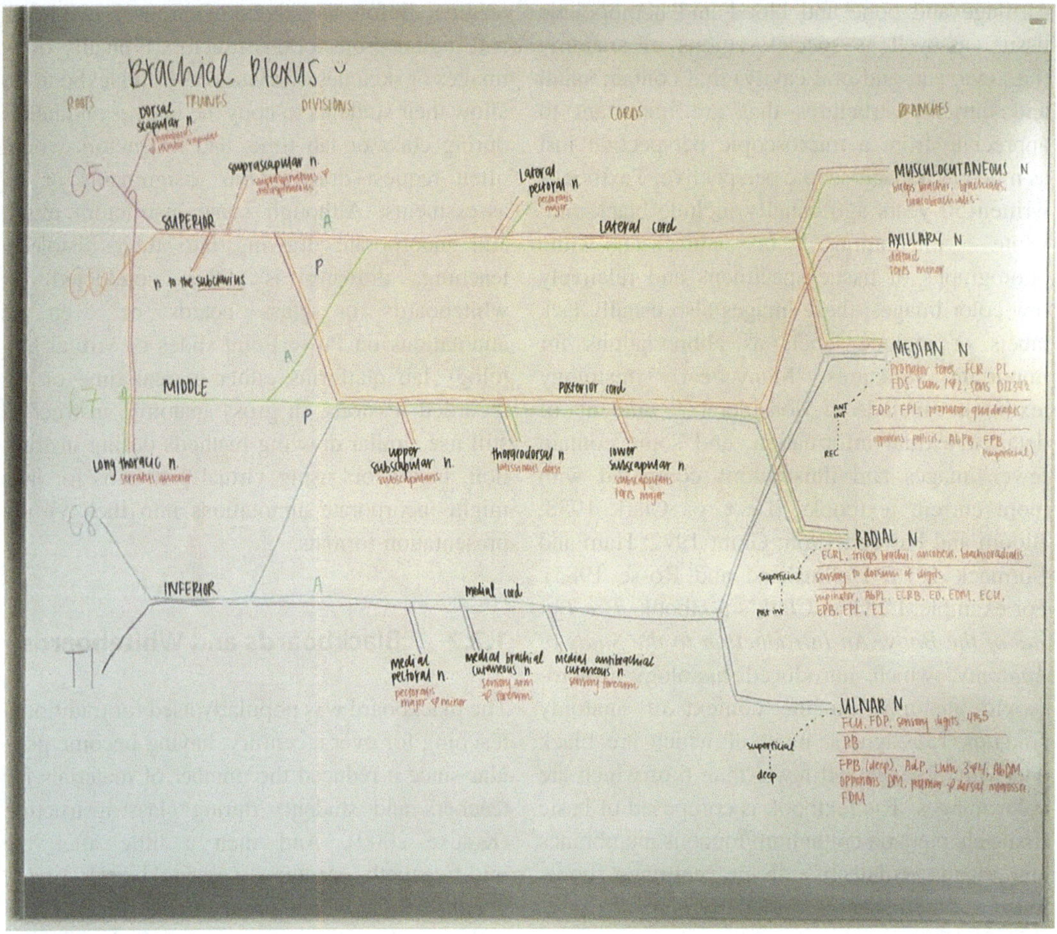

Fig. 1.1 An example of a whiteboard, with which students drew the brachial plexus and studied nerve branches, their distribution and innovation

boards today, aside from occasionally writing the password before exams through Examsoft or drawing and labeling critical anatomical structures such as the brachial plexus which is most ideally learned through the drawing process. However, our students often use these whiteboards in the small classrooms, drawing and labeling components of the brachial plexus and other nerve branches as well as blood supply networks to study and to test each other.

1.2.3 Overhead Projectors

Before the advent of computer-based projection, overhead projectors were commonly used beginning as early as the mid-twentieth century in the United States for displaying notes or illustrations written or drawn on transparent sheets onto a large screen, allowing students in small or large groups to view the material during class sessions. These transparent sheets were pre-prepared by the instructors, and sometimes, these sheets were color coded using ink pens. Up until 20–30 years ago, the overhead projectors used in traditional teaching were very simple, only requiring a single power cord connection (Fig. 1.2a). Overhead projectors have been improved significantly with the expansion of computer technology in terms of the heightened resolution of images displayed with advanced optics. However, the setup of projection systems

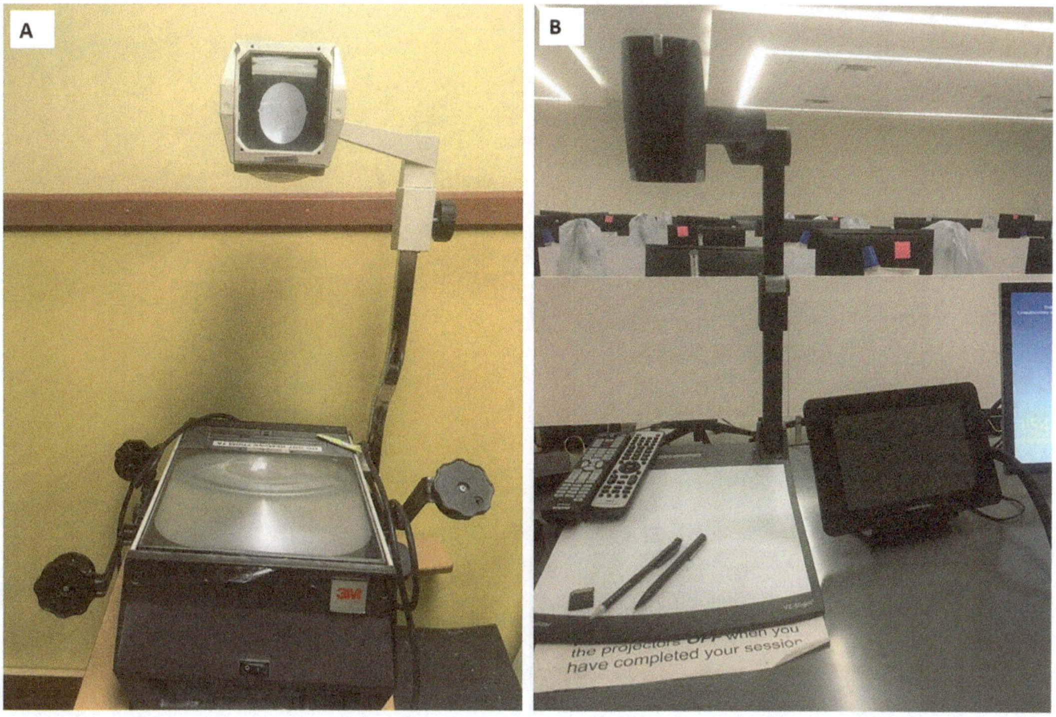

Fig. 1.2 (**a**) An example of an old overhead projector with simple and basic parts, which was used in traditional teaching 20–30 years ago. (**b**) An example of a new overhead projector with network connection, which is used in our current Histology computer laboratory

is more complex now as they are not only connected to power but also to the local network, a computer processing unit, a sound system, and sometimes multiple monitors. In our new histology laboratory within the medical school, the new overhead projectors are used to project transmission and scanning electron micrographic images on large screens in front of the classroom, and they are also connected to the computers and monitors of each station shared by a pair of students so that views or photographs from individual light microscopes can be displayed to the entire class (Fig. 1.2b).

1.2.4 Carousel Slide Projectors

Carousel slide projectors were popularly used in many professional schools and required pre-made Kodachrome slides to project images. These devices have the ability to project images onto a large screen without the reliance on a connection to a computer. The carousel slide projector was used for lecture presentations and slideshows, and it was composed of a base topped by a circular tray. The base part was a projector body that contained a motor that could rotate the circular tray. Each circular tray could hold about 100 slides, and each slide had to be placed in the tray upside down and backward so that images could be projected in their normal orientation. These Kodachrome slides were usually made by the Audio-Visual Department at the university and required considerable time to prepare. Instructors had to prepare their lectures several months before the presentations, and if one slide needed to be changed, they would have to wait for it to be re-made. Thus, any necessary significant changes could perhaps delay the delivery of presentations from year to year. Carousel slide projectors have become history in most institutions of higher learning, and computer

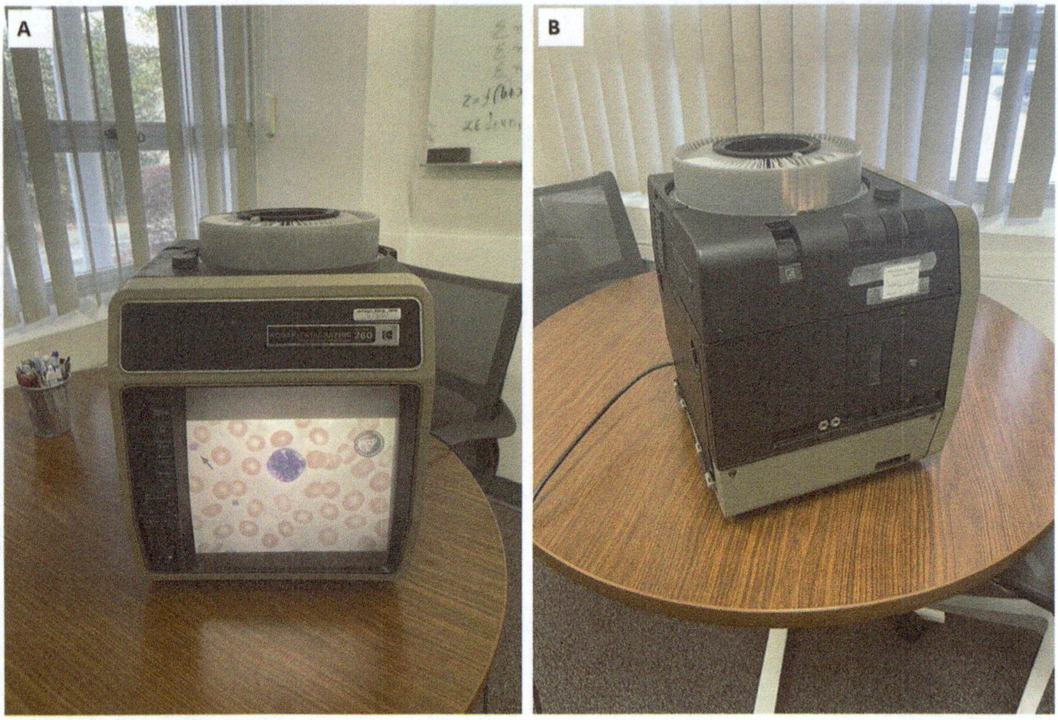

Fig. 1.3 (**a**) An example of carousel slideshow projector front view with a monitor and carousel slide tray. (**b**) Back view of the carousel slideshow projector

projectors in conjunction with Microsoft PowerPoint slides have pretty much replaced them. Here is an example of carousel slide project (KODAX EXTAGRAPHIC Audio Viewer Projector, Model 260, USA), with a monitor on the button and the movable carousel slide tray on the top. It was used for small group training (Fig. 1.3a, b).

1.3 Current Teaching Resources and Platforms

1.3.1 Integrated Textbooks and Virtual eBooks

Many more modern textbooks have been updated with more color images and some new features, such as accompanying videos and/or compact discs (CDs) containing additional resources (Netter 1998; Mescher 2013), flashcards (Gould 2008; Cui et al. 2013), and exam practice items (Haines 2011; Loukas et al. 2021). Textbooks in the anatomical sciences (among other disciplines), for instance, have also been integrated with clinical applications. An example of a textbook that has integrated content from histology, corresponding relevant gross anatomy, pathology, and clinically correlated cases into one resource, *Histology From a Clinical Perspective* (Cui et al. 2023). This textbook was also written from a systems-based perspective with a chapter featuring organ systems, but it is also composed of cell biology and basic tissue chapters. The textbook begins with an integrated chapter on histology and pathology, consisting of an Illustrated Glossary of Histological and Pathologic Terms which includes descriptive terms for normal cells with color images and descriptive terms for abnormal cells and tissues with color images. The second chapter includes a detailed description of cell organelles, specializations, and their functions, beginning with an introduction and an overview of key concepts of cellular components and including an integrated collection of histological and pathological images throughout the chapter along with clinically correlated cases. The chapter concludes with an integrated section titled "From Histology to Pathology" followed by a section with clinical vignette questions. Each consecutive chapter concludes with these two sections which feature a histological image of a cell or group of cells, a tissue or group of tissues, an organ, or some other structure and a comparative adjacent image depicting a pathological state of the featured histology and which include clinical vignette-style questions associated with the corresponding chapter's content, respectively. Answers to the clinical vignette questions are included in the back of the textbook. Each consecutive chapter also follows a format similar to the first chapter in that each chapter begins with an introduction and overview of key concepts, discusses the featured histological structures of the tissues (basic tissues unit) or the organs (organ system units), and features synopses of the functions of structures as well as integrated histological and pathological images and correlated clinical cases. In addition, the basic tissue chapters incorporate tables that include the tissue types, their main locations, functions, and clinical correlations. The organ system chapters also include an overview of each system along with color illustrations. Moreover, three-dimensional (3D) stereoscopic model images and computed tomography angiography (CTA) images are used to indicate anatomic structures and organs in anatomical and clinical orientation. Furthermore, all chapters include illustrations of cells, tissues, and/or organs alongside many of the featured histological images in order to convey their conceptual attributes more effectively (Cui et al. 2023). Another type of integrated textbook discusses histology with correlated cell and molecular biology (Pawlina 2018). This type of textbook for teaching histology was written based on a systemic approach, and it focuses on more detailed levels of anatomical integration. Each chapter contains detailed descriptions of histology with cell biology and molecular biology approaches, such as through the clinical correlation of diseases, cytogenetic testing, and structures targeted by pathogenic agents (Pawlina 2018). Most textbooks for teaching gross anatomy are still written based on a regional approach. Examples of integrated gross anatomy textbooks include Clinical Oriented Anatomy (Moore et al. 2017), *Gray's Anatomy for Students* (Drake et al. 2023); and Clinical Anatomy: Applied Anatomy for Students and Junior Doctors (Ellis and Mahadevan 2018). These textbooks integrate anatomy illustrations, radiological images, clinical correlations, and surface anatomy. Given the adoption of systems-based curricula at many medical schools in the United States, textbooks with integrated multidisciplinary content (i.e., gross anatomy, histology, physiology, biochemistry, and neuroscience) have also been published (Ward 2021).

Many textbooks which have been published by international publishers, such as Wolters Kluwer, Elsevier Churchill Livingstone, McGraw Hill, and Springer, have printed editions and virtual eBook editions or online access that enables users to have easy access to read the books online via computers, iPad, and iPhone. Some publishers only publish books online in order to reduce costs. There are many online resources available and free of charge, such as YouTube (YouTube, San Bruno, CA) videos and integrated linkable web resources and databases, many of which are published by institutions and universities, professional societies, and/or individuals or groups.

1.3.2 PowerPoint Slide Presentations

The PowerPoint slide presentation has been a gold standard for lecture-based instruction and presentation delivery for nearly 20 years due to its ease of use and its multiple, customizable features. It includes a variety of presentation templates, allowing users to choose best-fit options, such as Blank Presentation, Atlas, Gallery, Parcel, Ion Boardroom, and Quotable. Instructors can create their own presentations, starting with a Blank Presentation or another template, create a lecture title slide, add subtitle slides, and add content slides featuring lecture concepts and details. Instructors can also insert pictures, images, tables, audio, and videos in their presentations. In addition, there are multiple design formats for the background of the presentation, such as the dimensions of the presentation slides, which include standard, widescreen, and custom slide size. Instructors can add animations to attract students' attention, to reduce cognitive load, and/or to provide just-in-time teaching moments in which answers to questions appear only when the instructor clicks after students answer posed questions correctly. Flexible slide show options are also available, allowing presenters to start a presentation from the beginning or from a current slide or to provide a presentation file in the form of an outline or custom slide show. The rehearsal and record slide show features are options that enable instructors to practice timing during practice presentations or set fixed timings to slides for actual presentations. Review features for PowerPoint presentations include spelling check and accessibility check, and anyone editing a presentation can add comments or translate the text into different languages, thus allowing global use of the presentation. Moreover, PowerPoint presentations can be portrayed in several different views, such as normal view which shows slides and notes, outline view which shows slides and notes in an outline format, notes page view which displays lecture slides and notes in one page, reading view which displays all slides on a single screen, making sorting and organization of slides easier. Other features include slide master, handout master, notes master, zoom-in and zoom-out, and fit to window. Furthermore, during the presentation, instructors can choose to use a pointer to indicate structures on slides or a pen or highlighter drawing tool to annotate slides with text, sketches, or markings for emphasis. Therefore, this diverse repertoire of PowerPoint tools and features has made the software program a favorable and fairly intuitive technology for many users, especially since it has greatly improved utility, flexibility, and originality of presentation files and styles. Most importantly, Microsoft has included PowerPoint in their Microsoft Windows package along with other programs, allowing users to use it in different parts of the world and on different devices without interruption. In other words, Microsoft has dominated the market, and users are predisposed to gravitate toward using PowerPoint when designing and delivering lectures and presentations.

1.3.3 Online Learning Management Platforms

Blackboard and Canvas are two commonly used online learning management (OLM) platforms in the United States. Our institution used Blackboard initially and then switched to Canvas due to university contract changes. Both programs are good online teaching platforms, but Canvas may have more features than Blackboard. These teaching platforms allow instructors to post lecture files, notes, videos, and other teaching materials for continuous access or just-in-time access by students and to create and organize course content into modules for each block or section of a course. Considering Canvas as an example, the following features are available: (a) The Home page provides an introductory, welcome page that typically includes the course name and ID number, the academic term and year, and a list of faculty members teaching in that course. (b) The Announcements page allows faculty to post a welcome message to students before or at the beginning of the course and additional announcements and reminders about specific events (e.g., assessments, lab sessions, and lectures) or any necessary changes or updates to course content or format throughout the course, such as communications of pre-exam review sessions and notifications regarding the posting of old exam files and their associated answer keys. (c) The Syllabus page allows instructors to post content of the course syllabus, such as the course schedule, course description, and course objectives, and allows instructors to link events like lectures, lab sessions, and assessments to the course calendar in Canvas. (d) The Modules page is a space commonly used by many instructors. It allows faculty members teaching in a course to post various documents and other resources pertinent to the course, such as lecture and laboratory materials, classroom seating charts, lab group, and/or small-group team rosters. Instructors also have the ability to create separate modules for the different blocks or sections in a course and to create subpages for each day's lecture and laboratory materials, allowing better organization so that students can easily navigate the page and find materials that will be utilized in instruction on a given day. Instructors can also create subpages for assignments, quizzes, pre-lab videos, lecture recordings, or other materials for students. (e) The Assignments page allows instructors to create questions and/or items for online tests, quizzes, homework assignments, lab worksheets, etc., which can, in turn, be linked to subpages on the Modules page. The questions and items can be drafted in a single or mixed format, using single-best answer multiple-choice, multiple-choice, true-and-false, matching, fill-in-the-blank, fill-in-multiple-blanks, essay, text entry, and document submission options among others. (f) The Files page allows instructors to store their lecture, laboratory, and video files so that they can early upload these files in correlated folder, but it also allows instructors to store files that will not be visible to students until the day and time of the learning session events. (g) The Grades page compiles all assignment scores for each student in the course into a spreadsheet and calculates their weighted or point-based averages. As instructors create assessments (e.g., Exam 1 and Exam 2), worksheets, or other assignments on the associated Assignments page, they appear as titled columns in the Grades page spreadsheet. This page also includes a Total column for final grades whose settings, along with all the other grade columns, can be adjusted so that when students' assignments are graded, their grades will automatically be reported on the Grades page, and students will receive notifications based on the dates and times in the settings. (h) The People page contains the roster of all faculty members, teaching assistants, coordinators, reviewers, peer instructors, students, and observers participating in the course, including information such as their name, email address, activity time in Canvas, and/or student ID numbers. Instructors assigned as teachers in the course can add people and assign them to roles at any given time as long as those individuals have an institutional email address. (i) The Studio page is not visible to students, but it allows instructors in the course to record lectures in real time and upload the recordings or to upload pre-recorded lectures from another recording platform. This process creates an archived storage page that reduces the storage space utilized by large video files. These stored video files can then be linked as Arc videos to the Module page or any subpage. (j) The Virtual meeting platform pages, such as BigBlueButton, Microsoft Team meetings, and Microsoft Teams classes, allow instructors to insert virtual meeting links as subpages on the platform page. This centrally located collection of links allows students and instructors to meet via online lectures or other class session meetings, which is very useful for distance and virtual learning. (k) The Settings page allows instructors to set course details, course sections, and navigations for each page in the Canvas course and add external apps, feature options, and integrations as needed. Overall, these online learning management platforms provide students, faculty, and other participants in courses the opportunity to access course materials inside and outside of the classroom and provide instructors the opportunity also to post resources in the course at any time. Students can also take exams/tests and quizzes in the classroom or complete any other assignments anywhere when they have Internet access.

1.3.4 Using Meeting Apps for Virtual Teaching

Meeting apps have become more popular since the onset and aftermath of the COVID-19 pandemic. Since the coronavirus stopped face-to-face lectures inside the classroom, many schools moved their classes online, and while online instruction was not ideal in many ways for teachers and students, many people quickly learned some of the benefits of online, virtual learning. The meeting apps provide a bridge for instructors to meet students online without individuals having to take time out of their schedules to travel to a meeting or classroom location on a physical campus. These online sessions also reduce the costs associated with travel. While there are many meeting apps available on the market, the most commonly used meeting apps in the United States include Microsoft Teams (Microsoft Corporation, Albuquerque, NM, USA), BigBlueButton (Blindside Networks, Ottawa, Ontario, Canada), and Zoom (Zoom Video Communications, Inc., San Jose, CA, USA). In our university, BigBlueButton and Microsoft Teams are the main apps used in classroom teaching. BigBlueButton is an open-source virtual classroom application designed for online education. It was initially developed by the Technology Innovation Management Program at Carleton University in 2007 (Nettleton 2010). In 2020, BigBlueButton 2.2 was released and awarded by the President of Ente Nazionale Digitale Innovazione (ENTDI) as the best web conferencing system in Europe (ENTDI 2022). BigBlueButton 2.5 was released in 2022. It has multiple features that allow instructors to share audio, webcam views, slides, and computer screens with students in real time or asynchronously as a recording. It also has a whiteboard feature which provides a space for instructors to write and draw content virtually in the absence of a physical whiteboard. Breakout rooms allow instructors to divide students into small groups. The chat feature allows for text communication, and the polling feature allows instructors to receive responses from students during online class session or meeting. For example, if the instructor asked students the question "What type of epithelium covers the inner surface of blood vessels?" and provided the following answer choices: "(A) Simple squamous epithelium, (B) Simple cuboidal epithelium, (C) Simple columnar epithelium, (D) Stratify squamous epithelium, E. Stratify cuboidal epithelium," the instructor could open and close the poll at any time. After students answered the question, the instructor would be able to see what percentage of students answered A, B, C, D, or E and then post the percentages to the class anonymously. This polling option provides an opportunity for interaction between the instructor and students. A similar polling option is available in Zoom. Microsoft Teams allows instructors to join or create a team. Lectures can be presented during the team meeting via the presentation screen share or window share option. In each team meeting, there are Home, Class Notebook, Assignments, and Grades pages available for instructors. Instructors and students meet on the General Channels for the lectures. Instructors can also divide students into multiple small groups in sub-channels. Instructors can create test questions as assignments for students to complete online, and the Chat is available for communication between instructors and students during the meeting. Teams also include a calendar to which individual class or meeting sessions or sessions in a series on given days and times each week can be linked. Compared to Microsoft Teams, BigBlueButton as a platform for online teaching is relatively simple and easy to use while Microsoft Teams is more complicated with many hidden features. Nevertheless, Microsoft Teams has many more capabilities such as file storage and document sharing between various groups, or teams. The PDF file is typically the main lecture format used in BigBlueButton, and PowerPoint is the main lecture format used in Microsoft Teams. However, when an entire screen rather than a window is shared using BigBlueButton, PowerPoint is also compatible. BigBlueButton can have up to 10 small groups during a meeting, and Microsoft Teams can have up to 85 channels for small groups.

1.3.5 Recording Lecture Videos Using Canvas and Microsoft Teams

Lectures can be recorded synchronously or asynchronously using a variety of software programs and online learning management platforms. In Canvas, lectures can be recorded in real time during a live lecture or before a lecture is given in the Studio page, and in Microsoft Teams, lectures can be recorded during a meeting either with students present in real time or with no one present purely for the sake of procuring a recording to share later with the students. PowerPoint lectures need to be prepared by instructors before the presentation delivery and/or recording time. These presentations can be presented in the classroom or in the office, given the necessary equipment provided. A Logitech Camera (Carl Zeiss HD 1080p), a USB Condenser Microphone (CMKTECK), dual monitors, and a Dell Precision computer were used to record our lectures at our institution. For Canvas Studio recordings, the PowerPoint lecture needs to be displayed on one monitor, and the Canvas Studio recorder frame needs to be opened to select the recorded frame of view of the presentation on a different monitor so that the presentation view is recorded for and/or visible to the students while the presenter view with associated notes is visible only to the presenter The recording begins with the selection of the "record a new video" icon. For the first-time user, one must add or record new videos to build a video library. When a recording is complete, the video can be downloaded and saved with the option to add a title and description. The video lecture can be added to the course module for students to study on their own time. For a Microsoft Teams recording, the PowerPoint lecture needs to be displayed on the presentation screen. Instructors can open the team within Microsoft Teams associated with the course and create a team meeting or simply join a team meeting in real time without having to create a specific session for a given date and time and then screen share the presentation in the General Channels with students if the meeting event is pre-established. Alternatively, if a meeting is started in real time without having been scheduled for a specific date and time, the link can be shared with students via email and made viewable to anyone within the institution who receives the link. By selecting the "Record" at the beginning of the lecture, an instructor can initiate a recording for an entire lecture without pausing. When the lecture is finished, and the recording is stopped or the Teams session ended, the video of the recording is automatically saved in the associated team chat, and it can be downloaded as an MP4 file and/or as a shareable link that can be copied and pasted into an online learning management platform such as Canvas. Compared to the lecture recording process in Microsoft Teams, Canvas recording on the Studio page is more flexible, and it has the capability of pauses during the recording, the provision of post-production editing, and the option to re-record over mistakes. Moreover, the recordings are easily downloaded onto the computer, and all videos can be subsequently uploaded into the Canvas Studio all within the same platform. The lecture video recorded in the Microsoft Team will automatically upload into Teams, but serval steps must be taken to download the video onto the computer and/or retrieve the shareable link. Recordings cannot be paused during the lecture or meeting session, and users cannot re-record over mistakes. Any necessary editing must be completed on the MP4 file itself.

1.3.6 Immerse Technology in Biomedical Education

The approaches and technologies utilized in biomedical education are changing to accommodate the rapid advances in all biomedical fields, as well as the generational characteristics of learners (Ryan et al. 2022). One of the new approaches in biomedical education is the utilization of immerse technology, which includes virtual reality (VR), augmented reality (AR), and mixed reality (MR) (Maresky et al. 2019). The introduction of immersive technology in the learning process increases understanding of invisible concepts providing visual aids and promoting neural links for spatial perception in the cortex of learners (Ryan et al. 2022). The collection of hardware, tracking sensors, and delivering an immersive experience simulating reality is defined as VR (Hussein and Nätterdal 2015). The addition of more interactive experience of real-world scenarios augmented by computer-generated perceptual information is defined as AR (Pantelidis et al. 2017). And the amalgamation of virtual and real-world computer-generated new environments with visualizations of physical and digital coexisting and interacting in real time is defined as MR (Nicola and Stoicu-Tivadar 2018). These technologies are relatively new and still in development, therefore the outcomes of the learning process are under evaluation. However, reports from preliminary studies show that the use of Immerse Technology in the learning process increases learner satisfaction, self-confidence, and engagement (Ryan et al. 2022).

These approaches are worth to introduce in the learning process as innovative tools to face the rapid changes of content in all biomedical fields and to close the generational gap often present between educators and learners.

1.4 Discussion

Medical curricula have evolved from traditional course formats to longitudinal integrated formats at many schools in the United States and across the globe. Teaching resources and platforms have changed and improved from the mid-nineteenth century to current times. Textbooks have reduced textual content and increased colored images, integrated content, and additional features to adapt new curricula for the newest generation of learners. Textbook pages have been reduced due to reduced lecture and laboratory hours (Drake et al. 2009; Drake 2014); increased integration contents and features results from current integrated curriculum and quick learning tools needed for students. PowerPoint presentations have largely replaced lectures incorporating mostly written or drawn content on blackboards or whiteboards and presentations using overhead transparency projectors and carousel slide projectors. More teaching and learning technologies have been developed via institutional and educational research or made available on the market, having been released by software or communications corporations and companies. Online learning management platforms, such as Blackboard and Canvas, have allowed instructors to create their own courses and post their own course materials, content, and assignments more easily for students to access quickly and at any time. Meeting apps have made distance teaching and virtual teaching possible. Especially during COVID-19, when most of the schools in the United States and across the world had halted in-class teaching, meeting apps contributed to communications between instructors and students, thus allowing schools to continue functioning and serving students. Pre-recorded videos or recorded lectures during online meetings have made teaching and learning more flexible. Students can watch or re-watch lectures at the places and times of their choosing.

Although lectures can be easily transferred to virtual or online teaching, laboratory teaching for anatomical courses has largely remained inside of the laboratory for medical and other health professional schools since experiences involving human cadaveric dissection and/or prosection have been deemed highly valuable for learners by both educators and health professional students (Kochhar et al. 2022; Asante et al. 2021). Virtual teaching and learning have not completely replaced dissection or prosection study in many medical schools. Although some schools moved their anatomy teaching completely online due to the physical distancing policies of the COVID-19 pandemic (Memon et al. 2021), many of them reopened their dissection laboratories after the pandemic. Computer-Assisted Learning (CAL) and virtual learning can be useful tools if they are well-designed and integrated into current anatomy teaching methods and curricula (Tam et al. 2009). The first chapter in this book expounds upon the various virtual learning modalities in anatomy across a continuum. There is some evidence that suggests 3D stereoscopic models are superior to traditional two-dimensional (2D) illustrations such as PowerPoint slides and textbook figures in teaching complex anatomical structures (Luursema et al. 2006, 2008; Nicholson et al. 2006; Nguyen et al. 2014; Cui et al. 2017). However, there is a growing need to study 2D versus 3D visualizations in education, undoubtedly to maintain equality (Wilson 2015) and validity (Vorstenbosch et al. 2013) across anatomy education's evolving visualization requirements. Understandably then, the creation of virtual environments within the classroom or learning laboratory environment remains challenging both technically and logistically (Hoyek et al. 2014). Nevertheless, the exponential speed at which technological advances unfold provides some comfort and potential assurance that such seamless environments that are fully and partially immersive will be a common reality for all learners both inside and outside the classroom and laboratory. These advances will hopefully apply to hardware and software used in the laboratory and classroom spaces but also to software used on laptops and handheld devices. For histology laboratory teaching, most of the schools use virtual slides instead of using conventional glass microscope slides, and many schools have moved their histology laboratory from face-to-face sessions to online teaching (Blake et al. 2003; Krippendorf and Lough 2005; Saverino et al. 2021), especially during and after the COVID-19 pandemic.

1.5 Conclusions

Teaching and virtual learning resources and platforms have dramatically changed and improved in the last 20–30 years. In this chapter, we have compared and summarized several traditional teaching and current teaching resources and platforms from textbooks to presentation styles and from non-virtual to virtual teaching and learning tools available for the given time period and commonly used at our institution and at other institutions across the United States and beyond globally. We also provided some detailed information on utilized virtual teaching apps with comparisons of the pros and cons of their general usage and the utility of some of their features. We hope this information will be useful for students and educators.

References

Alpern RJ, Belitsky R, Long S (2011) Competencies in premedical and medical education: the AAMC-HHMI report. Perspect Biol Med 54:30–35

Asante EA, Maalman R, Ali MA et al (2021) Perception and attitude of medical students towards cadaveric dissection in anatomical science. Ethiop J Health Sci 31(4):867–874. https://doi.org/10.4314/ejhs.v31i4.22

Attardi SM, Harmon DJ, Barremkala et al (2021) An analysis of anatomy education before and during Covid-19. Anat Sci Educ 15:5–26. https://doi.org/10.1002/ase.2152

Barbeau ML, Johnson M, Gibson et al (2013) The development and assessment of an online microscopic anatomy laboratory course. Anat Sci Educ 6:246–256

Blake CA, Lavoie HA, Millette CF (2003) Teaching medical histology at the University of South Carolina School of Medicine: transition to virtual slides and virtual microscopes. Anat Rec (New Anat) 275:196–206

Bloodgood RA (2012) Active learning: a small group histology laboratory exercise in a whole class setting utilizing virtual slides and peer education. Anat Sci Educ 5:367–373

Bloom W, Fawcett DW (1968) A textbook of histology, 9th edn. W.B. Saunders Company, Philadelphia, 858 p

Boutonnat J, Paulin C, Faure C et al (2006) A pilot study in two French medical schools for teaching histology using virtual microscopy. Morphologie 90:21–25

Campbell G, Demetriou LA, Arnett TR (2010) Virtual histology in the classroom and beyond. Med Educ 44:1124–1125

Caruso MC (2021) Virtual microscopy and other technologies for teaching histology during Covid-19. Anat Sci Edu 14:19–21

Cottam WW (1999) Adequacy of medical school gross anatomy education as perceived by certain postgraduate residency programs and anatomy course directors. Clin Anat 12:55–65

Cotter JR (2001) Laboratory instruction in histology at the University at Buffalo: recent replacement of microscope exercises with computer applications. Anat Rec 265:212–221

Cui D, Moxham BJ (2021) A core syllabus for histology within the medical curriculum-the cell and basic tissues. Clin Anat 34:483–495

Cui D, Daley WP, Yang G (2013) Histology flash cards with clinical correlations. Wolters Kluwer/Lippincott Williams & Wilkins, Philadelphia, 475 p

Cui D, Wilson TD, Rockhold RW et al (2017) Evaluation of the effectiveness of 3D vascular stereoscopic models in anatomy instruction for first year medical students. Anat Sci Educ 10:34–45. https://doi.org/10.1002/ase.1626

Cui D, Daley WP, Yang G et al (2023) Histology from a clinical perspective. Wolters Kluwer, Philadelphia, 546 p

Drake RL (1998) Anatomy education in a changing medical curriculum. Anat Rec 253:28–31
Drake RL (2014) A retrospective and prospective look at medical education in the United States: trends shaping. Anatomic Sci Educ. J Anat 224:256–260
Drake RL, Lowrie DJ Jr, Prewitt CM (2002) Survey of gross anatomy, microscopic anatomy, neuroscience, and embryology courses in medical school curricula in the United States. Anat Rec 269:118–122
Drake RL, McBride JM, Lachman N (2009) Medical education in the anatomical sciences: the winds of change continue to blow. Anat Sci Educ 2:253–259. https://doi.org/10.1002/ase.117
Drake RL, McBride JM, Pawlina W (2014) An update on the status of anatomical sciences education in United States medical schools. Anat Sci 7:321–325. https://doi.org/10.1002/ase.1468
Drake RL, Vogl W, Mitchell AWM (2023) Gray's anatomy for students, 5th edn. Elsevier Churchill Livingstone
Ellis H, Mahadevan V (2018) Clinical anatomy: applied anatomy for students and junior doctors, 14th edn. Wiley-Blackwell
ENTDI (2022) ENTD - Ente Nazionale Digitale ed Innovazione (in Italian, translated to English using Google translate). https://entd.org/. Accessed 3 Mar 2023
Gould DJ (2008) Clinical anatomy flash cards. Wolters Kluwer/Lippincott Williams & Wilkins, Philadelphia, 350 cards
Grant JCB (1972) Grant's an atlas of anatomy, 6th edn. Williams & Wilkins, Baltimore. 665 F. (Figures)
Haines DE (2011) Lippincott's illustrated Q&A review of neuroscience, 1st edn. Lippincott Williams and Wilkins, New York
Ham AW, Cormack DH (1979) Histology, 8th edn. J.B. Lippincott Company, Philadelphia, 966 p
Hollinshead WH, Rosse CR (1985) Textbook of anatomy, 4th edn. Harper & Row, Philadelphia, p 1041
Hopkins R, Regehr G, Wilson TD (2011) Exploring the changing learning environment of the gross anatomy laboratory. Acad Med 86:883–888
Hoyek N, Collet C, Di Rienzo F (2014) Effectiveness of three-dimensional digital animation in teaching human anatomy in an authentic classroom context. Anat Sci Educ 7:430–437
Hussein M, Nätterdal C (2015) The benefits of virtual money in education- a comparison study. Göteborgs Universitet. University of Gothenburg, Gothenburg. http://hdl.handle.net/2077/39977
Kochhar S, Tasnim T, Gupta A (2022) Is cadaveric dissection essential in medical education? A qualitative survey comparing pre-and post-COVID-19 anatomy courses. J Osteopath Med 123(1):19–26. https://doi.org/10.1515/jom-2022-0016
Krause SD (2000) "Among the greatest benefactors of mankind": what the success of chalkboards tells us about the future of computers in the classroom. J Midwest Modern Lang Assoc 33:6–16. https://doi.org/10.2307/1315198
Krippendorf BB, Lough J (2005) Complete and rapid switch from light microscopy to virtual microscopy for teaching medical histology. Anat Rec 285B:19–25
Laliberte M (2017) The surprising reason why most blackboards are actually green and not black. Insider Business Insider. Available via internet https://www.businessinsider.com/why-most-blackboards-are-actually-green-and-not-black-2017-11. Accessed 16 Mar 2023
Le Gros Clark WEL (1958) The tissues of the body—an introduction to the study of anatomy, 4th edn. Oxford University Press, London, 415 p
Loukas M, Tubbs RS, Abrahams PH et al (2021) Grays' anatomy review, 3rd edn. Elsevier, Philadelphia
Luursema JM, Verwey WB, Kommers PAM et al (2006) Optimizing conditions for computer-assisted anatomical learning. Interact Comput 18:1123–1138. https://doi.org/10.1016/j.intcom.2006.01.005
Luursema JM, Verwey WB, Kommers PAM et al (2008) The role of stereopsis in virtual anatomical learning. Interact Comput 20:455–460. https://doi.org/10.1016/j.intcom.2008.04.003
Maresky HS, Oikonomou A, Ali I, Ditkofsky N et al (2019) Virtual reality and cardiac anatomy: exploring immersive three-dimensional cardiac imaging, a pilot study in undergraduate medical anatomy education. Clin Anat 32(2):238–243
Memon I, Feroz Z, Alkushi et al (2021) Switching from face-to face to an online teaching strategy: how anatomy and physiology teaching transformed post-Covid-19 for a university preprofessional program. Adv Physiol Educ 45(3):481–485
Mescher AL (2013) Junqueira's basic histology: text & atlas. Mc Graw Hill Lange, 544 p
Moore KL, Dalley AF, Agur AR (2017) Clinical oriented anatomy, 8th edn. Wolters Kluwer, Philadelphia, 1134 p
Moxham BJ, Plaisant O (2007) Perception of medical students towards the clinical relevance of anatomy. Clin Anat 20:560–564
Netter FH (1998) Atlas of human anatomy: combination package (book & CD-ROM), 2nd edn. Novartis (Ciba Pharmaceutical Co.), Cambridge
Netter FH (2006) Atlas of human anatomy, 4th edn. Saunders Elsevier, Philadelphia, 548 p (Plate)
Nettleton R (2010) BigBlueButton. In: Educational development Centre (EDC) blog. WayBackMachine. Available via internet archive. https://web.archive.org/web/20100814003302/http://edc.carleton.ca/blog/index.php/2010/06/04/bigbluebutton/. Accessed 3 Mar 2023
Nguyen N, Mulla A, Nelson AJ et al (2014) Visuopatial anatomy comprehension: the role of spatial visualization ability and problem-solving strategies. Anat Sci Educ 7:280–288
Nicholson DT, Chalk C, Funnell WR et al (2006) Can virtual reality improve anatomy education? A randomized controlled study of computer-generated

three-dimension anatomical ear model. Med Educ 40: 1081–1087

Nicola S, Stoicu-Tivadar L (2018) Mixed reality supporting modern medical education. Stud Health Technol Inform 255:242–246

O'Rahilly R (1986) Anatomy—a regional study of human structure, 5th edn. W.B. Saunders. Philadelphia, 809 p

Pantelidis P, Chorti A, Papagiouvanni I, Paparoidamis G, Drosos C, Panagiotakopoulos T, et al (2017) Virtual and augmented reality in medical education. In: Tsoulfas G (ed) Medical and surgical education-past, present and future. IntechOpen, London, pp 77–97

Pawlina W (2018) Histology: a text and atlas: with correlated cell and molecular biology, 8th edn. Lippincott Williams & Wilkins, New York

Ryan G, Callaghan S, Rafferty A, Higgins M, Mangina E, McAuliffe F (2022) Learning outcomes of immersive technologies in health care student education: systematic review of the literature. J Med Internet Res 24(2): e30082. https://www.jmir.org/2022/2/e30082. https://doi.org/10.2196/30082

Saverino D, Marcenaro E, Zarcone D (2021) Teaqching histology and anatomy online during the Covid-19 pandemic. Clin Anat 35:129–134

Tam MD, Hart AR, Williams S et al (2009) Is learning anatomy facilitated by computer-aided learning? A review of the literature. Med Teach 31:e393–e396

Verhoeven BH, Verwijnen GM, Scherpbier AJ et al (2002) Growth of medical knowledge. Med Educ 36:711–717

Vorstenbosch MA, Klaassen TP, Kooloos JG et al (2013) Do images influence assessment in anatomy? Exploring the effect of images on item difficulty and item discrimination. Anat Sci Educ 6:29–41

Ward PJ (2021) Netter's integrated musculoskeletal system: clinical anatomy explained! 1st edn. Elsevier, Philadelphia

Williams PL, Warwick R, Dyson M et al (1989) Gray's anatomy. Churchill Livingstone. 37 edition. Edinburgh, London, 1598 p

Wilson TD (2015) Role of image and cognitive load in anatomical multimedia. In: Chan LK, Pawlina W (eds) Teaching anatomy a practical guide. Springer International Publishing, Cham, pp 237–246

Yeung JC, Fung K, Wilson TD (2012) Prospective evaluation of a web-based three-dimensional cranial nerve simulation. J Otolaryngol Head Neck Surg 41:426–436

Developing a Flipped Classroom for Clinical Anatomy: Approaches to Pre-class Recordings and a Novel Approach to In-Class Active Learning

2

Stuart Inglis

Abstract

For centuries the established pedagogy for teaching in academia has been the lecture, in which an expert in a given subject speaks from a pulpit while a collection of students listens and takes notes. In recent decades, however, advances in technology have allowed educators to rethink the way in which they deliver course content. One approach that has been given attention is that of the flipped classroom, in which lectures are delivered outside of the class through video recording, and the class session is dedicated to more interactive activities. This chapter provides a rationale for this approach to learning, provides guidance in the development of pre-recorded lectures, and introduces a novel approach to the classroom sessions using audience response software.

Keywords

Active learning · Flipped classroom · Lecture videos · Lecture recording · Case-based learning · Teaching pedagogy

2.1 Introduction

A rite of passage experienced by the majority of scholars in academia is delivering their first lecture. Many times, the content and slide presentation are already developed, and the fledgling lecturer simply steps into the shoes of the senior lecturer, mimicking a presentation they have previously seen. In other instances, new lecturers may be encouraged to develop their own presentation, allowing more creative control over the content and delivery, provided that the content fits with the overall flow of the course. As these scholars progress in their careers, they are likely to deliver a set of lectures on a given topic or may have the privilege (or burden, depending on the perspective) of becoming the director of their own course. With this comes a greater sense of creative freedom over the content and how it is to be delivered: the topics that should be covered, the textbook that should be adopted for the class, and the number of tests that should be given over the term. In addition to these traditional questions regarding course design, another question that has started to be asked over the past decade is whether it is still necessary to have lectures at all in their traditional sense. And if the answer to that question is no, with what should they be replaced?

These were questions I began to ask in 2014, which led to a radical redesign to most of the classes that I presently teach. It was a long journey that got me to the point where I presently am,

S. Inglis (✉)
Department of Pathology and Anatomical Sciences and Surgery, Jacobs School of Medicine and Biomedical Sciences, University at Buffalo, Buffalo, NY, USA
e-mail: stuartin@buffalo.edu

D. Cui et al. (eds.), *Biomedical Visualisation*, Advances in Experimental Medicine and Biology 1431,
https://doi.org/10.1007/978-3-031-36727-4_2

filled with uncertainty and doubt. When I first decided to flip the classroom for our first-year medical students, I was visited in my office by another of the first-year instructors who happens to be a close personal friend.

"I hear that you want to do a flipped classroom approach," he remarked.

"That is the plan," I replied.

"Don't," he said, and spent the next 10 min presenting arguments for why this was a bad idea. By the time he left my office, I was feeling concerned that I might be making a mistake, but in the end, I followed my gut instinct, which was still telling me to move ahead with the flipped classroom approach. And I never looked back.

What follows is a narrative of how I came to the decision to adopt a flipped classroom approach to teaching and the lessons I learned along the way that shaped the approach I presently use in the majority of classes that I teach. Before proceeding, it is important to clarify that this narrative is not intended as an instruction manual or step-by-step guide to recreate an identical active learning experience. A number of different pedagogical methodologies inspired my approach to the flipped classroom teaching, but the thought of blindly duplicating a single established methodology felt like the fledgling lecturer mimicking a mentor's presentation; the shoes you step into never seem to fit exactly right. The hope is that readers take away ideas and, more importantly, inspiration from this narrative as they start to consider revisions to their own teaching methodologies. There is no single best practice that works for all instructors and all courses. The most important concept to impress upon readers is that they should feel comfortable with the approach they take to teaching.

2.2 My Epiphany

The transition to a flipped classroom style of teaching is fraught with uncertainty and second guessing and requires a great deal of effort in developing both the pre-class preparation material and in-class content. With all the work involved and the risk of failure, it is hard to understand why someone would bother with the transition in the first place. As with many large undertakings, the desire stems from a firm belief in an idea and the inherent motivation to make it a reality. I had previously read about the flipped classroom approach and was familiar with a variety of active learning strategies for the classroom, but for me, the transformative moment occurred while preparing a guest lecture for the fine arts department at the university where I was faculty at the time on the historical interplay between art and anatomy. As part of the talk was about the historical teaching of anatomy, I was searching through a collection of images of lecture theaters from the Middle Ages. As I worked, I started to notice how similar each of these images looked, with an instructor at a pulpit reading from a book, and students below transcribing notes (Fig. 2.1). These images did not look all that different from what would be seen in a modern classroom setting. This is when a thought occurred to me; could the reason that didactic lectures in academia are the norm simply be because that was the way it had always been done? After some brief research, I learned that the first modern university is generally considered to be the University of Bologna, founded around the year 1190, with other historic names such as Oxford and Cambridge appearing a few decades later (Verger 2003). On the other hand, Gutenberg's printing press would not be invented until 1436 (Meggs 1998). So, for the first 250 years of the modern university, books had to be transcribed by hand, making them a scarce and expensive item to own. This means that at the time the first universities were established, the only effective way to disseminate information to the masses was for an orator to read from a collection of notes or books owned by the university while students in the room transcribed what they heard for their own personal records. By the time textbooks would have been available to the masses, the practice of lecturing to a group of students would have been well established in academia, with professors teaching in the manner by which they learned from their mentors when they were

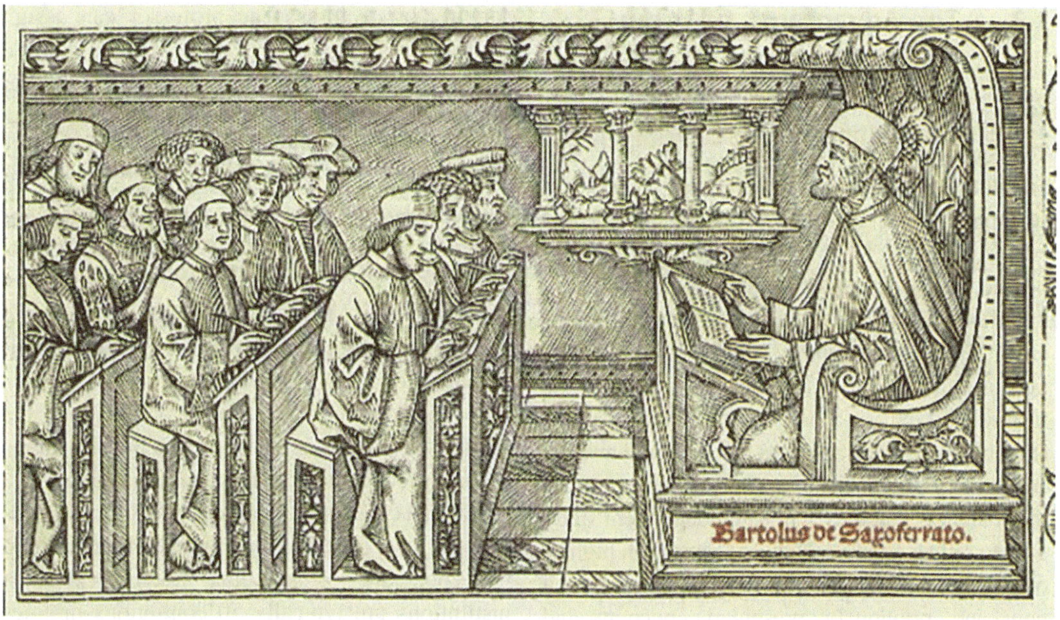

Fig. 2.1 A typical medieval classroom. Bartolo of Sassoferato, Repertorium super lecturis Bartoli, Lyon: 1523. Hiero von Holtorp Collection, Holtorp Box 6, John Rylands Library, The University of Manchester. Copyright of The University of Manchester

students. Even with the availability of mass-produced textbooks, there was little need to change the lecture format. The professor could expand on concepts beyond what was described in the course text or take time to explain the more difficult concepts for students to grasp.

It has only been within the past two decades that technological advances have provided opportunities to rethink the classroom in meaningful ways that were not possible before. The internet made it possible to share computer files with anyone with internet access. This allows professors to share notes and slides generated through word processing and slide presentation software with students, avoiding the need to transcribe everything that is said during a lecture. The development of screen capturing software combined with the high-performance computer processors allows professors to record their lectures in high-definition video and save it as a digital file. They can again use the internet to make these videos available to students for download to their personal devices. Instructors can also make use of video streaming services such as YouTube to post their videos, allowing the students to watch the video through a web browser on any device without having to download. Another recent trend is in open-access textbooks, which can be independently customized by professors to match the design of the course they are teaching. These types of technological advances allow for what feels like unlimited possibilities in designing curriculum and content delivery. While lecturing from a podium as students take notes is an effective and efficient teaching strategy in a world where printed text was not available, it is necessary to question whether it is still the most effective method of teaching in the new millennium. It was the question I asked myself as I put the finishing touches on my presentation for the fine art department, and my instinct told me there was a better way to deliver content to my students. This is what led me to adopt the flipped classroom approach to teaching my classes.

2.3 The Advantages of Using a Flipped Classroom Approach to Teaching

The concept of a "classroom flip" has been around since the late 1990s (Baker 2000) and was popularized by hedge fund analyst turned educator Salman Khan around 2010. The premise behind the flipped classroom has been detailed in the literature (Lage et al. 2000) and is well characterized by its name. It is based on a variation of a traditional lecture classroom, where the instructor stands at a podium and lectures while students take detailed notes on the fundamental concepts related to the course. The expectation of the student is that they will use independent study time to reflect on their notes and engage in higher-order assimilation, making connections between new and previously presented concepts and applying this newly acquired knowledge to higher-order application situations. The success of the design for a given class is unclear, as the opportunity for higher-order thinking occurs during independent study away from the instructor. With the flipped classroom, this process is reversed. The students are first exposed to the fundamental concepts through well-chosen readings or pre-recorded lecture sessions. They then arrive at class and expand on this fundamental knowledge through carefully designed classroom activities. One of the fundamental differences in this approach is that the professor is now present to encourage students to engage in higher-order thinking in regard to the material and can provide guidance to the students in this process. As students take a larger, more independent role in their acquisition of knowledge, the role of the professor changes from that of a "sage on the stage" to more of a "guide on the side" within the classroom (King 1993).

A number of advantages to using a flipped classroom approach have been described in the educational literature, both for the pre-class phase and for the in-class active learning portion. There are a number of options for approaching the pre-class phase, in which students are to attain the foundational knowledge that is traditionally acquired in lectures. The professor can assign textbook readings or carefully selected journal articles that cover the appropriate topics or refer students to publicly available videos on the topic. Probably the most common approach to pre-class foundational teaching is the development of pre-recorded lectures, in which the instructor records audio of themselves giving the lecture, typically through annotated slides in presentation software (e.g. Microsoft PowerPoint) or using screen recording software (e.g. OBS Open Broadcaster Software Project) to create a video media file. These videos are made available to the students to stream or download. A big advantage of this approach is in its asynchronous nature. In a traditional classroom approach, a lecture is scheduled for a set time and location which the student is responsible for attending. Lectures at most institutions are typically 50 continuous minutes in duration, or in the case of evening classes, 3 h in duration, typically with a break at the midway point. The pace is typically set by the lecturer, which requires the student to maintain their focus for the duration of the session. The concern is with the cognitive load that is placed on the learner. Previous work (Davis 1993) suggests that student focus will start to drift around 15 min into a continuous lecture, resulting in a loss of the narrative flow of the lecture. In situations where the content builds upon itself as the lecture progresses, it can be very difficult for the student to refocus their attention, having missed important portions of the dialogue. Even when a student can maintain their focus for extended periods of time, the pace set by the professor may be too fast to allow the student to fully process the information being communicated, and they find themselves taking down notes without really understanding their meaning. Conversely, the cognitive load on other students sitting in the same class may be low, and they find the pace of the lecture to be too slow for their liking (Francl 2014). Similar to a loss of focus, students may occasionally arrive late for the lecture, leave early to make an appointment, miss an entire class, or simply need to excuse themselves from a portion of the lesson to use the bathroom. All these scenarios result in students missing out on what could be critical curriculum content. There is also the occasional issue of the lecturer misjudging the time required to complete the lecture and either rushing through the latter portion of the lecture or leaving a portion of the content for the students to review independently. The issue of timing can be further complicated if the instructor does not accommodate the occasional questions that may arise from students in the audience. Conversely, as most students will not raise their hands during a lecture to ask questions, there is concern that students may leave the class with unanswered questions and uncertainty about what they have just learned.

The asynchronous viewing of pre-recorded lectures prior to the classroom session addresses a number of the issues outlined above. First, it addresses the issue of lecture scheduling. As there is no set time and location for the lecture, students can choose the time that is most convenient and the location most conducive to their learning. This flexibility allows students to work lecture sessions into their schedule around more rigid appointments. It also allows them to select a location more to their liking, such as a study room at a library, a coffee shop, or their couch at home. The use of pre-recorded lectures is also thought to address the issues related to cognitive load (Abeysekera and Dawson 2015). Unlike in-person sessions, each student controls the pace of the lecture. They can pause the recording any time they feel their concentration waning or need to take a break and can rewind and rewatch sections where they lost focus to avoid missing important details (Owston et al. 2011). This also allows students to take time to reflect on a concept that was just explained before moving on to subsequent content (Francl 2014). This may involve quiet reflection on the concept, making detailed notes they can refer back to at a later time, consulting additional resources for further clarification on the concept, rewatching a portion of the video to ensure that they understood the concept that was being explained, or a combination of these strategies. Students can also use this time to pose questions, either to the professor through email or to classmates through online discussion boards or social media platforms. This allows an opportunity for students who are normally reluctant to ask questions during an in-person lecture session an opportunity to do so using an approach with which they are more comfortable. The extra time it takes to complete a lecture raises a concern about overwhelming students, but it should be noted that this extra time is being used to assimilate and comprehend the material, a necessary process that is typically postponed until closer to a test, when the information is no longer fresh in the student's mind. As one student in a professional school gross anatomy course expressed on a survey of the flipped classroom approach, "Upfront time commitment is painful, but I felt very prepared for the unit exam even without reviewing the material." The use of pre-recorded lectures also addresses the other end of the cognitive load spectrum. Most video playback software programs have some form of speed control built into the system. This allows students who are familiar with the material and find the cadence of the lecture to be too slow for their liking to watch the videos at a faster speed.

With the development of live lecture streaming and recording that has taken place over the past decade, it is now common for students to forego attending lectures in person but instead watch the live stream from a remote location or the recording at a later time. This should not be confused with the flipped classroom approach to teaching. The critical element of the flipped classroom approach is the greater focus that is placed on guidance during higher-level cognitive activities. Using Bloom's taxonomy (Anderson and Krathwohl 2001), the process of learning can be broken down into six hierarchical levels of cognitive complexity, ranging from basic recall of information and comprehension of concepts and progressing through the application of knowledge to novel situations, making inferences about the concepts, adding novel ideas to the field of study, and evaluating the validity of the various elements of the topic. In a traditional approach, lecture is primarily dedicated to the lower levels in Bloom's taxonomy of knowledge and comprehension, and the student works independently on higher-order levels of cognition such as application, analysis, synthesis, and evaluation. With the flipped classroom, lower-level cognitive activities, which require little individual guidance, are accomplished outside of and in advance of class time, which can now be dedicated to higher-order cognitive activities with instructor and peer attention and support (Francl 2014). This allows the instructor to observe students as they work through complex thought processes, providing guidance and redirection where appropriate, and addressing difficult questions that may arise through direct discussion. The focus on active learning in the classroom is believed to increase levels of student motivation. In relation to the self-determination theory of motivation, an active learning classroom has the potential to provide both intrinsic and extrinsic motivational factors to the students (Abeysekera and Dawson 2015). If the active learning session is well designed and pertinent to what the student is learning, they will find inherent value in both the exercises and the pre-class learning necessary to prepare for the in-class session. In working through complex tasks with minimal assistance from others, the student gains self-confidence in their abilities and a sense of accomplishment, which motivates them to attempt additional learning tasks. This also provides students with a sense of contribution to the learning process, rather than being just a passive observer. They have a chance to offer their thoughts and ideas, which can affect the direction of a discussion or alter the direction in which the class is heading. With this autonomy, students develop a greater sense of responsibility for making meaningful contributions to the class and are more likely to come to class prepared and ready to participate (Lage et al. 2000). This may be particularly true in instances when students work in groups, due to concern with letting others in the group down if not well prepared for a given class session. Scheduling critical thinking activities during class time is also more likely to engage students who may otherwise not concern themselves with the same activities if assigned as homework exercises outside of class time (Mok 2014).

2.4 Designing Pre-class Learning Activities

It is difficult to provide concrete suggestions on how to design the pre-recorded videos, as a variety of styles and approaches have proven successful. This section is therefore more focused on general recommendations and common pitfalls that should be avoided when developing video lecture recordings. The first thing to mention is how surprisingly different recording a video lecture can be when compared to a live lecture. While public speaking can be daunting for some, experienced lecturers are used to speaking in front of a live audience of students and develop a flow in which they can look out and appraise the level of attention and understanding of those in attendance. Performing the lecture sequestered in a small room with no audience to feed off can be daunting at first for those unused to the experience and takes practice to perfect. Many educators gained a sense of this when presenting lectures through videophone software during the pandemic. Those planning to start generating a series of lecture videos should plan to take a few weeks to acclimatize to the experience, as their first few recording attempts are likely to appear amateurish and unnatural compared to later attempts, after a rhythm is established.

The first question that the course instructor should ask is if it is necessary to record these lectures in the first place. The process of lecture recording can be intimidating and time consuming, especially for those with little to no experience with the process. Instructors may, therefore, assign textbook and/or journal article readings to take the place of the traditional lecture approach or may provide the students with links to publicly available videos as substitutes for in-house video recordings. The advantage of this approach is that it saves time in developing course content and, if properly selected, provides the students with high-quality resources from which to learn from. The disadvantages include the difficulty in finding and compiling enough resources available to cover all the necessary content, which may prove more time consuming than creating novel lecture recordings. External readings or recordings are also likely to contain superfluous details unrelated to the course that create unnecessary work for the students and confusion regarding what they are responsible for learning for the class. Selecting from a variety of resources can also create disjointedness with the class, as the pre-class delivery style changes from one class to another.

A number of research articles (e.g., Guo et al. 2014), Lange and Costley (2020), as well as a variety of internet websites (e.g., Borer 2021; Myers 2016) offer their perspectives on strategies for creating effective lecture videos. Some general advice appears to be universally accepted across this multitude of platforms:

1. Videos should be short.
2. Videos should be informal.
3. Videos should be interactive.
4. Videos should be novel.

2.4.1 Videos Should Be Short

A common mistake that I made the first time I developed pre-recorded lectures was to regard them as a carbon copy of the in-person lecture. I therefore hit the screen record button and just started speaking, as if I was talking to a lecture hall full of students and hit stop when I came to the end of the lecture I was used to giving. The advantage of pre-recorded lectures, I thought, was that I was not bound by the 50-min time limit placed on live lectures, where students may have another lecture to get to, and another class is waiting for the room. I would therefore have recorded lectures on a topic that would go well over an hour, which I felt was offset by other shorter lectures that were well under the 50-min time frame. In reading through student feedback in course evaluations, I realized that students were not fond of this approach, as they found the lectures long and exhausting to view. While students did have the option of pausing at any point in time, most would push through out of concern of losing their train of thought and having to rewatch a portion of the video to remind themselves of what they had just seen. Several students also called out the fact that many of the lecture recordings went beyond the allotted time mandated by the credit hours for the course. In rethinking and redesigning the lecture recordings for the following class, I found natural break points in most lectures that could result in breaking the traditional 50 min lecture into a collection of three or more videos that run between 10 and 20 min each. Using my discipline of anatomy as an example, a 50-min session on the forearm can be broken down into a 10-min video of the bones and joints, a 20-min video of the muscles, and an 8-min video of the vessels and nerves. Importantly, at the end of each session, the students are encouraged to take a break before starting the next segment. By segmenting the lecture according to subtopics, natural breaks are created that allow students an opportunity to pause, reflect on the portion of content they just watch, and reestablish their focus before moving on to the next subtopic. Instructors should also ensure that the combined time for the video segments for a given lecture does not exceed the time that would be allotted to the lecture if given in person. Students will be paying attention to this measure and will be quick to point out if the instructor consistently runs overtime. Strategies include trimming unnecessary content and/or being more scripted in the approach to avoid unnecessarily long pauses in the recording where the instructor is thinking about what they wish to say next (discussed in the next section).

2.4.2 Videos Should Be Genuine

In making lecture recordings, the creator is encouraged to be themselves and allow their natural personality to shine through. Instructors who use humor and anecdotes in a live lecture session should use the same principles in a recording of that lecture. The danger is in trying to be something you are not in order to impress the camera. Recording a video lecture introduces options that are not possible for live lectures, including script development, recording in segments, and post-production editing. There are certain advantages to having these options available. For example, there is probably not a lecturer reading this who cannot recall a time that they misspoke during a live lecture and had to backtrack to make a correction. With video recording, if the lecturer makes a mistake, they can repeat the segment as if nothing happened and cut out the portion containing the error in post-production. That being said, a number of sources caution against making the recording too polished, arguing that it allows for the instructor's genuine personality to shine through in a way that is lost if the video is overly scripted and overly edited. While this may be the case, the use of scripts and recording in planned segments allows for a concise delivery that decreases the overall run time for the video, as discussed in the previous section. As previously stated, it is critical to ensure that the overall run time for a pre-recorded lecture does not exceed the amount of class time that would have otherwise been used for a live lecture. My personal preference is to utilize a script and post-production video editing to remove segments where I may have misspoken and corrected myself, stumbled over my words, or left too long of a pause between sentences. One concern with reading from a script is that the dialogue becomes artificial and robotic. To avoid this pitfall, I closely monitor the cadence of my delivery as I record, including inflections, dramatic pauses, and the same anecdotes that are included in my in-person lectures.

2.4.3 Videos Should Be Interactive

Earlier we discussed the importance of keeping the video segments short to maintain student focus. Another way to keep focus is to include interactivity in the videos with the inclusion of segments that promote viewer interaction. This could be as simple as including reflection exercises that ask the viewer to pause and consider a concept and how it may apply to various scenarios or multiple-choice questions based on what the viewer has just seen. A potential downside to this simple approach is that it requires a willingness on the student's behalf to be an active participant in the process. Less committed students can bypass the pause function in order to complete the lectures as quickly as possible. For some educators this is not of concern, as the exercises are considered optional rather than mandatory. In instances where the instructor requires participation, a discussion forum can be established in the school's Learning Management System where participation in the discussions is built into the student's grade. In addition, software programs exist that can embed multiple-choice questions into the video at specific checkpoints that require the student to select an answer before being able to move on to the next section of the video.

2.4.4 Videos Should Be Novel

This can be thought of as a "don't reinvent the wheel" suggestion. The thought being that pre-existing videos can be adopted into the class, saving the instructor time in trying to develop identical content. Although efficient, caution should be taken in adopting this approach. As previously mentioned, selecting from a variety of pre-existing videos can be time consuming in itself, may contain unnecessary details, and disrupt the flow of the class if numerous sources are used. Students may also question the value of the class if they are constantly being referred to external resources rather than receiving direct instruction from the course director. It should also be noted that, while better quality sources may be available, students will often prefer the instruction provided by the instructor, as it is someone that is familiar to them. A colleague confided in me that she once had to use an external video she had found to cover a class while she attended a death in the family. She was impressed by the quality of the video and remarked that it was much better than anything she could have produced on her own, but in the course evaluations many students recommended she take time to create her own recording for the incoming class, as they preferred learning from someone they were familiar with rather than from the stranger in the video she had found.

2.5 A Novel Example of an In-Class Learning Activity

As difficult as it is to give concrete advice on generating pre-class lecture videos, it is more difficult for the in-class active learning phase. A variety of well-established strategies exist for approaching in-class learning, such as problem-based learning (PBL), team-based learning (TBL), and process-oriented guided inquiry learning (POGIL), to name a small few. Many of these approaches are highly structured from start to finish and for this reason, varying the design is not recommended. Other approaches are not as rigidly structured, and modifying elements of the approach is acceptable, so long as certain critical components are included in the approach.

What follows is an outline of a novel approach that I use for active learning in the classroom. Having attended a number of workshops for different active learning pedagogies I was eager to incorporate a number of the principles I had learned about into my own class sessions, but also found that there were elements in each that I did not feel would work in my classroom design. Through trial and error and with the assistance of feedback from students in the course evaluations, I developed an approach that has come to be known as a Facilitated Active Learning (FAL) exercise. Putting a name to the approach allows students to know what to expect when they see it on a calendar and, to the best of my knowledge, is not currently being used to characterize any other pedagogical approaches to active classroom learning.

The approach I have developed utilizes a "think-pair-share" (Prahl 2017) approach to answer a series of clinical cases that the student should be able to solve based on the content covered in the pre-recorded lecture videos. The FAL approach bears certain similarities to the Peer Instruction (PI) approach developed by Eric Mazur in the 1990s (Mazur 1996), in that both utilize questions that students work on both independently and in groups to answer. Differences include the use of pre-recorded videos rather than textbook readings as a pre-class activity, the selection of pre-determined questions carefully designed to extend student knowledge rather than in-class questions based on student responses to pre-class questions, and the use of gamification to incentivize and motivate student participation.

The session makes use of audience response system (ARS) software to collect student responses. A variety of both free and subscription-based programs are available on the market. What is critical in my approach is the ability to collect individual responses for each question in order to examine which response each student selected and how many questions each student answered correctly. This type of functionality is absent from a number of the free programs that are available.

Each FAL session can be broken down into the following stages:

1. Pre-class set-up
2. Readiness assessment quiz (RAQ)
3. Class announcements and questions
4. Facilitate Active Learning (FAL) cases:
   A. Case presentation
   B. Independent polling
   C. Poll results
   D. Polling with group discussion
   E. Class discussion
   F. Answer reveal
   G. Case summary and follow-up questions

2.5.1 Pre-class Set-Up

As with a live lecture, this is the time used to set up the room computer and projector. This set-up time is important, as there are a number of moving parts to the session. I typically use my personal laptop plugged into the external port of the podium, as it allows me to utilize the extended screen function that is not possible with the classroom computer. I have found that operating both the audience response system and PowerPoint as separate programs work best, and as both programs have a "presenter mode" function, I can display the presentation for each through the projector, while using my laptop screen to see what slides and questions are coming next, how many students have answered a poll question, and the distribution of selections before the question closes. Importantly, it gives me an opportunity to set the mood for the class. After plugging my laptop into the projector system, I will open some sort of music program and play music for the class to hear. I have found that this creates a welcoming environment and can be stimulating for students, particularly in a morning class where they may have only woken up a few minutes ago.

The pre-class set-up ends with a segue into the next phase that begins each class: the readiness assessment quiz described next. From experience, I have found that from the moment I announce that a quiz is beginning it can still take a couple of minutes to get every student's attention and eliminate talking, which eats into valuable class time. To address this, I developed a PowerPoint slide with a countdown time that runs from 60 s and a distinct melody that starts at the appearance of the slide and ends when the timer hits 0 (Fig. 2.2). I use the first minute of the song "Intro" by the musical group "The XX," as it creates an appropriate sense of urgency, builds in intensity with a percussion line that mimics a beating heart, and has a natural break at precisely the 1-min mark. A less intense option would be a 30-s timer that coincides with the "Think!" jingle that is used during final jeopardy in the popular game show. Since introducing this practice, I have found that I can start the slide when I am just about ready for class to begin and will have the students in their seats and ready to go in exactly 1 min without delay. This includes students rushing into class to find their seats, having heard the countdown music in the hallway.

2.5.2 Readiness Assessment Quiz

Each class begins with a brief quiz consisting of five multiple-choice questions based on basic information from pre-class lecture material. This serves the purpose of motivating students to complete the pre-class activities which provide the foundational material that will be applied to the active learning session. The questions should test major concepts to encourage students to focus on the big picture and key points and should be of low difficulty such that students should do well having been recently introduced to the material. Overly difficult questions at this phase will be discouraging and undermine the purpose of the quizzes, which is to engage students in the pre-class material.

While giving a quiz at the start of each class increases class engagement in the material, it also takes up a portion of the allotted class time, which is particularly problematic when class duration is an hour or less. The quizzes must, therefore, be delivered efficiently to avoid unnecessary loss of class time. The system I have developed takes advantage of the audience response software. As the quiz begins, the first question is projected onto the classroom screens from a PowerPoint slide. Students read the question and use the audience response software on their phones to submit their responses. Importantly, the question is not displayed in the audience response software, which only instructs the student to select between options A through E. This ensures that the student must be present in the room in order to view the question on the projection screen. In regard to timing, students are instructed to enter their responses as soon as they decide on an answer. Once 75% of the class has answered the questions, the remaining 25% are given 15 s to decide on an answer before the question is closed and the next question appears. This method of timing was established through trial and error, giving students in the audience ample time to respond without feeling overly stressed and without allowing the question to drag out overly long. It also matches the time allotted to each question with its difficulty. Simple questions with short stems are answered quickly, whereas more time can be given to more difficult questions with longer stems. To ensure nondiscrimination on the basis of disability in higher education, students with documented accommodations should be provided with an alternative quiz format if they prefer. This typically involves offering the quiz in the instructor's office just prior to class. From personal experience using this

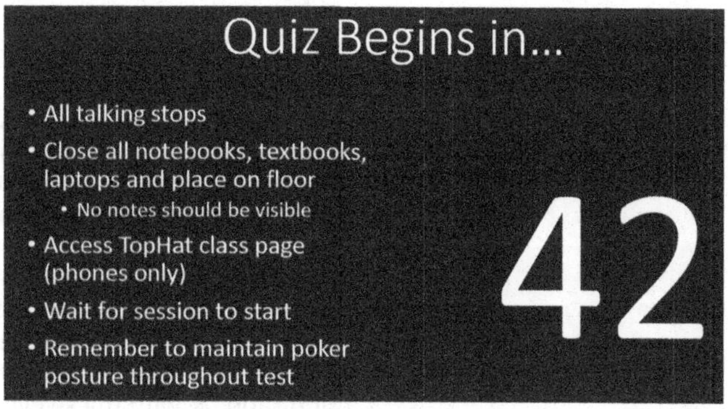

Fig. 2.2 Example of the countdown slide used to start the readiness assessment quiz, frozen at 42 s before the start of the quiz

approach, most students with accommodations opt to take the quiz with the remainder of their classmates, as they find they perform well using the format explained above. Once all questions are presented, the answers are given so that students will immediately know how they performed.

2.5.3 Announcements and Questions

Immediately following the quiz review is a short period for making announcements to the class. These can be formal announcements such as events or review sessions, details about an upcoming test, and changes to the course calendar. It can also include some less formal announcements. For many of my courses, students are encouraged to notify me of any classmates having a birthday, and we take a few moments during the announcements to sing "happy birthday" to these individuals. Students are then encouraged to ask questions about class material or the course itself. Often, I will have received poignant questions through email regarding the pre-class lecture recordings and will address these questions anonymously for the benefit of the class as a whole.

2.5.4 Facilitated Active Learning (FAL) Cases

What follows for the remainder of the class are a series of FAL cases for the class to work on in a synchronous fashion. It is typically possible to cover anywhere from 3 to 5 cases in a 50-min classroom session, depending on the amount of time taken up by the quiz and announcements as well as the difficulty and depth of discussion the cases produce.

2.5.4.1 Case Presentation

Each case typically begins with a clinical vignette displayed on a PowerPoint slide (Fig. 2.3). The vignette is read out loud to ensure the details are presented to the class in unison. Images and videos can also be incorporated at this point to provide students with additional details. For an anatomy course, this can include video of a musculoskeletal injury caught on camera as it is occurring, video of a patient during a hospital visit, and any medical imaging that may be associated with such a case.

2.5.4.2 Independent Polling

Once the case has been presented to the class, students are given a multiple-choice question related to the case (Fig. 2.4). For the anatomy class, this typically takes the form of a stem such as "damage to what anatomical structure is responsible for the patient's condition?" although more elaborate questions are commonly used. The distractor list includes as many plausible options as can be imagined for the students to discuss. For the quizzes at the start of the class and the tests and exams associated with the exam, questions are normally limited to five multiple-choice options which allows the student to select their response in an appropriate amount of time

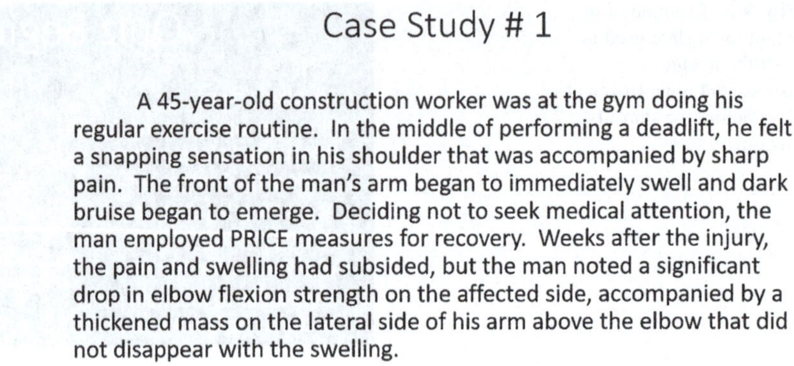

Fig. 2.3 Example of a case used during an FAL session

based on the test duration. For the FAL sessions, the point of the exercise is for students to thoroughly discuss and debate all feasible choices in deciding on their answers.

When the question is presented and all answer choices are read out loud, students select from the answer options independently. This stage coincides with the "think" portion of the "think-pair-share" approach to active learning. Students are instructed to treat this period as if they were in a test and are not permitted to talk or look around at others as they submit their answers. Importantly, after the question is fully read, students are given a 25-s countdown timer in which to submit their answers. This encourages students to select the option that first jumps out at them and does not give them time to consult additional resources such as class notes or the internet in deciding on an answer. This also speeds the process along, allowing as much time as possible for the group discussion phase.

2.5.4.3 Poll Results

Once the 25-s countdown timer completes, students are shown the response frequencies for each of the multiple-choice options (Fig. 2.5 top). Importantly, the correct answer is not yet disclosed to the students. Showing the polling results helps students to narrow their focus to the more common answer choices for the next

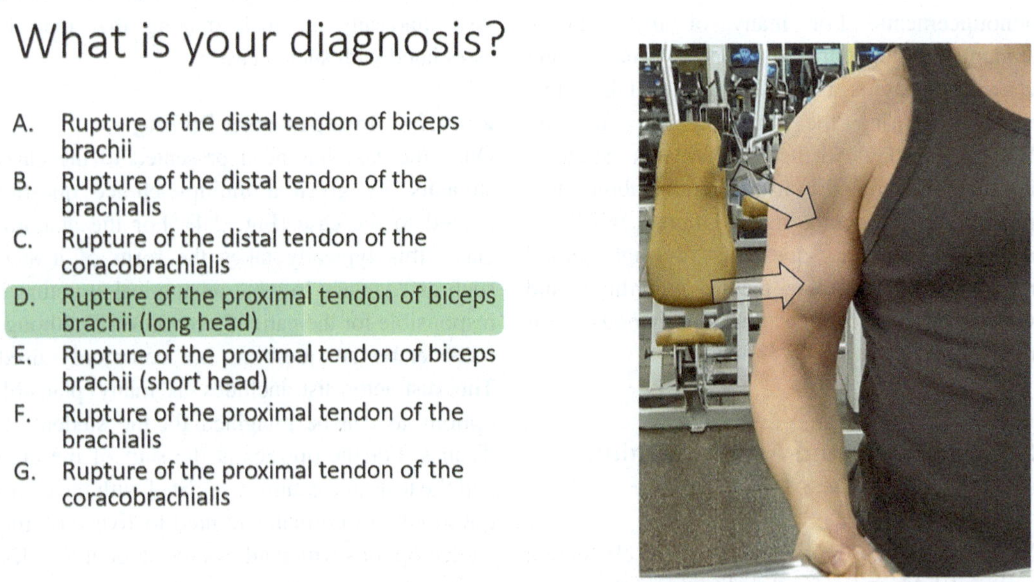

Fig. 2.4 The multiple-choice question associated with the case depicted in Fig. 2.3. Note that correct answer is highlighted in the image, which is kept hidden until the answer reveal phase.

Fig. 2.5 Polling results taken after individual polling (top) and group discussion (bottom). Options F and G have been omitted from view as neither had any selections during either phase. Graphic generated through TopHat audience response software

stage of the process. In most instances, student instincts are right, and the correct answer has the greatest number of selections. Occasionally, however, this is not the case, which makes for interesting discussions in later phases.

2.5.4.4 Polling with Group Discussion

Once the polling results are shown, the question is reopened, and students discuss the question in their small groups. This phase is closely associated with the "pair" portion of the "think-pair-share" approach and is critical to the active learning element of the class. Students take this time to discuss the answer options, offer their reasoning for the options they chose during the independent polling phase, and decide on a correct approach. Importantly, each student in the group still submits a response through the audience response software. This gives students who disagree with the remainder of the group the autonomy to make their own decision after hearing the group discussion. It also allows groups that can decide between two or more responses to strategically split the responses between these options, to increase the chances that a portion of the group will land on the correct response.

Students are also instructed to use this time to consult additional resources through their computers or phones. This is considered an important aspect of becoming a lifelong learner, as they will need to learn how to further their knowledge of a subject beyond their university education. This also opens the door to frank discussions of credible sources of information and using critical thinking skills in evaluating the validity of information being provided. Students will use this time to search for symptoms on the internet or read further into textbook descriptions of conditions to determine if they match the case being discussed. Students are also encouraged to look up the meaning of medical words used in the stem of the question that they may not presently understand.

This phase also involves the instructor walking around from group to group and listening to the discussions that are taking place. Instructors may also engage the students in conversation through questions such as "Is there an answer choice that looks good to you?" followed by "What are your reasons for focusing on this choice?" This makes for two-way conversations that are mutually beneficial. For the students, the instructor may point to additional information from the pre-class instructional material or other resources that the students may have overlooked. It also gives the instructor important insight into the student's rationale for making an answer choice. Instructors often suffer from the "curse of knowledge" (Wieman 2007), in which they may fail to remember the complexity in first learning the subject they are so well versed in and not appreciate the thought processes of novices working through the material for the first time. Listening to student discussions and explanations offers insight into student perceptions and misconceptions that can be addressed in later phases of the case.

With regard to the time spent during the group discussion phase, I use a similar approach to that used during the quiz portion of the class. Students are instructed to submit their responses to this second attempt at the question as soon as they are certain of their answer choice and the instructor monitors the number of submissions as they move between groups. Once the response frequency reaches 75%, the remaining students are provided with a 60 s countdown to when the question will close. This allows students reasonable time to wrap up their discussion and decide on their answer choices.

2.5.4.5 Class Discussion

This phase is associated with the "share" portion of the "think-pair-share" approach to active learning and begins with revealing of the new polling frequency following group discussion (Fig. 2.5 bottom). In most cases, the most commonly selected choice during the individual polling phase has an even greater selection frequency. It is also common for an initially less popular choice to overtake the other options in popularity, which tends to be met by surprise and interest in the class.

Students are now given the option to offer their reasoning for the answer they selected. This can be done by simply asking for volunteers from the class to raise their hands and offer their thoughts. Two setbacks to this approach are long pauses with students hesitant to offer their thoughts and the discussion period being dominated by a small number of enthusiastic students. Long pauses may feel awkward but showing patience and looking expectantly toward the class creates something akin to a staring contest, in which the student should always blink first. These pauses usually subside after the first few classes, as students learn that the only way to end the pause is for a student to speak. It is also important for the instructor to express appreciation for any sort of student participation to encourage further contributions from the class as a whole. I personally make a point of thanking students who offer incorrect explanations for their bravery in speaking up and stressing that the class often benefits more from the discussions that come out of incorrect answers than from explanations of the correct answer, itself. The instructor can also ask for input from a student that has not yet spoken during that class or students that have not yet contributed to class discussions in the course to encourage more students to participate in the discussion. Another strategy that can be used is assigning each group a distinct number and using a random number generator to select which group will speak first during the group discussion, although this approach could increase student anxiety and foreboding during the group discussion phase, as students are unsure of when they will be selected to speak.

Often the first to speak gives a clear and thorough explanation for the answer choice, and little needs to be added to the discussion. Other times the student may offer incomplete thoughts or erroneous explanations. In these instances, the instructor should offer probing questions for the class to answer that lead them to a full understanding of why a particular answer is correct for a given case. The key element is allowing the students the opportunity to discover the correct answer for themselves without having to be told.

2.5.4.6 Answer Reveal

Once dialogue and discussion have run their natural course, the correct answer can finally be revealed to the class. In most instances this is anticlimactic, as the students have already determined the correct answer by this point. There are occasions, however, where students will have missed a critical element to a case, and the correct answer is one that only a small collection of individuals selected. This generates surprise in the majority of the class, and elation in the small number of students that answered the question correctly, and they are typically happy to share their reasoning now that they realize their instincts were correct despite being in the minority. On one memorable occasion, a student jumped out of his seat, pumping his fist and pointed at his group, exclaiming "I told you guys!" These create memorable moments for the class, and students are more likely to remember these cases particularly well.

2.5.4.7 Case Summary and Follow-Up Questions

The final stage of the case is to provide summary slides with the critical details related to the case as well as further details such as treatment and prognosis (Fig. 2.6). Most of the details about the case have usually been identified and discussed by this point, but there may be a few additional details to bring to students' attention. This would also be the opportunity for the instructor to clarify some aspects of the case based on conversations that may have come up during the group discussion phase. Students should also have the opportunity to ask additional questions to clear up any lingering confusion related to the case or to extend their understanding of other situations. Once the summary is completed, a new case can be introduced, and the process repeated as many times as allowed according to class duration.

2.6 Gamification as Motivation for the Active Learning Session

One lingering question that remains for setting up and implementing an active learning session in the classroom is student motivation for performing to the best of their ability. In other words, what reason does the student have for trying to perform well during the active learning session? This is, in part, addressed through the five-point quiz at the start of the class. Students have already studied the material to perform well on the quiz and are already present in the class, so there is no reason not to try to do well in the group exercise. The instructor may also choose to assign grades for performance during the active learning session, but this is something that I would discourage against. These sessions provide students with an opportunity to use creative thinking, group discussion, and Internet research to explore answers to meaningful questions, which may lead to tangents that stem from student curiosity. The ability to think freely may be hindered by constant concerns regarding the effect an incorrect response may have on a student's performance

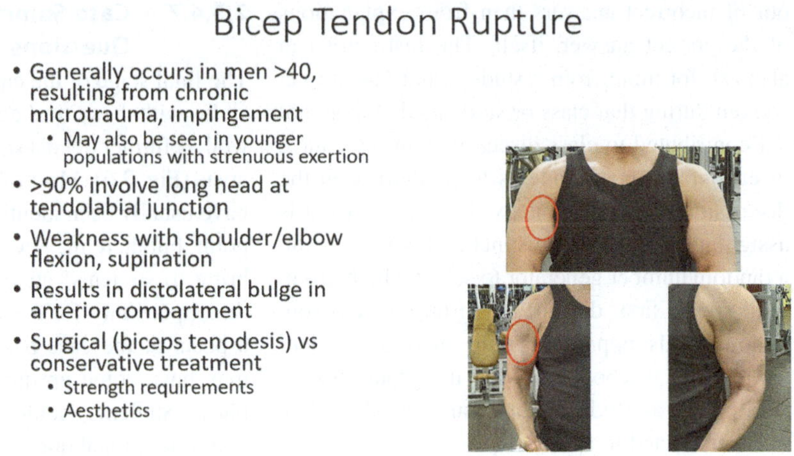

Fig. 2.6 Summary slide for the case depicted in Fig. 2.3

and may discourage the exploration of other topics to which a grade is not attached.

An emerging paradigm in education is the gamification of learning (Kapp 2012), which incorporates elements of game play into the classroom setting to motivate students to learn. Gamification can be used as a positive reinforcement to promote participation in the active learning classroom. A key element of the facilitated active learning sessions that was just described is monitoring both individual and group scores for each student and averaging these scores with the other members of their group to derive a group total for each session. The top three groups are revealed during the announcement portion of the following active learning session with the top group being presented with a trophy for their efforts (Fig. 2.7). In developing this strategy, I established the "Teacher's Trophy," which was created years earlier when performing the silver bottle experiment for a high school chemistry class I taught (Fig. 2.8). The winning group gets to keep the trophy at their seats for the duration of the class, can sign the bottle with a felt pen, and take photos of themselves with the trophy. As described to the class on the first day, "this trophy has absolutely no significance, has no inherent value and no meaning, and all of you will fight tooth and nail to win it." The trophy presentation

Announcements

Winners	Table 21
2nd Place	Table 17
3rd Place	Table 10

Fig. 2.7 Slide used during the announcement for the top performing group from the previous class

Fig. 2.8 The "Teacher's Trophy" that is presented to the winning group from the previous class. Note the signatures on the bottle from previous winners

has become a highlight of each class, with groups celebrating their first-time victories and winning streaks, and friendly rivalries developing between groups who consistently finish in the top three places. As an added incentive, we will often play music during the lab sessions, and the winning group is allowed to choose the music playlist for the day. This provides students with the motivation to try and perform well but does not create negative consequences should they perform poorly.

2.7 Closing Remarks

This chapter presents what I hope is compelling evidence for adopting a flipped classroom approach to learning, as well as a blueprint for a method by which it can be implemented. While such a blueprint can serve as a helpful starting point in developing an approach to active learning, the Facilitated Active Learning (FAL) approach should not be considered absolute and unalterable. It emerged as inspiration from other approaches to active learning I had studied and took form as I discovered what worked and didn't work for my class structure and teaching style. Similarly, those that may choose to adopt the model described above should be encouraged to modify the approach as their instincts guide them to develop through trial and error an approach that works for them.

It is also not necessary to make wholescale changes to a class all at once, which can be a daunting task. Those interested in the flipped classroom approach but unsure of where to start can select one or two lessons over the course of a semester to present in the flipped classroom manner. They can then reflect on their experience of what seemed to work well and not work well, as well as solicit feedback from the class by means of a questionnaire. As the model progresses, they may find the motivation to take the flipped classroom to the next step and prepare classes to be conducted entirely as a flipped classroom, or to take their class in an entirely different direction. What is important is for educators to not be satisfied with the status quo, but always ask themselves if there is a better way that their course can be taught. This allows education to break free of an institutionalized dogma that stagnates our ability to do better.

References

Abeysekera L, Dawson P (2015) Motivation and cognitive load in the flipped classroom: definition, rationale and a call for research. High Educ Res Dev 34(1):1–14. https://doi.org/10.1080/07294360.2014.934336

Anderson LW, Krathwohl DR (2001) A taxonomy for learning, teaching, and assessing: a revision of Bloom's taxonomy of educational objectives, 1st edn. Longman, New York

Baker JW (2000) The "classroom flip": using web course management tools to become the guide by the side. Selected papers from the 11th international conference on college teaching and learning, Jacksonville, April 2000, 9–17

Borer GJ (2021) Engaging video lectures in 4 easy steps. https://screencast-o-matic.com/blog/engaging-video-lecture/. Accessed 14 Feb 2023

Davis BG (1993) Tools for teaching, 1st edn. Jossey-Bass, San Francisco

Francl TJ (2014) Is flipped learning appropriate? J Res Innovative Teach 7:119–167

Guo PJ, Kim JK, Rubin R (2014) How video production affects student engagement: an empirical study of MOOC videos. 41–50. https://doi.org/10.1145/2556325.2566239

Kapp K (2012) The gamification of learning and instruction: game-based methods and strategies for training and education, 1st edn. Pfeiffer, San Francisco

King A (1993) From sage on the stage to guide on the side. Coll Teach 41:30–35

Lage MJ, Platt GJ, Treglia M (2000) Inverting the classroom: a gateway to creating an inclusive learning environment. 31:30–45. https://doi.org/10.2307/1183338

Lange C, Costley J (2020) Improving online video lectures: learning challenges created by media. Int J Educ Technol High Educ 17(1):1–18. https://doi.org/10.1186/s41239-020-00190-6

Mazur E (1996) Peer instruction: a user's manual, 1st edn. Prentice Hall, Upper Saddle River

Meggs PB (1998) A history of graphic design, 1st edn. Wiley, Hoboken

Mok HN (2014) Teaching tip: the flipped classroom. J Inf Syst Educ 25:7–11

Myers S (2016) 6 tips for creating engaging video lectures that students will actually watch. https://teaching.temple.edu/edvice-exchange/2016/03/6-tips-creating-

engaging-video-lectures-students-will-actually-watch. Accessed 14 Feb 2023
Owston R, Lupshenyuk D, Wideman H (2011) Lecture capture in large undergraduate classes: student perceptions and academic performance. Internet High Educ 14:262–268. https://doi.org/10.1016/j.iheduc.2011.05.006
Prahl K (2017) Best practices for the think-pair-share active-learning technique. Am Biol Teach 79:3–8
Verger J (2003) Patterns. A history of the university in Europe, vol 1, Universities in the middle ages, 1st edn. Cambridge University Press, New York
Wieman C (2007) The 'curse of knowledge', or why intuition about teaching often fails. APS News 16: backpage

An Overview of Traditional and Advanced Visualization Techniques Applied to Anatomical Instruction Involving Cadaveric Dissection

3

Kenneth Hisley

Abstract

This work provides an overview of the role, basic concepts, significance, and instructional contributions of current and evolving digital visualization technologies being applied in first-year anatomy curricula. These are visualization methods that have been and are being used to support both basic science and clinical applications of gross anatomical teaching and learning to the health professions (i.e., medical, dental, physical therapy, chiropractic and nursing students). It first presents a foundation as to how this discipline has been and is being taught within the professional school environment using visualization and illustration: aspects of learning, the format of the first-year anatomy curriculum, the nature of the visual information network in support of educational excellence and newer opportunities afforded by advanced technologies placing the student at the center of the learning experience. Then, the nature of each of these new methods is presented with their individual unique characteristics, and the results that anatomy faculty running cadaveric dissection courses had with the evaluation of the new technologies.

The Conclusion section lists key points found in the literature as reported. Finally, the Future Work section proposes investigations into standardizing the presentation and assessment of anatomical concepts using prominent in situ structures of viscera, their enclosures and resident compartments for more precise and reproducible measurement of then instructional effectiveness of the new techniques.

Keywords

Medical education · Anatomy education · Visualization techniques · Augmented reality · Virtual reality

3.1 Introduction

This chapter provides a basic overview of traditional and newer advanced instructional visualization methods supporting and augmenting human cadaveric dissection during the implementation of first-year professional gross anatomy instruction. Additionally, it reports on academic pursuits attempting to evaluate the effectiveness of these advanced technologies using semi-immersive and fully immersive methods. Instructional materials comprising both pre-digital visualizations using the principles of perspective and digital methods applying the principles of stereopsis have been in use for some time. As

K. Hisley (✉)
Pre-Clinical Studies – Anatomy, William Carey University College of Osteopathic Medicine, Hattiesburg, MS, USA
e-mail: chisley@wmcarey.edu

D. Cui et al. (eds.), *Biomedical Visualisation*, Advances in Experimental Medicine and Biology 1431,
https://doi.org/10.1007/978-3-031-36727-4_3

described below, even 3D tools portraying 3D spatial perception of clinically oriented structure and function have been available in both the nineteenth and twentieth centuries using the principles of stereopsis. Current digital applications provide not only remarkable 3D spatial representations of complex anatomy but support the direct student interrogation of these perceived models with supporting textual annotations, in-depth information, and motion.

Traditional printed and digital instructional resources are not new in the medical school curriculum given the common existence of digital documents (handouts, slide sequences, laboratory instructions, and visual materials); however, the evolving digital methods of student-controlled 3D visualization used in conjunction with cadaveric dissection extend these rigorous and confirmative learning experiences. Examples of these newer digital methods are: (1) 2D screen-projected 3D anatomy (2D-VR), (2) augmented (AR—spatially semi-immersive), and (3) virtual reality (VR—spatially fully immersive) software applications. These instructional tools create remarkable and compelling visual scenarios that draw student learning-oriented attention and interest inspiring direct interaction and exploration leading to greater instructional success. To set the stage for understanding these effects and the literature summarized, Sect. 3.2 provides a description of the anatomical curriculum based on the first-year non-integrated curriculum within our experience. Additionally, Sect. 3.2 provides relevant definitions of learning, the range of visual instructional materials available, descriptions of the currently available advanced visualization methods and a basic description of the anatomical data formats required to design and produce these student experiences.

Finally, Sect. 3.2 includes reports from several articles describing the principles of stereopsis (as opposed to those of perspective) that support understanding of these new technologies and how their instructional effects might be better measured, assessed in terms of student curricular success.

Section 3.3 provides example papers from the literature comparing the instructional results of dissection with those of the virtual technologies described above on student outcomes. The cited literature gives insight into the application of advanced methods of visualization that both support and augment cadaveric dissection in unique ways.

Thus, it will be shown that these new techniques can achieve student empowerment by giving them the technologies that draw them perceptually into 3D worlds where the students actively manipulate complete anatomical configurations to achieve their assigned learning objectives. Thus, by direct manipulation of these real-time interactive learning environments promoting active learning modes, they can be drawn into their own focused instructional environments via their spatial perception enhanced by rotation and individual structural manipulation (fading, hiding). Additionally amplified by on demand user rotation, zooming, and annotations, they are able to interact with medical school-level gross anatomy that reinforces their dissection experience increasing the probability of academic success. Ultimately, when combined within appropriate curriculum design, these new and quickly evolving instructional technologies should represent the next step in the evolution of clinical anatomy instructional tools.

3.2 Background

3.2.1 The Importance of Representing Spatial Representation in Anatomical Instruction

As cadaveric dissection represents the student-based confirmation of didactic regional and systems knowledge with which students are presented in lecture materials, the question must be asked, "Why is the spatial exploration of gross human male and female bodies through cadaveric dissection important?"

Moore (1998) attempted to answer this question. He noted the challenges related to dissection in the medical school curriculum due to its expense and large laboratory space requirements, its requirement for highly trained faculty and the shifting of curricula to problem-based learning supported by computer-based methods. He made the point that anatomists strongly support the introduction of new methods that emphasize and complement dissection without replacing dissection. He felt that this traditional learning activity was central to anatomical education due to the fact that (1) each cadaver is physically real, providing direct volumetric sense and haptic feedback during this laborious, directed exploration. Thus, he felt this activity was instrumental for students to acquire a three-dimensional internal conceptual model of human form that supports the experience of understanding the body as a spatially integrated configuration of tissues with different functions and tissue strengths, and (2) dissection offers an appreciation of human variability and the logical and critical thinking that each student will apply to the broad array of patients he or she will treat.

3.2.2 Passive Versus Active Learning Approaches

The work of Bui et al. (2021) defined the two major types of learning: passive and active. *Passive learning* was considered to be a teacher-centered approach where the delivery of information is mostly one-way to students and the acquisition of knowledge by students often occurs without their making a conscious effort. On the other hand, *active learning* is a student-centered approach that not only involves listening and taking notes but also requires students to make conscious interactive efforts in the presented instruction to build internal models of the assigned information underlying retention and recall. It encourages students to ask questions and engage in discussions through interaction with presented learning models. Therefore, Bui and colleagues saw the advantages of active over passive learning as consisting of active thinking and applying problem-solving skills. Their work is expanded below.

3.2.3 Sequence of Learning Steps for any Given Assignment

From my years of experience in anatomy courses designed around dissection laboratory exercises, I have found that the exploration of anatomical form using cadavers and dissection ideally involves four steps.

3.2.3.1 Didactive or Conceptual Preparation Before Dissection

This first step usually involves text and visual lecture/laboratory instruction materials carefully designed within the directives of specific structured learning objectives. Supporting learning materials usually include large anatomical charts and systems tables, dissection overview videos, articulated skeletons, individual bones, physical models, and cross-sectional radiological image sequences. In this step, the attempt is made to endow each student with an internal cognitive model preceding dissection that represents the basic form and function assigned (e.g., heart within the thorax, mediastinum, and pericardium with its superior mediastinal vessels). This prepares the students for the anticipation of the form that he/she should be expecting to see during the dissection experience. This cognitive model is intended to serve as a guide to physical confirmation, feature identification, and correspondence to clinical correlations that illustrate its anatomy and allow the student to remember the visceral complexities and its structural naming. Ultimately this initial exercise sets the stage for the following step–the corresponding confirmation of this internal model by guided physical exploration and inspection. During this phase, these didactic textbooks focus on a list of anatomical feature terms, usually emphasized in dissection manuals—both emphasized by boldface fonts. These terms, which we call the Anatomy Study List (or ASL) represent key features that are fair game for written examination and practical examinations.

3.2.3.2 Dissection of the Assigned Cadaver

This second step involves the estimation of surface anatomy to be removed; the location of the assigned viscera within its natural context, boundaries, and landmarks; identification of specific surfaces and gross features; and direct cutting, separation, and cleaning of surfaces and internal organ chambers and structures. The assigned dissector guidebook provides detailed instructions for these events, again, guided by each term in the ASL. Ideally, each student uses both the simple diagrams in the dissector and the more detailed and compelling paintings in the assigned atlas as problem-solving tools. It is hoped that this exercise confirms the initial cognitive model laid out didactically and assists understanding and memory with the shape characteristics acquired in a haptic sense (or related to the sense of touch sensation and perception)—a feature only available in cadaveric dissection (see Fig. 3.1).

3.2.3.3 Experiencing Anatomical Variation

Most dissection laboratories contain multiple cadaveric specimens, each within their own tank. These specimens include both male and female bodies, many with some form of pathology. We run a lab with 24 specimens—50% male and 50% female. Thus, each cadaver is somewhat different from the others in terms of stature, external features, and in situ internal organ systems—yet sufficiently alike in their anatomy to be supported by all of the assigned instructional materials. If successful, this feature of the laboratory experience introduces each student to the idea of human variability, which clinicians view as a core factor in the successful practice of medicine.

3.2.3.4 Laboratory Assignment Review of Prosected Cadavers

In addition to the above activities, we have found great value in the preparation of four completely dissected prosected cadavers (two males and two females) and in encouraging student access to them–particularly before a given dissection session but certainly during and afterward. These prosected cadavers are produced by our prosection graduate course held each Spring. This course was designed to not only produce the detailed dissections of all regions by our advanced first-year students but also digital images, videos, and instructional PowerPoints as a requirement for learning to think as an instructor during dissection and capture the sequences as completed. Additionally, these prosections are used to prepare students for practical examination review via mock examinations (see Fig. 3.2).

3.2.4 The Range of Visualizations Applied in First-Year Anatomy Courses

The ultimate academic intention of these four steps of instruction for students is to build, confirm, and strengthen the resulting internal cognitive anatomical model so as to allow successful participation in the clinical courses of physical diagnosis treatment. This should provide a firm foundation for the clinical rotations that follow in the third and fourth year.

The required anatomy activities and requirements of medical students as described above are designed to lay out a systematic consideration of student experiences that form the design requirements for anatomy instructional materials. This includes visual and illustrative exhibits involving direct exposure, perception, manipulation, and free rotation of viscera as complex spatial objects within the bodily context and after extraction. This is usually achieved by lifting, rotating, cutting into visceral chambers and parenchyma and examining all ASL features with the goal of simultaneously understanding their basic form and function.

These course/lesson design requirements may be extended to advanced technologies such as VR and AR that may reproduce many of the same capabilities that are meant to be an adjunct but not to replace the cadaveric component in the same manner that other visual exhibits do. It should also be emphasized that the VR and AR

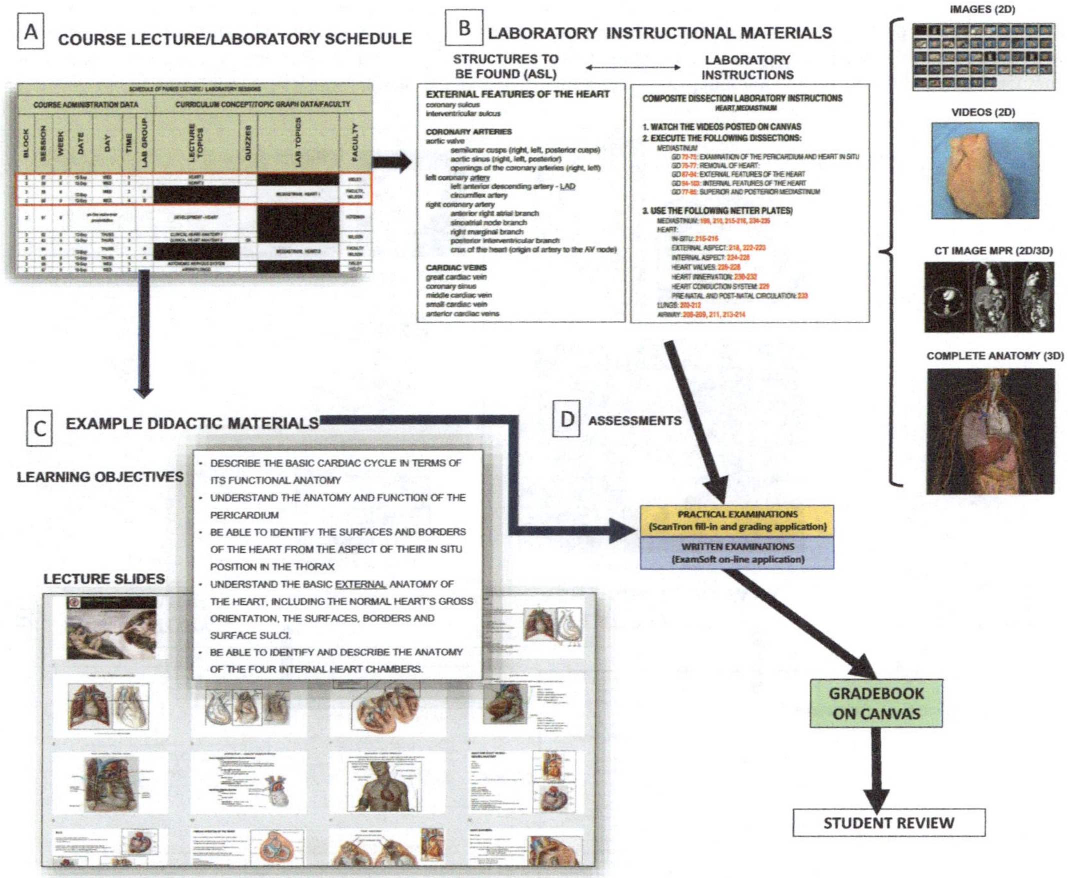

Fig. 3.1 The organized activities for didactic and practical of the heart using the ASL as the organizing core. This organization is a network of text, visual and tabular materials that support visual exhibits of many types. (**a**) course schedule assignments, (**b**) laboratory instructional materials, (**c**) didactic instructional materials and (**d**) practical and written assessments and their grade captures. From the integrative work of K. Hisley, 2023

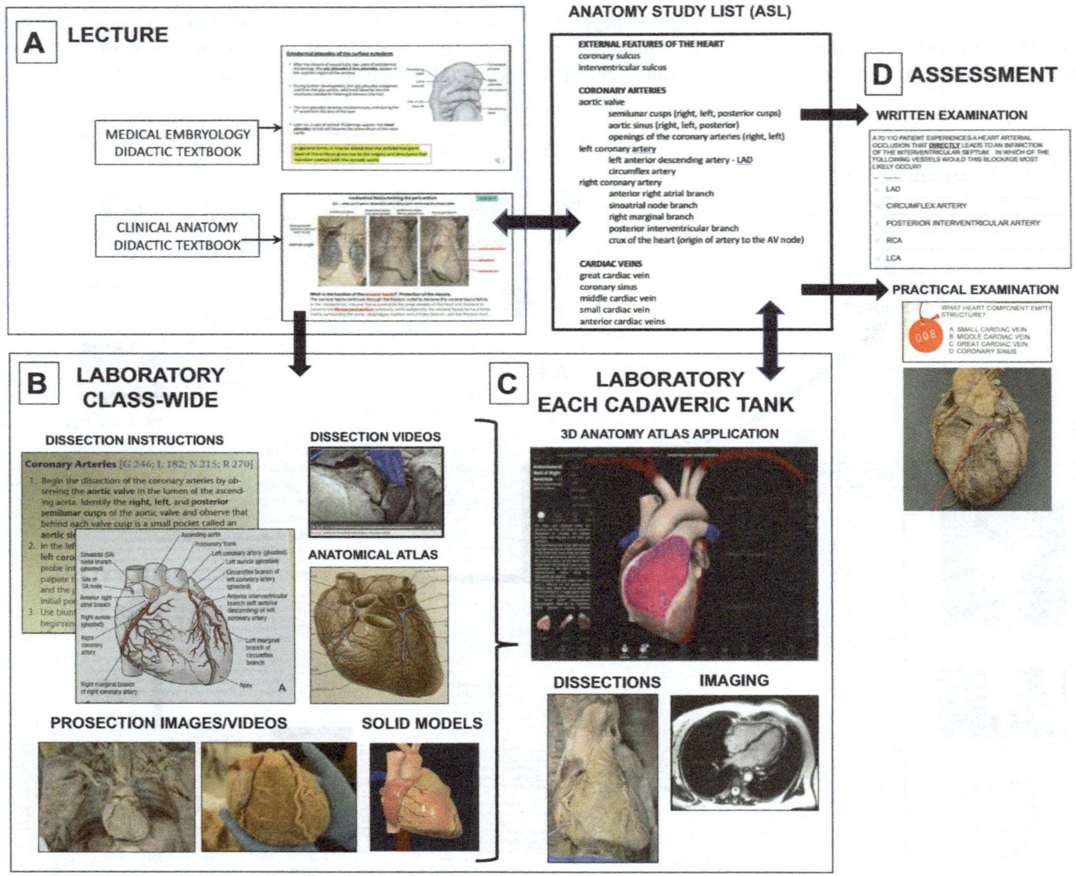

Fig. 3.2 This graphic is an extension of the information network shown in Fig. 3.1 emphasizing the locations of visual materials in the instructional stream of information—identifying those points where augmented and virtual reality applications could amplify the lessons, again, *synchronized with the Anatomy Study List (ASL) for any given lesson*. (**a**) represents didactic figures used to illustrate the required concepts as defined by the learning objectives; (**b**) the laboratory instructional materials for the entire class; (**c**) laboratory resources supporting each student tank group's cadaveric dissection exploration: 2DVR model exploration, sequential dissection observations, and corresponding radiology exhibits; (**d**) written and practical examination specimens tagged with corresponding ASL structural name and distractor entries. From the integrated work of K. Hisley, 2023

experiences allow the student to more easily perform spatial manipulations from any viewpoint and view associated ASL annotations quickly and easily, both inside and outside of the laboratory–functions not easily performed with a cadaveric specimen. Additionally, these learning environments appear capable of generating compelling interest and exploratory focus when used with all other faculty-supplied resources.

Ultimately, we wish to define and describe these more advanced digital applications that seek to extend the student's perception of this discipline as it relates to the didactic and practical activities required of students studying basic and clinical anatomy at the first-year level.

This includes the instructional assignments and materials in a two-semester first-year anatomy course sequence with a foundation of rigorous whole-body cadaveric dissection. It seeks to describe the evolution of current printed and physical instructional media we use including printed 2D diagrams and paintings found in the assigned anatomical didactic textbooks, the dissectors used to guide invasive exploration of the entire human body, and the supporting indexed sets of full-color paintings in atlases. All of these instructional materials are applied to each body region and their contents as laid out in the course schedule (see Fig. 3.1).

I believe that the successful study of human anatomy requires the simultaneous perception of forms and their related functions that may be understood through visual means. Thus, although the discipline of anatomy—both basic and clinical—rests on ordered sequences of adjacent *regions* (localized functional collections of viscera, spaces, and protective enclosures that perform specific functions) and *systems* (those transport and signaling entities that support and synchronize each region as a working integrated entity), successful learning requires a basic understanding of each regional or system-based structure's function obtained didactically to truly understand cause and effect relationships in the human individual, especially as it applies to the successful application of the clinical arts and sciences.

It makes sense that most in-depth learning requires *visualization* (the formation of internal visual mental cognitive models from perceived displays) and *illustration* (visualizations + text *annotations + sometimes animation*) to visually understand the dynamic structure and function of the subjects of study defined by the learning objectives. Real-time user interaction allows direct manipulation and interrogation of these perceived models and appears to aid in active learning and recall.

At the current time, visualization and illustration are applied in various information formats covering the above operating whole with *specifics* (text), *ordered collections* of related anatomical structures (tables), and *related visuals* (pictures, images, diagrams) that, together with structural text annotation, try to provide a conceptual, learnable cognitive model to the student that allows reasonable understanding and success in basic and clinical science assessments, including Boards. Additionally, for some years, each topical didactic lesson and its corresponding laboratory sessions in most schools have been posted for students on Learning Management Systems (LMS) where all instructional materials are posted in digital format, including lecture slide collections, text documents, images, and videos. These online systems are critical in the current medical school curriculum for the distribution of learning materials, including video segments, currently with the exception of VR and AR experiences. Examples are shown using the heart as an example in Fig. 3.3.

As commonly understood, these instructional methods have recently been extended with 3D anatomical application designs based on technologies supporting digital gaming applications. Some of these applications are remarkable 3D anatomical atlases that allow students to view extremely realistic human anatomy (regions and/or systems) as projected on 2D screens with intuitive man-machine interfaces requiring only a reasonable computer/tablet and some practice to become proficient. Additionally, some of these applications are so inexpensive that each student can buy their own license, or the school can purchase a site license that supports

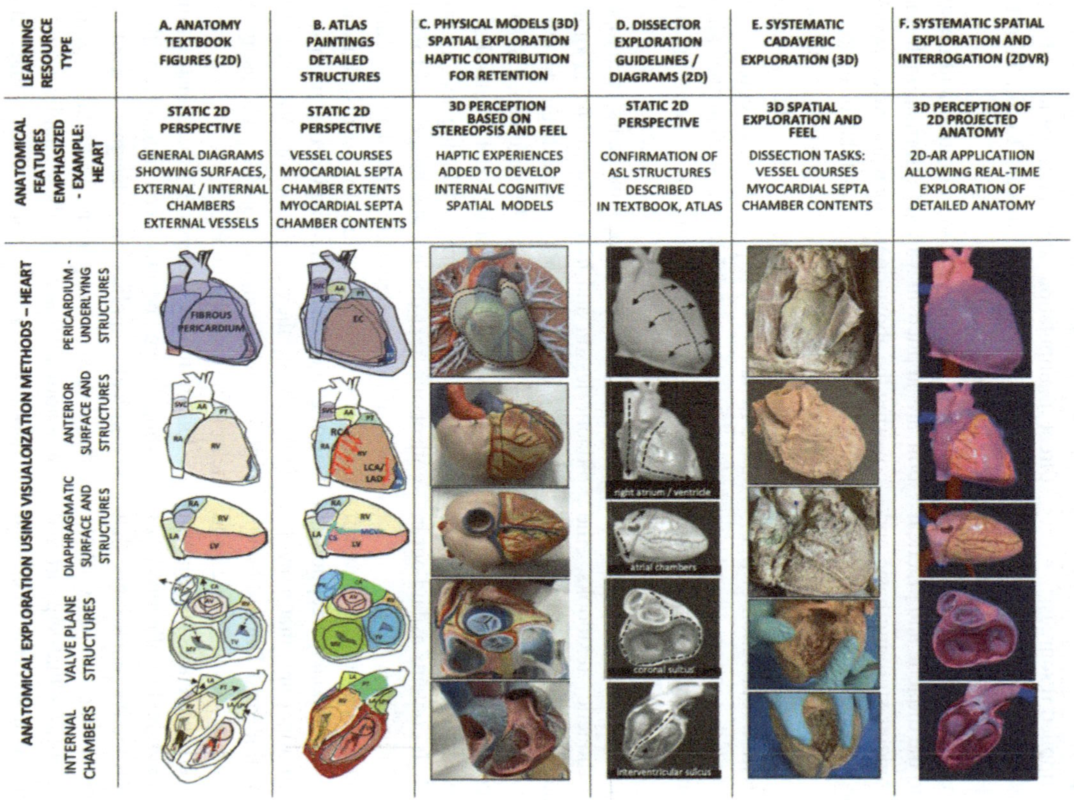

Fig. 3.3 This exhibit shows a logical sequence of traditional figures, diagrams, paintings, physical models, and dissection videos/prosections/actual dissection learning activities widely used in the current medical school anatomical curricula. The internal heart is used as an example. Note the progression from figure and diagrammatic perspective representations, to painted perspective plates, physical models, actual prosected specimens to finally 2DVR screen-projected digital models generated and interrogated by the student. Note that there are two rows in this figure: the top row describes the type of knowledge representation (**a**–**f**) and the bottom row describes the basic anatomical features represented in each (**a**–**f**). From the integrated work of K. Hisley, 2023

its general addition to the curriculum for students *and* faculty. Three-dimensional computer graphics are computer-based mathematical data structures that use a representation of geometric data within a standard defined space (often Cartesian) that is stored in the computer for the purposes of performing calculations and rendering digital images.

The objects in 3D computer graphics are often referred to as 3D models. Unlike the rendered image, a model's data is contained within a graphical data file. A 3D model is a mathematical representation of any three-dimensional object; a model is not technically a graphic until it is displayed.

A model can be displayed visually as a two-dimensional image through a process called 3D rendering, or it can be used in non-graphical computer simulations and calculations. With 3D printing, models are rendered into an actual 3D physical representation of themselves, with some limitations as to how accurately the physical model can match the virtual model.

Important to the application and design of future anatomical instructional applications is some sense of what is actually going on in a given dissection laboratory session given the stresses of these modern times with decreased teaching time windows and many more students. Recognizing the heterogeneity of the learning experiences going simultaneously in dissection sessions must be recognized and used to design effective instructional interventions with the new technologies.

In this light, Winkelmann et al. (2007) investigated student activities during dissection laboratories. I found that this paper had important aspects for planning and supporting future dissection laboratory activities given decreased time and higher student loads. They grouped and gauged the time percentages in three classes of activities: (1) direct dissection (33%), (2) prosection exploration (27%), and (3) and non-cadaveric learning (31%). They also noted the wide variability in direct dissection. In our school, we have observed similar results. The question to be asked here is if individual students were able to experience compelling 3D and 3D VR and AR anatomical visualizations supporting conceptual illustrations (immersive or semi-immersive), would the overall time percentages for direct dissection and prosection examination significantly increase? Their key observation was that there was not a single anatomy course dissection laboratory, but a series of separate pursuits based on individual student capability and willingness to apply the guidance that they receive. My point here is that if the newer visualization technologies can provide enough compelling experiences in a direct manner, they may be able to pull students with disparate learning abilities and desires into more of a homogeneous group. This will certainly be important for the design and implementation of these new methods in the future.

3.2.5 The Classes of Anatomy Applications Using Advanced Technologies

3.2.5.1 Extended Reality

Extended Reality (XR) is an umbrella term encapsulating 2D-VR, Augmented Reality (AR), Mixed Reality (MR), and Virtual Reality (VR). AR and VR offer a wide range of revolutionary instructional experiences in real-time anatomical spatial exploration and click-based information interrogation.

With the advent of head-mounted displays and control methods, the 3D atlas concept has been extended to perceptual immersion and semi-immersion. More advanced visualization methods include instructional technologies that have now been extended to fully immersive Virtual Reality (VR) and semi-immersive Augmented Reality (AR). For either, the student must wear a headset—different for each type of technology.

Taylor and Soneji (2022) summarized the VR, AR, and MR visualization modalities. VR is a fully artificial environment while AR technologies overlay artificial objects on the real world, but users can only interact with the projected objects themselves. In MR, the participant can interact with both the artificially projected world and the real world upon which it is projected. In fact, real and projected objects can affect each other, such as in terms of lighting, depth, and interrogation (see Fig. 3.4).

3.2.5.2 Projected 3D Models on 2D Screens (2D-VR)

The first of these visualization modalities is non-immersive and is referred to as 2D-VR application. These highly detailed anatomical models comprise an extensive database of polymesh surface models, supported by on demand anatomical information about each structure. These models allow extensive spatial manipulation including full rotation, panning, and zooming. Additionally, the user can search for structures individually and "build" the desired model. Finally, each structure may fade to semi-transparency or hidden completely. These capabilities allow the user to investigate complex anatomical structures, such as the mediastinum/heart in great detail from many different angles. Because these models are based on polymesh mathematical surfaces comprised of dense networks of vertices and connectors, these models respond immediately to digital programming resulting in efficient spatial manipulation on the screen. This is true of the other application classes described below. Finally, the licensing costs for both faculty and students are low. This 2DVR visualization method uses non-immersive experiences. This class of application, an example of which is COMPLETE ANATOMY (Elsevier) does not allow true stereopsis to simulate object three-dimensionality but simulates this perception through the principles of perspective as displayed on a flat screen. This type of 2DVR application is mouse-driven and the range of all body regions and systems may be activated at different levels of visibility. Additionally, the user may interrogate anatomical structures directly by clicking on them providing in-depth information sets describing the selected features.

3.2.5.3 Virtual Reality (VR)

VR is the creation of an artificial environment that is experienced through sensory stimuli (such as sights and sounds) provided by a computer and in which one's actions partially determine what happens in the environment. This generated artificial environment completely immerses or surrounds the individual within a scenario where all experiences he or she perceives and controls with handheld game controllers arise from the projections, and the outside world does not exist—to the extent that the individual reacts automatically to the stimuli within the scenario. Virtual reality (VR) is the first alternate reality that came on the mass market and can be experienced by wearing a VR headset that is directly tied spatially to handheld controllers similar to those used in most video games. Wearing VR glasses is like having a screen in front of your eyes. Physically you cannot see anything else other than what is presented by the headset, which means that you are completely immersed in the given digital environment. One example of VR applied to anatomical learning is Oculus Quest 2 running an application running the SHARECARE YOU web-based anatomy application service. This application uses stereopsis to simulate a full sense of object/environmental three-dimensionality. Active learning is supported by on demand direct control usage and model manipulation by a handcontroller-based beam supported by a simple but effective user interface. The anatomical structures, functions, and pathologies are visually compelling which may aid interest, absorption of information into knowledge, and the retention of that knowledge.

VR and AR "fool" the mind into thinking that it is completely or partially within a different real environment where the individual contributes to this artificial reality with his/her head motions and direct interaction with the perceived objects using handheld game controllers (VR) and specific hand gestures (AR). Describing this technically, Parisi (2016) in his textbook describes VR as the ability to be fully immersed in experiences and to feel like you are really there. He described VR as a collection of technologies: 3D stereoscopic head-mounted display (HMD), motion tracking hardware, input devices such as handheld game controllers, software applications frameworks, and developer tools. The Oculus HMD creates the illusion of depth by generating separate

	Virtual reality (VR)	Augmented reality (AR)	Mixed reality (MR)
What is it	Fully artificial environment 	Virtual objects overlaid on a real world environment 	Virtual environment combined with real world 
Experience	Full immersion in a virtual environment 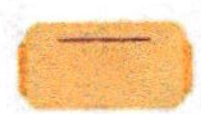	The real world enhanced with digital objects 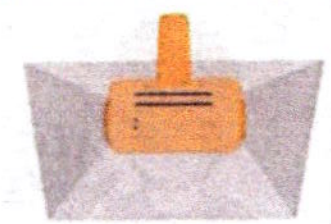	Interact with both the real world and the virtual environment 
Models	HTC Vive, Valve Index, Meta Quest 2, Meta Rift, PiMax, HP Reverb, Varjo, Pico Neo 2	Mobile phones, Snapchat glasses	(a) Projection based: Hololens 2, NReal (b) Camera Based Varjo, Quest 2
Pros	Most mature technology, least expensive Wide field of view to increase immersion	User is less isolated from environment; useful for training by overlaying instructions on the display; user has existing hardware	User can see the environment and can engage with other items e.g., mouse, keyboard and colleagues (b) has wide field of view
Cons	Difficult to interact with people or objects in the same room; some versions have complicated tracking set- up; may get VR motion sickness; may need connection to a computer with high powered GPU	Limited interaction options; limited field of view	Objects look slightly see- through in the display in (a) Has narrow field of view (52 degrees) On cheaper solutions of (b) pass through quality is poor (e.g., Quest) (b) can exhibit artefacts due to the real time processing

Fig. 3.4 Summary of semi-immersive and immersive interactive visualization technologies (XR) and their advantages and disadvantages. From Taylor S, Soneji S (2022) Bioinformatics and the Metaverse: Are We Ready? Front Bioinform. Open Access—this article is licensed under a Creative Commons Attribution 4.0 International License, which permits use, sharing, adaptation, distribution, and reproduction in any medium or format, as long as you give appropriate credit to the original author(s) and the source, provide a link to the Creative Commons license (https://creativecommons.org/licenses/)

images for each eye—each slightly offset from the other. This simulates "parallax"—the visual phenomenon where the brain perceives depth based on the apparent position of objects in the perceived visual field. Rapid head motion mimics how we see our surroundings, tracking head position with the projected 3D scene in real time using "inertial measurement functions" within the HMD. In this way, the perceived objects and changes in their perceived positions constantly synchronize with head motion adding to this perceived "reality." Handheld controllers common in video gaming equipment and software provide direct real-time interaction with application operational menus on demand and the real-time ability to directly interact with scenario objects, allowing selection, rotation, sectioning, and annotation. These AR/VR experiences and the equipment/software that supports them are complex undertakings requiring expert levels of data acquisition, digital object modeling, scenario construction, and interaction capabilities. There are specific software engineering packages that support these creative endeavors.

3.2.5.4 Augmented Reality (AR)

AR is an interactive experience that combines real world and computer-generated content. In this technology, the user relies on specifically learned pre-defined standard hand gestures to manipulate the environments that are sensed and interpreted by the headset. The content can span multiple sensory modalities, including visual, auditory, haptic, somatosensory, and olfactory. AR can be defined as a system that incorporates three basic features: a combination of real and virtual worlds, real-time interaction, and accurate 3D registration of virtual and real objects. The primary value of augmented reality is the manner in which components of the digital world blend into a person's perception of the real world, not as a simple display of data, but through the integration of immersive sensations, which are perceived as natural parts of an environment. One commercial example of this is example is Hololens 2 running the HoloHuman application (GigXR). This application uses stereopsis to simulate a full sense of object/environmental three-dimensionality and a standard set of learned hand/finger positions.

3.2.5.5 Mixed Reality/Extended Reality (MR)

Mixed reality (MR or XR) combines the best aspects of both VR and AR. As the name implies, mixed reality combines virtual content with the real world in an interactive, immersive way. In Mixed reality, virtual objects appear as a natural part of the real world, integrating with real objects. This is similar to augmented reality as it will not remove you from your viewable surroundings, but rather enhance your surroundings with potentially interactive digital objects added to these surroundings.

Real objects can also influence the shadows and lights of virtual objects. This natural interaction between real and virtual perceptions potentially creates new methods within this artificial space for interaction with other individuals and objects—especially in collaborative activities more so than with standard virtual or augmented reality methods. Mixed Reality gives the ability to see yourself and interact with your colleagues while (for example) designing a virtual object or environment. For this modality to be valuable to professionals, it has to be convincing–blending real and virtual content to the point that it is impossible to tell where reality ends and the virtual world begins (see Fig. 3.5).

[Note: Given the above information and the relative paucity of literature on MR applications in the anatomical curriculum, this work will emphasize the relevant literature covering the application of AR and VR technologies.]

3.2.6 Understanding Two Basic Types of Visual Data Object Formats Used in All Classes of Digital 3D Modeling and Presentation

The understanding of the formats of digital anatomical structural datasets, where they originate and how they may be used to illustrate complex anatomical configurations is key to obtaining, processing, and implementing

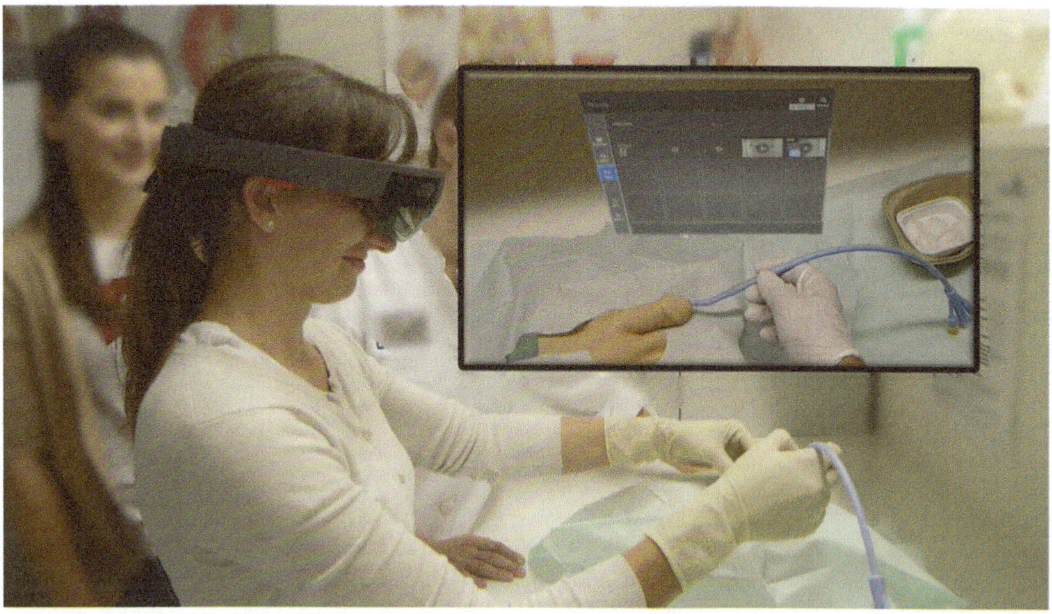

Fig. 3.5 Example of a mixed reality (MR) semi-immersive experience allowing teaching through simulation of catheter placement to medical students. In mixed reality scenarios, there is the capability for one or more participating users to collaboratively act on the artificial world within an environment, implementing complex, cause-effect reactions from participants to simulated objects, changing characteristics of those objects, then supplying feedback back to the user. From Schoeb et al. (2020). Open Access: This article is licensed under a Creative Commons Attribution 4.0 International License (https://creativecommons.org/licenses/)

anatomical models in visualization applications. The objects and contexts required in VR and AR applications to represent visual and illustrative anatomical information sufficient to teach and learn at the professional school level may be *volume projection models* or *polymesh models*. Software applications supporting volume projection model generation are a relatively common function in radiological workstations, especially in the area of 3D CT cerebral angiography. This representation is three-dimensional only in the sense of its representation of object surfaces at a given angle, and each rotation requires a re-projection of the entire scene at each angle of rotation. Figure 3.6 shows the alternate projections of several mediastinal structures. Each discrete structure is segmented separately, and a separate model of a contiguous set of pixel values in 3D space is formed. Each separate model may be made to appear or not. They may, however, be annotated in graphics after printing. These objects, if sectioned, will show no internal detail–only the exterior surfaces are rendered as colored surfaces. See Mori et al. (2019).

On the other hand, polymesh surface projections require detailed segmentation of object edges through a sequential space of cross-sectional images. These results provide the computed definition of solid object boundaries from which the generation of mathematical datasets representing tessellated polymesh surface models may be generated. These models may be quickly and easily rendered on the screen, freely rotated, sectioned, and annotated by the user on demand in real time allowing manipulation of the rendered model in real time, thus amplifying the active learning requirement of interactivity. This class of anatomical models is used in 2DVR applications, such as Complete Anatomy (Elsevier) and Anatomy.tv (Primal

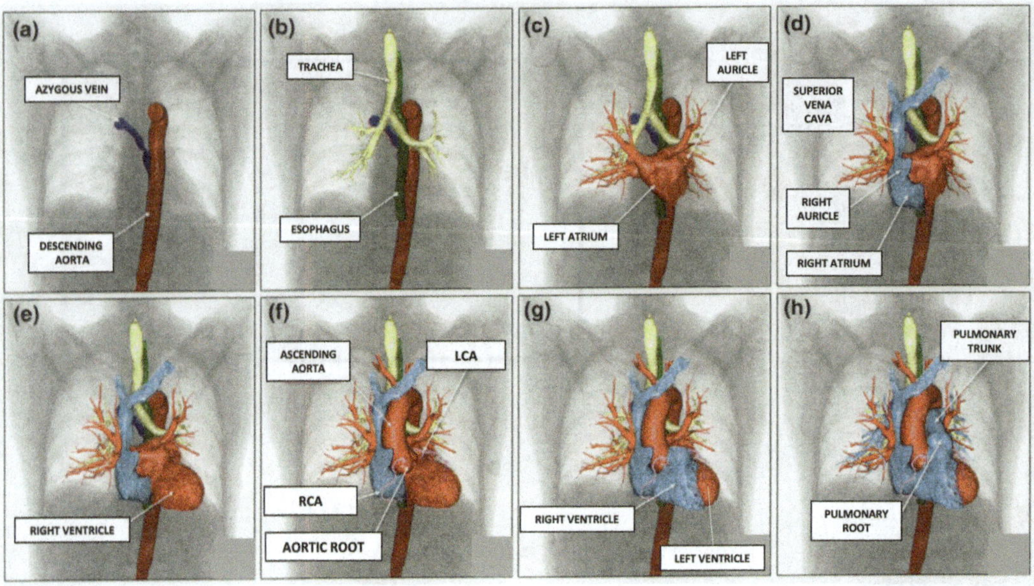

Fig. 3.6 The stepwise virtual dissection shows the contents of the middle and superior mediastinum as viewed from the frontal direction. By tracking from the left upper (a) to the right lower panels (h), the components constructing and supporting the cardiac contour are demonstrated in order from the posterior to the anterior direction. Anatomical structures are contiguous sets of pixels of the same CT intensity and must be re-rendered at different angles to show different viewpoints. Different structures may be made invisible by not including those pixel sets in the re-rendering. From Mori et al. (2019). Open Access: This article is licensed under a Creative Commons Attribution 4.0 International License (https://creativecommons.org/licenses/)

Pictures) providing excellent and detailed annotated 3D anatomy surface models that may be easily used by students and faculty. Ultimately, this method for the capture of 3D spatial structures allows the real-time manipulation of complex structures providing smooth perception of rotating curvilinear surfaces in the same manner that a student would view a static physical model (see Fig. 3.7).

Zilverschoon et al. (2017) describe the technical applications and detailed process of generating polymesh model datasets of the hand which run on their own application in this fashion. This work demonstrates the level of effort in generating such models, especially those that represent anatomical detail at the professional student level.

Accordingly, it is this type of polymesh surface-based anatomical modeling that forms the backbone of commercially available VR and AR software applications as shown in Fig. 8 (columns F, G and H).

3.2.7 The Importance of Stereopsis to Semi-Immersive and Immersive Visualization Methods

The following two articles laid a good technical foundation for understanding the effects of stereopsis in the new semi-immersive and immersive visualization technologies as applied to anatomy education, as opposed to the principles of perspective. Since binocular depth perception is critical to identifying detailed structures and their interrelationships, the role of stereopsis should be a key factor. Wainmen et al. (2020) were interested in understanding the role of stereopsis defined as 3D depth perception, occurring when your brain combines the two images received

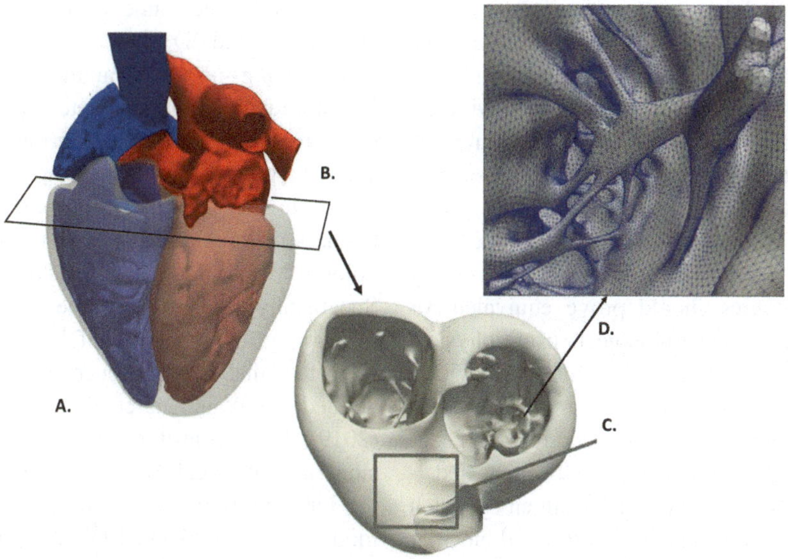

Fig. 3.7 This figure shows the steps to generate surface polymesh models of solid pixel-based volume segmentation. Polymesh surfaces are mathematical descriptions of <x,y,z> coordinate vertices and their interconnectivity using arcs. Based on the resolution specified before mesh generation—the higher number of vertices and arcs extracted from the pixel volume configuration, the more similar in detail is the surface representation. The reason for generating detailed polymesh surfaces is that being mathematical constructs projected by computer graphics algorithms, they may be manipulated in space in real time by the user with mouse or hand controller keys. This process makes cause-effect human exploration of complex curvilinear object configurations possible (in this case anatomical regions and viscera) with minimum delay and gaps in cognitive apprehension. On the left (**a**), an example of heart geometry, elaborated from the solid pixel volume-rendered heart model of the solid myocardium and major vessel walls and their enclosed spaces, (**b**) the plane around the coronary sulcus showing the view after rotating to display the model from above, (**c**) demonstrates the surfaces generated from the solid forms in the form of tessellated polymesh surfaces of vertices and arc interconnections, and (**d**) the internal view of the right ventricle demonstrating how these detailed surface meshes are rendered into surfaces via computer graphics algorithms. From Fedele and Quarteroni (2021). Open Access: his article is licensed under a Creative Commons Attribution 4.0 International License (https://creativecommons.org/licenses/)

from each eye and creates one single 3D image. Also, Eroline 2019 reported from the literature on the findings of the use of stereopsis in anatomy education (see Sect. 3.3). These separate research efforts recognized that there is an appeal to current visualization technologies that support dual eye projections simulating stereopsis, including fully rotatable 3D anatomical models projected on screens (2D VR) as well as fully semi-immersive and fully immersive AR and VR presentations over prosections, illustrations, photographs, and physical models. They asked if these newer visualization technologies actually enhanced learning of the bony pelvis as an example. They noted that in educational research studies at that time, learning names and locations of anatomical structures using these screen-based 2D VR applications had no advantage over flat "key view" representations of structures (i.e., top, bottom, front, back, and sides).

Additionally, these studies reflected disadvantages for learners with poor spatial ability. In contrast, they found literature that reported the superiority of physical models for learning structural names over all other materials including key view images, lectures, 2D presentations, and computer-based programs. These studies have shown that advantages of the physical models (PM) may be based on stereopsis. Wainman showed the significance of the phenomenon of stereopsis by doing comparative tests during learning tasks on PM and 2D VR presentations of the female pelvis, using both eyes on each subject, and then covering one eye with an eyepatch (creating monocular vision) and repeating the test. Their main hypothesis was, assuming that VR and MR (AR applied within a closed artificial environment) simulations included true stereoscopic projections, testing students using these technologies should prove equivalent or exceed the test results of subjects using physical models for the same tasks. A group of key views of the pelvis presented as a sequence of PowerPoint slides served as a control. Four student groups of 20 participants each were tested—key views, MR, VR, PM—half with an eyepatch (monocular vision) and half without (binocular vision capable of stereopsis).

Their results showed that the MR and VR learning modalities were significantly inferior to the physical model. The VR modality was superior to MR, and MR was not superior to learning from the key views. They suggested that the reason why the PM outperformed its digital counterparts is that MR and VR technologies cause a vergence-accommodation conflict that inhibits performance by creating a performance-inhibiting fatigue. Interestingly enough, this work showed that the removal of stereopsis by covering the dominant eye significantly decreased performance on the PM but not the 2D VR presentation. This demonstrated that stereopsis is critical to performance in anatomical learning and that stereopsis is not present when perceiving 2D VR images. Finally, the removal of stereopsis significantly decreased performance on the PM and VR presentations, but not the AR modality. This would confirm that stereopsis assists in PM and VR perception and their instructional capabilities and that the AR headset image projection capabilities are more like pictures on a page and may not incorporate true stereoscopic effect, limiting its instructional value in the exploration of complex spatial objects. Finally, this work provides evidence supporting the pivotal role of stereoscopic vision in efficient anatomy learning and that the current status of XR/AR technologies may need improvement before they may be functional in the same anatomical instruction class as VR and 2D VR.

Following up on Wainmen et al. (2020), Bogomolova (2020) completed a meta-analysis of stereoscopic 3D visualization technologies applied to anatomy learning. There were identifying features that contribute to the effectiveness of three-dimensional visualization technology (3DVT) in teaching anatomy. Specifically, they reviewed the role of stereopsis in learning anatomy with 3DVT. Their findings indicated that the presence of stereopsis, as defined by Wainmen et al. (2020) and Erolin (2019) above, contributes to a better comprehension of anatomical knowledge especially when the binocularly-perceived 3D digital anatomical models are accompanied by active manipulation and dynamic exploration–much like navigating through a neighborhood and becoming familiar with discrete structural landmarks which you can then re-navigate with less difficulty. The benefit of these two interlocking capabilities (stereopsis and interactivity) should provide effective anatomical learning to most students, particularly those with low visual-spatial abilities (VSA). These observations support the hypothesis that the stereoscopic view contains spatial information (not found in a corresponding monoscopic view) that assists in generating an internal cognitive 3D mental representation of an anatomical scene. Students with low VSA may attempt to generate this internal model when they monoscopically memorize a series of key views of a complex configuration leading to cognitive overload and poorer learning. Thus, in good 3DVT presentations with user-centered dynamic interactivity, the binocular depth cues are automatically generated and may be perceived with less difficulty. Therefore, active manipulation, including dynamic exploration of the perceived model through rotation, zooming in and out, and panning provide a rich basis for constructing and understanding the internal model—especially when corresponding annotations are present.

Ultimately, this work found that stereopsis is an important element of 3DVT that has a significant positive effect on the acquisition of anatomical knowledge when implemented within an interactive framework. In terms of anatomical instruction as supported by this meta-analysis, stereoscopic 3DVT contributes to a better comprehension of anatomy and is preferred over monoscopic 3DVT, such as screen projections of even excellent 3D models as each eye receives the same view. This is especially true when user-centered dynamic interactive capabilities are provided.

In Fig. 3.8 is a summary grid of the traditional visualization methods (to the left of the vertical dotted line), what I feel should be considered *transitional methods* (within the dotted lines) and the more advanced modalities that still await full exposure within anatomy curricula (to the right of the dotted line). I mention the middle 2D-VR applications as transitional as they are highly affordable and feasible to integrate into current anatomical coursework—certainly supporting the first year of medical school. The more advanced technologies are impressive and compelling but, in my opinion, still await general integration and assessment as to their teaching design/delivery and learning effectiveness by both faculty and students. Note that the methods of depth perception underlying each are shown.

3.3 Example Literature Relating to the Application of AR and VR to Anatomical Instruction

Erolin (2019) provides an excellent overview exploring four aspects of the application of 3D digital models for anatomy and medical education: (1) the history and development of virtual 3D anatomical resources; (2) an overview of some of the current methods of producing 3D models suitable for instruction; (3) methods of distribution and interactivity characteristics of these models within the curriculum, including virtual learning environments, websites, interactive PDFs, standalone AR and VR applications, and 3D physical printing; and (4) specific methods of instructional application of these resources. The questions she addressed included how they may be used to enhance student learning, application in the classroom, and self-directed study and how suitable they may be for replacing cadaveric dissection in practical learning sessions (i.e., laboratory dissection).

As in other references, she describes anatomy as an inherently spatial subject whose core activity is learning the 3D interrelationships of structures. Importantly she emphasizes that student achievement in certain medical disciplines may be directly related to the individual's spatial ability. Importantly, she reviews references that describe the instructional importance of direct, active student interaction with 3D models vs. passive presentation viewing (i.e., kinesthetic and visual learning as opposed to visual presentation alone).

From antiquity, artists, from the battlefield dead and the deceased, produced 2D sketches and drawings of gross normal and distorted anatomy using the principles of perspective to represent depth–both for classical reasons of beauty and the clinical interest of clinicians. Bones and skeletons were also produced by maceration to provide 3D information on osteology for the detailed examination, identification, and naming of specific features. Beginning in the Renaissance, this osteology and artwork were then used to produce fine, life-size wax models using wax-based methods representing both regional viscera and systems.

Students using stereoscopic tools relying on the principle of stereopsis experienced semi-immersive 3D depth perception from the paired anatomical scenes captured from slightly different angles as if viewed in life. These tools appeared in the mid-1800s with the creation of stereoscopes which could demonstrate anatomical dissections visualizing organs, structures, and medical conditions within their surrounding context. See Wainmen et al. (2020) and Bogomolova (2020). Cards containing parallel photographic images from slightly different left and right views with annotations and instructional information in the area above them were inserted into wireframes and then viewed through paired lenses through an enclosure that blocked peripheral vision and, therefore, forced the mind to focus on the presented scene.

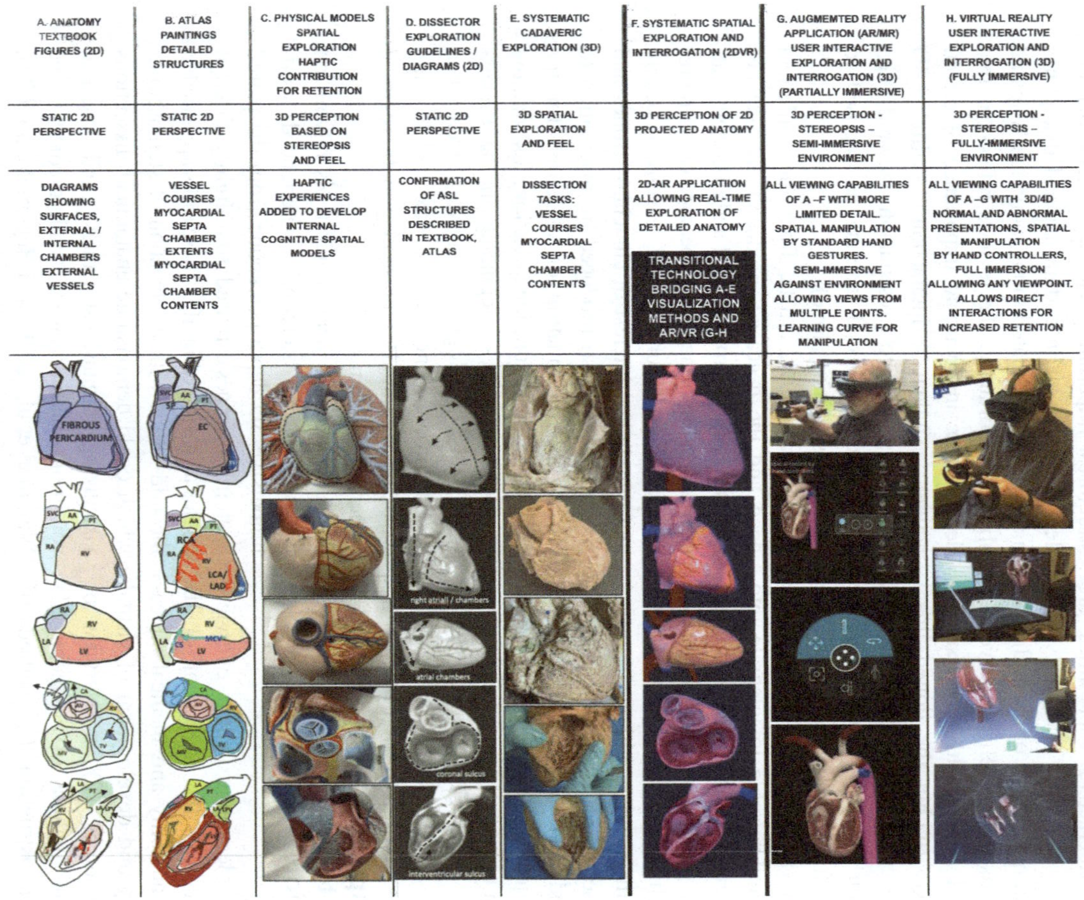

Fig. 3.8 Using the same format as that used in Fig. 3.3, this exhibit shows the extension of the logical sequence of traditional figures, diagrams, paintings, physical models, and dissection videos/prosections/actual dissection learning activities in Fig. 3.3 extended to AR (partially immersive) and VR (fully immersive) applications. The internal heart is used as an example. Note the progression from figure and diagrammatic perspective representations, to painted perspective plates, to physical models to actual prosected specimens, and finally to the advanced visualization technologies allowing stereopsis and real-time manipulation and structural information interrogation. Note that, as experienced in our work, the 2DVR application proves to be an inexpensive "transitional application" allowing the fast design and assessment of different anatomical models applied to a specific instructional task. These initial learning models may then be used to design AR and/or VR learning experiences more efficiently. From the integrated work of K. Hisley, 2023

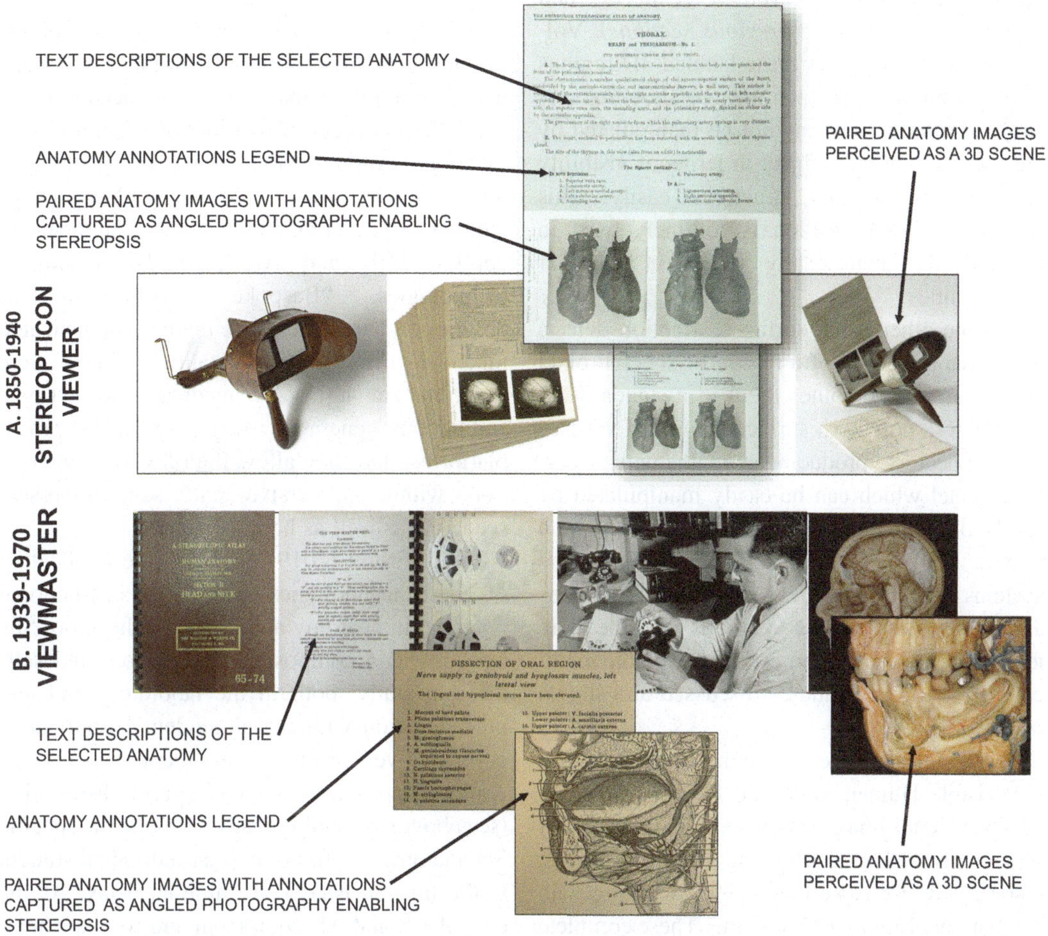

Fig. 3.9 Examples of early 3D anatomical spatial exploration tools. (**a**) Stereoscope—Stereoscopic Studies, Edinburg Anatomy; (**b**) Viewmaster—A Stereoscopic Atlas of Human Anatomy, The Thorax, Williams, and Wilkins. Viewing in three dimensions in both of these viewing tools uses the phenomenon of stereopsis. From Hisley 2023, and the Smithsonian Institute images found using Google search. Open Access: this material is licensed under a Creative Commons license (https://creativecommons.org/publicdomain/zero/1.0/)

The Viewmaster commercial device, originally developed for children's stories, was applied to gross anatomy with complete sets of regional dissection disks taking the place of the earlier stereoscope cards (see Fig. 3.9).

The evolution of radiological imaging methods to spiral X-ray computed tomography (CT) and magnetic resonance imaging (MRI) allowed sequential cross-sectional anatomical sets of body regions to be routinely generated. In this manner, diagnostic radiology evolved into a process where the radiologist routinely scrolls through the sets in the standard orientation planes looking for abnormalities and unusual contrasts in intensities. This process created another method for anatomical instruction for students who are now introduced to the cross-sectional anatomy image sequences they will see in their clinical rotations. Computer applications allowed the segmentation of different areas of intensity through these images, allowing contiguous areas of pixels of a specified intensity to be remembered and internally labeled. With this additional information from the image sets, two new post-processing methods were possible: *volume rendering* and *surface extraction. (See previous sections).* Volume rendering portrays these labeled 3D regions as projected objects which may be shown at any angle of rotation but may not be manipulated by the user as a discrete object in terms of visibility and viewing internal structures easily. This method of 3D projection is routinely used in cerebral CT angiography to detect vascular abnormalities.

On the other hand, surface extraction detects and saves the edges of a region of specific pixel intensities and combines them in a standard 3D space into a polymesh surface dataset of vertices and connectors, producing a mathematical surface model which can be easily manipulated by levels of visibility, color/tone assignment, and intersection with other anatomical objects and systems which become visible with sections and/or visibility. These methods are the core techniques of most commercially available medical school-level 3D atlases discussed and shown in previous sections.

It is of interest here to note the development of the Visible Human Male and Female cadaveric cross-sectional image sets where actual cadavers were sectioned in 1-mm and 0.33-mm slices, respectively, by Ackerman (1995) working with the National Library of Medicine. These complete head-to-foot sets, released in 1994 and 1995, provided the foundation of several anatomy atlases and modern XR applications.

Of particular importance, given the existence of the above modern technologies applied to anatomical instruction, is the concept of *distribution*: feasible and cost-effective provision of access to interactive, highly defined anatomical 3D models to the student as close as possible to the time of instructional/problem-solving need. This provides effective user access to these learning models.

As an example, if minimal user perceptual delay in manipulating and exploring a complex 3D object configuration, efficient distribution sometimes requires the downloading of the model's polymesh surface data and rendering characteristics (i.e., light sources, color, tonal qualities, texture, and semi-transparency). This process may be seen as a component of distribution systems as the model datasets source, through network connectivity, to the user's internal application database may be understood as a part of the concept of distribution. Erolin (2019) mentions several methods: online web pages, digital 3D PDF, Oculus and HTC Vive (VR), Hololens (AR, MR), and "bespoke" applications applying VR and AR methods to gaming technologies. "Bespoke" applications as described here are essentially sophisticated mixed reality (MR) applications. These were described in terms of medical and surgical simulations generated using game development platforms, and they allow high levels of interaction within immersive and semi-immersive scenarios as well as haptic feedback.

Also, Erolin (2019) describes the application of these technologies with high instructional potential in anatomy and medical education. She saw this as a composite of classroom and self-directed study, both in the laboratory and any other location within each student's personal or collaborative learning environment. This consideration emphasizes the concept of distribution (see above), providing ready access to the specific 3D anatomy information to an individual student at the time of need with minimal delay. As the cost of VR and AR equipment and software anatomy applications are expensive, they will probably continue to be available at most school locations; however, she felt that 3D PDFs can be used in place of traditional handouts. Finally, Eroline reiterates the current realities of decreasing time for anatomical instruction, increasing emphasis on other curricular demands as well as some schools' turning completely to digital dissection which would provide the rationale for the adoption of new proven technologies as described above.

I would suggest that desktop/phone-based 3D anatomy atlases using 2D projection can provide a multitude of exhibits, both in captured still images, video creation from the screen projections using their extensive structural feature highlighting and annotation, and high-structured assignments for students to explore the assigned anatomy using these applications themselves. We assign each first-year student their own license for the year, so this is quite feasible if faculty will collaborate on standard presentation designs and if students will complete such assignments.

Estai and Bunt (2016) reviewed the best teaching practices for anatomy education. They noted a significant and continuing decrease in traditional cadaveric dissection anatomy education time, even though their data showed less than one-third of new residents in surgery had the required anatomical knowledge. Importantly, from the viewpoint of both anatomy faculty and their students, many learning anatomy resources were useful for learning objectives understanding, but that no single teaching tool met all aspects of the instructional process. They also mention that moving to a systems-based anatomy curriculum has contributed to this decreased time in the laboratory. This group mentioned several classes of learning resources: cadaveric dissection and prosection review, plastination, computer-based learning (CBL), medical imaging, living anatomy, lecture-based teaching, integrated and systems-based curricula. I found that the most significant items mentioned that were related to the application of VR/AR technologies were CBL and systems-based curricula. They reported that software applications supporting VR anatomical representations had importance in addressing this problem of decreased time and emphasis. Supporters of dissection state that only cadavers can provide tactile manipulation of tissue, 3D interaction, and engagement of multiple senses. In rebuttal, opponents of dissection appear to argue that VR technologies offer advantages over cadaveric reality, including the ability to explore body anatomy in ways not possible with human tissue: different viewing angles, portability, and standardization. They felt that students interested in surgery could dissect whereas others could use prosections. Finally, this work described the transformation of region-based approaches to systems-based orientations where each system is presented with all relevant pre-clinical sciences, including anatomy (gross, histology, embryology, and neuroanatomy). I find positive aspects of both approaches, and I was surprised to find that no mention of the integration of the two methods was suggested as a holistic solution–combining the best of both worlds.

In their work, Bui et al. (2021) reviewed, compared, and contrasted four major modalities that have direct applications in medical education: slide-based lectures, 3D printed models, and the current AR and VR experiences. They considered the importance of distinguishing between the two main types of learning: passive and active (see above) and felt that the application of the above-listed anatomical exhibit types, with proper design and implementation, supported active learning which is seen as the more desired instructional method.

Slide-based instruction, according to students surveyed, helped sustain attention and allowed information organization along with added notes. They also felt that with the many slides presented during the course of a week in the medical curriculum, they became overwhelmed and engaged passively with the presented materials.

The addition of 3D printing uses physical materials in the place of ink that have different physical qualities (density, elasticity, color) allowing the creation of detailed 3D anatomical models.

This process is accomplished by printing sequential physical layers using different chosen material properties to represent different components of a 3D anatomical model. These physical models are created from polymesh contours generated from sequential cross-sectional image sets (e.g., high spatial resolution CT axial sets, the National Library of Medicine's Visible Human Male and Female cadaveric sections). This process requires specialized printing devices and supporting application software. Students, including residents, felt that the ability to see and touch these physical models from various viewpoints greatly aided their understanding—emphasizing the importance of haptic apprehension of 3D anatomical configurations. Limitations of this method are the training and expert labor required to select appropriate image sets and segment the regions, compartments, viscera, and systems desired.

Bui et al. (2021) also reported that students felt that AR and VR technologies were more interesting and fostered long-term knowledge retention. This work included evolving methods to allow haptic feedback from AR models as well as grouped AR experiences with distant instructors involving anatomical cases with pathologies. Using the AR method is interesting in that students must learn a standard set of hand expressions to guide the perceived functions rather than relying on controller button activation. Overall, the AR method was gauged to be an effective and reliable tool for educational functions as it helped students to learn actively via interaction while allowing remote supervision when required.

Fully immersive scenarios presented to the students had the ability to change the instructional environment from passive to active, promoting in-depth engagement. Additionally, this method of instruction, through this active engagement, significantly reinforced memory retention on tedious subjects. Therefore, VR was found to promote active engagement for both students and faculty. Limitations in VR's immersive 3D environment caused adverse effects in some students, including headache, dizziness, motion sickness, and eye fatigue.

Zhao et al. (2020) used a meta-analysis to examine the effectiveness of VR instructional technology in anatomy teaching. Importantly, they found that VR applications could be used effectively in conjunction with human dissection when laboratory time was reduced in the curriculum. They noted that medical students often had trouble acquiring an adequate understanding of three-dimensional anatomy from graphic images (2D representations in textbooks and lecture slides). Therefore, there is importance in creating modern strategies concentrated on efficient and high-quality anatomy education. They saw that VR methods allow the exploration and manipulation of immersive complex detailed anatomical structural configurations in real time. In their view, these applications promise to provide compelling presentations of information that must be learned within a curriculum in minimal time. They found examples of students' performing significantly better with the use of VR simulation when learning anatomy and resident-level surgical simulations. This group performed a meta-analysis to explore the educational effectiveness of VR applied to anatomy education in comparison with conventional or 2D digital methods as used in lecture. This analysis included consideration of medical student or resident learners, courses involving course type (skeletal anatomy and neuroanatomy), VR applications (3D interactive models and VR simulations), and teaching methods (conventional vs. digital methods). They found that only 2 of 15 studies compared cadaveric dissection to corresponding VR assignments. As an emerging new technology, they felt that there is evidence that VR applications have the potential to transform medical teaching. This study showed that when compared with conventional teaching methods, VR can improve the effectiveness of anatomical instruction but, as they found a lack of standardized measures and a high heterogeneity of research projects, these results may not be certain. Additionally, appropriate modes for the inclusion of these new technologies into the anatomy curriculum have not been systematically explored.

This paper is important as it suggests the need for standardized spatial criteria in the stepwise exploration of oddly shaped viscera, compartments, and cavities. I believe that such standardized methods of student-perceived spatial manipulation of viscera are necessary to move forward to more accurate teaching and assessment of learning objectives (e.g., the coronary sulcus and the interventricular sulci as initial landmarks for the external surface of the heart—see Future Work below Sect. 3.4.3.

Bergman et al. (2014) investigated eight factors that had a negative influence on the quality and amount of anatomical knowledge attained by their medical students. They cited the great influence that basic science knowledge had on successful clinical reasoning which should produce a working knowledge of normal anatomy. Of the factors noted in this work, they looked for potential deficiencies that could be useful using dissection coupled with the new VR and/or AR technologies. The significant reduction of dissection time allotted to it in anatomy courses was felt to be a key factor. Also, they reported that anatomy should be presented in the context of applied surface anatomy, imaging, and physical examination of patients. The new technologies, both non-extended and extended (VR, AR) technologies support these functions and, thus, if properly designed into each laboratory assignment in correspondence to actual dissection, could amplify student learning within the decreased scheduling.

Aziz et al. (2002) reported their experiences and thoughts about the human cadaver in the digital bioinformatics age and computer-based learning. They felt that when connected to dissection, medical informatics (digital databases and atlases) can enhance and expedite the preparation of students for a patient-based profession. Their work listed several learning objectives for any student: the establishment of the primacy of the patient, apprehension of the multi-dimensional body, touch-mediated perception of the human being, anatomical variability, acquisition of the language of anatomy as it relates to medicine, competence in diagnostic imaging and computer-assisted learning. Their core thought that I felt is primary here is that "despite the convenience of store anatomical or medical information in the computer, it is the physician who must conceptualize." This information could possibly provide a framework for directed laboratory learning objectives used to design structured immersive and semi-immersive digital experiences.

Riva (2003) did early systematic work on the role of virtual environments (VR) in medicine and medical education. They foresaw that digitizing massive amounts of physiological and anatomical information and representing them in spatial contexts would allow students and clinicians to better understand the principles and interrelationships of both. In the spatial sense, they understood that VR could be used by students to explore organs and organ systems by "flying" around, behind, and through them to note their structural interrelationships in ways that would be difficult or impossible to do with cadaveric material only. He noted the significance of the generation of VR anatomy textbooks using the high-resolution axials image sets of the Visible Human Male and Female datasets in 1991. These datasets were adopted by the United States National Library of Medicine for the development of digital tools for 2D and 3D visualization. He also foresaw the development of digital anatomy applications capable of visualizing normal and abnormal/pathological anatomy. Riva (2003) also saw that VR techniques and digital spatial anatomy datasets would be applied to surgical simulation and planning.

Birbara et al. (2020) performed a study with anatomy students using a digital model of the skull generated from serial CT sections in a fully immersive VR presentation room and the same skull in 3D views projected on a desktop screen—the latter being characterized as a "less immersive" version of the same anatomical data. This effort collected student demographic data and qualitative responses. There were three cohorts: (1) stereoscopic full immersion student participants, (2) "less immersive" desktop projection student participants, and (3) trained anatomy tutor participants who experienced both methods. All participants recorded their experiential perceptions in a survey with items that were rated using a 10-point Likert scale. Participants, in one of the three presentation methods, experienced the same presentations: first, a clinical scenario video was presented, followed by a series of 10 "pit-stops" around the rotating skull–each stop asking the student to answer a question about the individual bone or feature that involved the student spatially navigating through a foramen or canal and to answer additional questions along the course of movement. Additionally, tutorial guidelines were available along the way, including an announcement for the next stop on the skull.

Results showed that most anatomy students liked aspects of both methods. Spatial visualization of the skull and the exploration of individual features were positive, especially feature learning in the desktop method. Students using the immersive method particularly liked its level of interactivity and the sense of the 3D perspective in exploration of the overall skull and its individual features. The least liked perceptual quality of both methods was the sense of disorientation. This response was emphasized by participants in the fully immersive stereoscopic method coupled with a feeling of physical discomfort.

An important outcome of this effort was the consideration of student cognitive load in learning anatomy as a new subject in the design of virtual systems. Birbara et al. (2020) felt that virtual systems delivery of anatomy instruction could be more effective if a given student had a preceding basic foundation of anatomy using traditional methods.

A study by Chen et al. (2020) compared learning of the skull among third-year undergraduate medical students between (1) viewing atlas material for the skull, (2) examination of human skulls, and (3) a VR model of a skull generated from CT sections and manipulated using hand controllers. Pre- and post-tests consisted of theory (18 multiple-choice questions) and identification tests (25 write-in answers based on numbered features on the cadaveric skull). An additional five-part perception survey using a five-point Likert scale addressed participant enjoyment, learning efficiency, attitude, intention to use, and the tool's authenticity. Finally, student demographic data was collected, including GPA, age, sex, self-reported VR headset experiences, and video game experiences.

The results from this work were that the three test groups had no significant data differences for post- and pre-test differences—including total result, theory, and identification parts. Additionally, the theory/identification test within-subject scores showed overall improvement for all three groups, but the magnitude of improvement between the two tests was significantly different. Most importantly, the VR group performed somewhat better on the identification test than the physical skull and printed atlas groups, although with no significance. One explanation Chen et al. gave for the equivalent results across the three experimental groups was that their experimental design included a preliminary introductory lecture identical for the three separate groups followed by the three different self-directed intra-group training. This preliminary training phase may have provided sufficient identical foundation material for the three cohorts that might have overridden potential differences as experienced by the three presentation methods. This emphasizes the importance of experimental training designs in emphasizing subject experiences directly related to the differences in visualization methods experienced.

The results of the perception survey asked for responses regarding discomfort including headache, blurred vision, dizziness, and nausea. There was evidence that members of the VR group reported these effects somewhat more frequently; there was no significant difference between the groups in this part of the survey. The VR and cadaver groups did report a significantly higher intention to use the skull instructional material they experienced in their general studies.

Even though Chen et al. (2020) did not find significant differences between the instructional methods, they felt that 3D VR methods could provide rapid and feedback-based reinforcement of repetitive task learning, allowing students to receive instant visual feedback based on a scripted sequence of navigations. Ultimately, this group felt that VR methods may best be used to augment both lecture-style and dissection laboratory preparation exercises.

It is not difficult to imagine more sophisticated VR system designs that demonstrate normal and abnormal embryology of the skull and the consequent causality of abnormalities as well as conditions of skull trauma and the visualization of destructive forces acting on the contents of the cranial cavity and the resulting clinical outcomes, including signs and symptoms.

Hisley et al. (2008) completed a research effort that compared and contrasted two conceptually different methods for the exploration of human anatomy in the first-year dissection laboratory by accomplished students: "physical" dissection using an embalmed cadaver and "digital" dissection using three-dimensional volume-rendered modeling of whole-body CT and MRI image sets acquired using the same cadaver. The goal was to understand the relative contributions each method makes toward students' intuitive acquisition of practical anatomical knowledge gained during "hands-on" structural exploration tasks. The main instruments for measuring anatomical knowledge under this conceptual model were questions generated using a classification system designed to assess both the visual presentation manner and the corresponding response information required. Students were randomly divided into groups based on exploration method (physical or digital dissection) and then anatomical region. The physical dissectors proceeded with their direct methods, whereas the digital dissectors generated and manipulated indirect 3D digital models. After 6 weeks, corresponding student anatomical assignment teams compared their results using photography and animated digital visualizations. Finally, to see whether each method provided unique advantages, a visual test protocol of new visualizations based on the classification schema was administered. Results indicated that all students, regardless of gender, dissection method, and anatomical region dissected, performed significantly better on questions presented as rotating models requiring spatial ordering or viewpoint determination responses in contrast to requests for specific lexical feature identifications. Additional results provided evidence of trends showing significant differences in gender and dissection method scores. Moro et al. (2017) express a common opinion among pre-clinical and, especially anatomy faculty teaching within the first 2 years of medical school, that anatomy remains a fundamental area of study—although schools have routinely decreased teaching time for this discipline, a majority of these effects reduce time spent in the dissection laboratory to make way for applied clinical topics in the curriculum. Without a proper understanding of anatomy, in any healthcare profession, practitioners can be unable to perform competent physical diagnosis and academic coursework required to remain current. Fortunately, instructionally powerful, commercially available educational technologies provide ready access to students to expand and amplify their rigorous anatomical experiences and spatial viewpoint/feature identification understanding. Thus, the release and application of VR and/or AR devices and their supporting software systems allow learning that approximates and amplifies the anatomical experience most effectively in conjunction with a well-designed and rigorous whole-body anatomy course sequence. Decreased laboratory time limits student cadaveric exposure.

Well-designed and sophisticated anatomy instructional systems are changing the way students perceive and dynamically interact with assigned biomedical information regarding form and function. These learning resources allow students to use their innate abilities to grasp complex concepts and transform presented information into their remembered and accessible knowledge. This technology is best when it integrates into the curriculum with instructional designs that mitigate passive learning and inspire students' natural problem-solving abilities.

Moro et al. (2017) make the point that learning resources in anatomy have traditionally included examination of static 2D images, 3D physical models, and cadaveric specimens. Due to the nature of this discipline, these resources present visualizations and illustrations of spatial structures and their corresponding annotations in both 2D and 3D formats.

They make an excellent point that however excellent 2D representations of 3D structural configurations are (figures, diagrams, atlas paintings), these resources are limited by their nature. Two-dimensional graphics require students to remember and transform what they see into the actual 3D structures they represent—a process, which in the best of worlds, is confirmed by careful dissection. Their key point here is that static images, even if mentally rotated, may generate errors in students' minds unless they are able to directly manipulate the represented structures. In this work, Moro et al. (2017) consider this technology in terms of virtual reality (VR), augmented reality (AR), and now ubiquitous digital tablet displays as supporting basic functions of anatomical learning where desired structures can be examined from all angles and within, I suggest, their regional and systems-related contexts.

Additionally, Moro et al. (2017) suggest that even though cadaveric specimen-based investigations are foundational, they do have their limitations in terms of easy rotation and repositioning–functions that the advanced visualization technologies support naturally. The purpose of this work is to compare the effectiveness of AR, VR, and tablet-based applications in anatomical education. Fifty-nine student participants representing the biomedical disciplines, including biomedical/health sciences and medicine participated in this comparative study involving the skull with VR, AR, and tablets displaying a 10-min instructional sequence designed and aurally presented by a specialist surgeon. After this, each VR/AR/tablet-using student was required to complete a 20-question multiple-choice test. There were no significant differences between these cohorts. Those with previous anatomy experience performed, as would be expected, significantly better than the others. The experience survey responses of 80% preferred to learn at their own pace rather than be paced by the lesson designer and deliverer. Of note, the VR cohort reported that "The Oculus VR experience could be used as a good learning tool as it lets you see all parts of the skull at any angle you wish. "Being able to visualize what we are learning in any way is helpful." In terms of adverse health effects during their experiences, participants in the VR group reported significantly more symptoms than the other groups (discomfort, nausea, headache, and disorientation). For eye-related symptoms, the VR group also reported greater frequencies of blurred vision, focusing difficulties, and double vision. One student stated that "I became very distracted by how cool the program was that I wasn't focusing on the learning aspect of it." This point may become important when considering future designs of the visual presentation and levels of interactivity allowed students to keep the learning sequence on track within a feasible timeframe.

3.4 Conclusion

3.4.1 Key Conclusions

Key conclusions from the above work, including the Introduction and Background sections, are the following:

- New methods of 3D visualization that emphasize and complement dissection without replacing dissection are strongly supported.
- During practical dissection activities—there was not a single anatomy course dissection laboratory, but a series of separate pursuits based on individual student capability and willingness to apply the guidance that they receive–requiring a range of instructional methods at different levels.
- Medical students often have trouble acquiring adequate understanding of three-dimensional anatomy from graphic images (2D representations in textbooks and lecture slides). Therefore, there is importance in creating modern strategies concentrated on efficient and high-quality anatomy education. Stereoscopic views contain spatial information not found in a corresponding monoscopic view and this assists viewers in generating better internal cognitive 3D mental representations of an anatomical scene.
- Applications of these technologies have high instructional potential for anatomy and medical education. Anatomical learning was seen as a composite of classroom and self-directed study, both in the laboratory and other locations on- and off-campus within each student's personal or collaborative learning environment. This potential emphasizes the institutional requirement for effective materials distribution—providing access to visualization technologies that represent specific 3D anatomy information to individual students at the time of need with minimal delay.

- AR and VR technologies were more interesting and fostered long-term knowledge retention.
- Immersive interactive scenarios had the ability to change the instructional environment from passive to active, promoting in-depth engagement.
- Methods of instruction incorporating the use of advanced visualization methods promoted active engagement that significantly reinforced memory retention on subjects experienced as tedious.
- VR's immersive 3D environment can cause adverse effects in some students, including headache, dizziness, motion sickness, and eye fatigue after extended usage.
- VR methods that allow the exploration and manipulation of immersive, complex, detailed anatomical structural configurations in real time were effective. These applications promise to provide compelling presentations of information that must be learned within minimal time windows.
- Compared with conventional teaching methods, VR can improve the effectiveness of anatomical instruction, but with a lack of standardized measures based on the actual viscera and regions studied as well as a high heterogeneity of research projects, their true instructional efficacy is still to be determined.
- The new semi-immersive and fully immersive technologies, if properly designed into each laboratory assignment in correspondence to actual dissection, could amplify student learning within decreased curriculum time assigned to anatomy.
- A core point in teaching patient interactions aligned with the required diagnostic information is that it is the physician who must conceptualize. AR and VR instructional applications could provide and emphasize this providing an essential instructional framework for directed laboratory learning objectives. Used to design structured immersive and semi-immersive digital experiences. These learning objectives would include: the establishment of the primacy of the patient, apprehension of the multi-dimensional body, touch-mediated perception of the human being, anatomical variability, acquisition of the language of anatomy as it relates to medicine, and competence in diagnostic imaging and computer-assisted learning.
- Aspects for both AR and VR methods that appeared to be especially liked by students were spatial visualizations of the skull and the exploration of individual skull bone features, especially feature learning in the desktop method. Students using the immersive method particularly liked its level of interactivity and the sense of the 3D perspective in the exploration of the overall skull and its individual features. The least liked perceptual quality of both methods was the sense of disorientation.
- VR and cadaver groups did report a significantly higher intention to use the skull instructional material they experienced in their general studies.
- perception surveys asking for responses regarding discomfort from using the advanced technologies including headache, blurred vision, dizziness, and nausea. There was evidence that members of the VR group reported these effects somewhat more frequently.

3.4.2 Final Conclusion

Graphical methods using the principles of perspective as created in drawings, diagrams, and paintings when applied to anatomy instruction have a long history extending to the current time with perspective projections on digital screens. The principles of stereopsis, now the basis for advanced visualization techniques we know as augmented and virtual reality, have, in fact, been known and applied to anatomy since the late 1800s. Modern applications of VR and/or AR devices and their supporting software systems to anatomical education, both didactic and laboratory phases, allow more instructionally effective active learning interactivity that approximates and amplifies the anatomical concepts and learning experiences in both the didactic presentation phase and the dissection laboratory phase of instruction. Our direct academic experiences and those of the literature reported here provide strong evidence that emphasizes the most effective and retained professional school-level anatomical knowledge lies in curricula that fully integrate all visualization and illustrative techniques discussed above with a well-designed and rigorous whole-body anatomy course sequence.

3.4.3 Future Work

One absence found in the AR/VR literature as applied to anatomical education was the lack of standardized, reproducible methods involving anatomical regions/sub-regions. In order to accurately measure instructional effectiveness using the new technologies, including their comparison with existing methods as described above, we are currently working on models of anatomical instruction based on each structure's and viscera's natural structural geometries represented as organizing annotations on their surfaces using overlying graphics (e.g., dotted lines, planes, and arrows).

These simple and immediately understood 3D annotations will indicate compartments, organizing features, sulci, and/or neurovascular courses through the anatomical space of interest and indicate those major features within the organ that align with the annotations. If these framework annotations were then applied to instructional effectiveness to cohorts of students, they would be manipulated in set patterns of consistent directions, rotations, and orientations. Instruction and questions to students would then be based on their understanding of these basic landmarks. For example, the heart's spatial investigation for reproducible assessment would involve dotted line overlays around the coronary and interventricular sulci representing the valve plane and interventricular septum, respectively. Student instruction and/or assessment would then involve this heart representation, first with this framework annotation and then without it. The model would thus be rotated around each of these axes with the coronary vessels being introduced relative to the framework, then anterior atrial and ventricular walls being hidden, the valves in the valve plane/cardiac skeleton indicated, their perfusion/drainage territories corresponding to each vessel indicated with color overlays and then the conduction system overlaid as well.

Accordingly, if the patterns of 3D rotations were directionally restricted to this natural framework, and all corresponding voiceovers were identical in content and timing, then we believe that this mechanism would provide one basis for reproducibly measuring the instructional efficacy within and between different methods of visualization, including printed, physical, 2D-VR, AR, and VR modalities.

We are currently working on an initial effort to apply this framework concept to the instruction and assessment of the heart and superior mediastinum.

References

Ackerman MJ, Spitzer VM, Scherzinger AL et al (1995) The visible human data set: an image resource for anatomical visualization. Medinfo 1995(Pt 2): 1195–1198

Aziz M, McKenzie J, Wilson J et al (2002) The human cadaver in the age of biomedical informatics. Anat Rec 269(1):20–32

Bergman EM, Verheijen I, Scherpbier A et al (2014) Influences on anatomical knowledge: the complete arguments. Clin Anat 27:296–303

Birbara N, Sammut C, Pather N (2020) Virtual reality in anatomy: a pilot study evaluating different delivery modalities. Anat Sci Educ 13:445–457

Bogomolova K, Hierck B, Looijen A et al (2020) Stereoscopic three-dimensional visualization technology in anatomy learning: a meta-analysis. Med Educ 3:317–327

Bui I, Bhattacharya A, Wong S et al (2021) Role of three-dimensional visualization modalities in medical education. Front Pediatr 9. https://doi.org/10.3389/fped.2021.760363

Chen S, Zhu J, Cheng C et al (2020) Can virtual reality improve traditional anatomy education programs? A mixed-methods study on the use of a 3D skull model. BMC Med Educ 20(1):395–405

Erolin C (2019) Interactive 3D digital models for anatomy and medical education. Adv Exp Med Biol 1138:1–16. https://doi.org/10.1007/978-3-030-14227-8_1

Estai M, Bunt S (2016) Best teaching practices in anatomy education: a critical review. Ann of Anat 208:151–157

Fedele M, Quarteroni A (2021) Polygonal surface processing and mesh generation tools for the numerical simulation of the cardiac function. Int J Numer Method Biomed Eng 37(4):e3435. https://doi.org/10.1002/cnm.3435

Hisley K, Anderson L, Smith S, Kavic S, Tracy K (2008) Coupled physical and digital cadaver dissection followed by a visual test protocol provides insights into the nature of anatomical knowledge and its evaluation. Anat Sci Educ 1(1):27–40

Moore N (1998) To dissect or not to dissect. Anat Rec 253(1):8–9

Mori S, Tretter J, Spicer D et al (2019) What is the real cardiac anatomy? Clin Anat 32(3):288–309

Moro C, Stromberga Z, Raikos A et al (2017) The effectiveness of virtual and augmented reality in health sciences and medical anatomy. Anat Sci Educ 10(6): 549–559

Parisi T (2016) Learning virtual reality – developing immersive experiences and applications for desktop. Web and Mobile. O'Reilly

Riva G (2003) Applications of virtual environments in medicine. Methods Inf Med 42(5):524–534

Schoeb DS, Schwarz J, Hein S et al (2020) Mixed reality for teaching catheter placement to medical students: a randomized single-blinded, prospective trial. BMC Med Educ 20(510):1–8

Taylor S, Soneji S (2022) Bioinformatics and the Metaverse: are we ready? Front Bioinform 2. https://doi.org/10.3389/fbinf.2022.863676

Wainmen B, Pukas G, Wolak L et al (2020) The critical role of stereopsis in virtual and mixed reality learning environments. Anat Sci Educ 13:398–405

Winkelmann A, Hendrix S, Kiessling C (2007) What do students actually do during a dissection course? First steps towards understanding a complex learning experience. Acad Med 82(10):989–995

Zhao J, Xu X, Jiang H, Ding Y (2020) The effectiveness of virtual reality-based technology on anatomy teaching: a meta-analysis of randomized controlled studies. BMC Med Educ 20(1):127–137

Zilverschoon M, Vincken K, Bleys L (2017) The virtual dissecting room: creating highly detailed anatomy models for educational purposes. J Biomed Inform 65:58–75

4 Technology-Enhanced Preclinical Medical Education (Anatomy, Histology and Occasionally, Biochemistry): A Practical Guide

Jian Yang

Abstract

The recent explosion of technological innovations in mobile technology, virtual reality (VR), digital dissection, online learning platform, 3D printing, and augmented reality (AR) has provided new avenues for improving preclinical education, particularly in anatomy and histology education. Anatomy and histology are fundamental components of medical education that teach students the essential knowledge of human body structure and organization. However, these subjects are widely considered to be some of the most difficult disciplines for healthcare students. Students often face challenges in areas such as the complexity and overwhelming volume of knowledge, difficulties in visualizing body structures, navigating and identifying tissue specimens, limited exposure to learning materials, and lack of clinical relevance. The COVID-19 pandemic has further exacerbated the situation by reducing face-to-face teaching opportunities and affecting the availability of body donations for medical education.

To overcome these challenges, educators have integrated various educational technologies, such as virtual reality, digital 3D anatomy apps, 3D printing, and AI chatbots, into preclinical education. These technologies have effectively improved students' learning experiences and knowledge retention. However, the integration of technologies into preclinical education requires appropriate pedagogical approaches and logistics to align with educational theories and achieve the intended learning outcomes.

The chapter provides practical guidance and examples for integrating technologies into anatomy, histology, and biochemistry preclinical education. The author emphasizes that every technology has its own benefits and limitations and is best suited to specific learning scenarios. Therefore, it is recommended that educators and students should utilize multiple modalities for teaching and learning to achieve the best outcomes. The chapter also acknowledges that cadaver-based anatomy education is essential and proposes that educational technologies can serve as a crucial complement for promoting active learning, problem solving, knowledge application, and enhancing conventional cadaver-based education.

Keywords

Preclinical education · Anatomy and histology · Educational technologies · Active learning

J. Yang (✉)
The School of Biomedical Sciences, LKS Faculty of Medicine, The University of Hong Kong, Pok Fu Lam, Hong Kong, China
e-mail: jianyang@hku.hk

D. Cui et al. (eds.), *Biomedical Visualisation*, Advances in Experimental Medicine and Biology 1431,
https://doi.org/10.1007/978-3-031-36727-4_4

4.1 Introduction to the World of Technology in Preclinical Education

Technological innovations have advanced explosively in recent years. The rapid and continuous growth of mobile technology, virtual reality (VR), 3D printing, AI, and other digital technologies have created many new possibilities and avenues for improving the way how we facilitate and support the learning process and outcomes of students in preclinical education, especially in the subjects of anatomy and histology (Santos et al. 2022). Integration of technologies for enhancing, or in some cases replacing human specimen/cadaver-based education has provided a brand-new area of research interest for preclinical education.

Anatomy, including gross anatomy and microanatomy (histology), is a fundamental component of medical education that teaches the structure and organization of the human body. Understanding the human body's structure is essential for medical and healthcare-related students to comprehend the organization of the human body, build up foundations for grasping concepts across different disciplines in medical sciences, develop clinical skills, and perform safe invasive procedures.

Although anatomy knowledge is considered to be most essential during clinical practice by both students and clinicians, it is also widely considered by students that anatomy is one of the most difficult subjects to study (Ahmad et al. 2020). It has been reported that a large proportion of medical students and junior clinicians have low confidence in their anatomy knowledge (Mitchell and Batty 2009; Singh et al. 2015; Farey et al. 2018). Researchers have been investigating what are the major challenges for studying anatomy, and several themes emerged: the complexity and overwhelming volume of anatomy knowledge; difficulties in visualizing body structures; lack of clinical relevance; limited time and exposure to anatomical learning material; lack of time for repetition in learning anatomy; overloaded curriculum (Smith et al. 2014; Javaid et al. 2018; Cheung et al. 2021a, b). On the other hand, the decline in teaching hours for anatomy and histology in medical institutions globally and the shortage in the supply of cadavers and tissue specimens further exacerbate the situation (Drake et al. 2014; Singh et al. 2015; McBride and Drake 2018; Zhang et al. 2020). In addition, due to the ongoing COVID-19 pandemic that has lasted for 3 years, the opportunities for face-to-face anatomy teaching sessions have been significantly decreased, and the availability of body donations for anatomical education has also been affected (Onigbinde et al. 2021).

Educational technologies appeared to be the perfect solution to many of these challenges, including enhancing 3D visualization and visual-spatial ability, increasing exposure to learning materials, improving knowledge application and clinical relevance, and stimulating motivation. To improve students' learning experiences and overcome these challenges, many educational technologies have already been introduced in preclinical anatomy education. Some of the technologies used include virtual/online classroom platforms, virtual reality, digital 3D anatomy apps, 3D stereoscopic videos, digital dissection tables, 3D printed models, 3D scanning, tablet computer and video recording device, AI chatbot, virtual microscopy, and social media (Hennessy et al. 2016; Iwanaga et al. 2021a, b; Izhakoff et al. 2022; Owolabi and Bekele 2021; Patra et al. 2022). Many studies have demonstrated the effectiveness and limitations of these education technologies in improving students' learning experience, and the understanding and retention of anatomical knowledge (Trelease 2016; Cui et al. 2017; Cheung et al. 2020, 2021a, b; Zargaran et al. 2020).

It is apparent now that educational technologies provide educators with brand-new tools, modalities, and possibilities for meeting the challenges and potentially enhancing preclinical education. However, it is crucial that the advancement of educational technologies need to be matched with corresponding pedagogical advancement. The roles of education technologies need to align with the education theories to help build up a multi-modality and student-centered learning paradigm. The logistics and pedagogy of integrating technology must be developed, tested, and improved. Appropriate, student-centered, clinical-oriented, and active-leaning-based pedagogies are essential for utilizing educational technologies in promoting active learning, problem-solving, and knowledge application instead of rote memorization.

It is worth pointing out that cadaver-based anatomy education is essential for medical education and plays a crucial role in medical humanity and professionalism training (Turney 2007; Fruhstorfer et al. 2011 and McBride and Drake 2015; Hajj et al. 2015; Chiou et al. 2017; Douglas-Jones 2017; Saw 2018). However, the controversial gradual shift from cadaveric-based learning toward technology-based alternative learning model, utilizing digital anatomy apps, 3D printed models, and VR dissection labs has been adopted in many institutions in the world to address the challenges such as the high expenses of maintaining the dissection lab, shortage of cadaver, and shortage of experienced teachers (Ghosh 2017).

We believe that, under the condition of sufficient resources, education technologies will serve as an essential component for enhancing conventional cadaver-based education. The integration of technologies into conventional teaching activities will be essential for educators to overcome the challenges of traditional anatomy education and enable healthcare professional students to achieve optimal learning experiences and outcomes.

One important message from the author is that every technology has its own benefits and limitations and is best suited to certain learning scenarios. Different technologies complement each other, and educators and students should utilize multiple modalities for teaching and learning to achieve the best outcomes. In this chapter, the author will provide some practical guidance and examples about the pedagogical approaches and logistics for effectively integrating technologies into anatomy, histology, and biochemistry preclinical education.

4.2 Features, Benefits, Limitations, and Potentials of the Technologies We Currently Deploy in the Preclinical Medical Curricula

4.2.1 Virtual Reality

VR technology has advanced significantly in recent years and has been gradually adapted into preclinical education, especially in the field of anatomy education, in tertial medical institutions (Stirling and Moro 2020; Moro et al. 2021; Taylor et al. 2022). Many types of VR devices are currently available on the market, ranging from low-tech cardboard VR glasses to standalone VR goggles and then VR suits with powerful PC workstations. Coupled with VR anatomy software, VR technology allows medical students to interact with the highly detailed 3D reconstructed human body in a simulated world, explore complex anatomy concepts and spatial relationships, and apply anatomy knowledge in solving problems. It provides medical students with a true 3D immersive experience for exploring and learning anatomy, which is impossible with traditional learning (Moro et al. 2017; Maresky et al. 2019; Cheung et al. 2020). With the emerging new technologies of augmented reality (AR) and mixed reality (MR), more options and potentials will be available for educators to enhance preclinical education (Bork et al. 2019; Chytas et al. 2020; Little et al. 2021).

Three major challenges for learning anatomy in the medical curriculum identified are the inability to visualize anatomical structures, lack of time and repetition for memorization and identification, and lack of clinical integration to facilitate knowledge transfer (Cheung et al. 2021a, b). With significantly decreased contact hours in anatomy and histology teaching, increased demand for higher-level cognitive learning, limited dissection opportunities, and exposure to human specimens, the need for applying technology-enhanced novel pedagogies to complement traditional learning interventions is

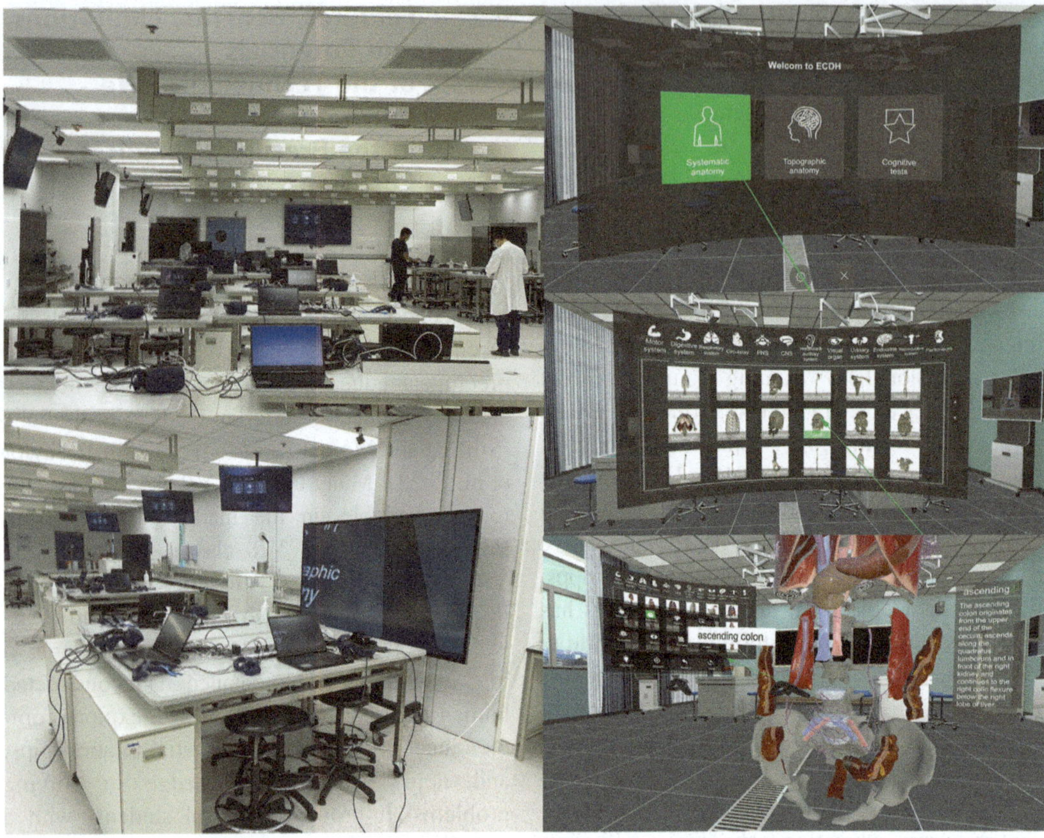

Fig. 4.1 VR anatomy equipment and VR dissection lab setup

becoming increasingly urgent. Combining a comprehensive VR anatomy learning software with a high-quality realistic digital human body, powerful VR equipment, clinical-oriented tasks, and active learning-based pedagogies, the application of VR technology is perfectly positioned to meet the imposing critical challenges in anatomy education (Fig. 4.1).

VR anatomy is best applied to specific topics involving complex and compact body regions, such as the head and neck region and pelvic cavity, where spatial relationships are challenging to comprehend, and the structures are intricate or occluded (Dobson et al. 2003; Parikh et al. 2004; Smith et al. 2007; Nguyen and Wilson 2009; Anderson et al. 2014; Ellington et al. 2019). From an educators' perspective, VR anatomy dissection stimulates the motivation in the new generation of learners, enables learning of complex anatomy concepts through problem solving, provides timely feedback, and encourages peer interaction and knowledge transfer (Stepan et al. 2017; Erolin et al. 2019; Chen et al. 2020). From students' perspectives, VR anatomy dissection helps them to visualize complex structures, comprehend spatial relationships, substantiate anatomy concepts, and prepare for the following cadaver dissection sessions (Codd and Choudhury 2011; Ekstrand et al. 2018; Cheung et al. 2020). VR technology makes anatomy education more engaging and fun. It provides the students with increased opportunities and enhanced freedom for exploring human anatomy.

Several VR physiology and molecular biochemistry software have also been developed and tested in preclinical education. VR has also been used for clinical education in medical and nursing courses to build a realistic virtual working environment. Students can carry out consultations with virtual patients, interact with colleagues, proscribe drugs and tests, perform routine medical procedures, and simulate invasive procedures (such as arthroscopic or endoscopic surgeries) (Nakai et al. 2022). These VR learning activities mimic clinical activities, allow students to gain more exposure to real-life tasks and problems, and have become especially convenient during the COVID-19 pandemic.

Benefits

1. *Enhancing the learning experience*: VR technology provides a highly immersive and interactive learning environment, allowing students to explore anatomy in a way that was impossible with traditional modality learning.
2. *Enhancing visualization and spatial understanding*: Unlike 2.5D digital anatomy modalities, VR technology provides a true 3D immersive learning modality. Students can manipulate and explore anatomical structures in a virtual environment, which helps them to better understand difficult anatomy concepts and complex spatial relationships. Students with lower mental rotation abilities may benefit more from VR learning.
3. *Enhancing motivation*: VR provides a fun, engaging, and immersive learning experience, catering to the learning habits of the new/future generations of learners.
4. *Promoting deep learning and knowledge application*: Coupling with task-oriented pedagogy, educators can integrate basic as well as clinical knowledge into the teaching sessions, promoting students' active learning, problem solving, and knowledge application abilities.
5. *Facilitating clinical integration*: VR technology provides students a safe and controlled environment to practice surgical procedures and gain experience. Without the risk of damaging the crucial structures in the cadavers, students can apply their anatomy knowledge to solve clinical problems.
6. *Enhanceable User interface (UI) and contents*: Although the initial development of the UI, digitalized body, and educational contents are considerably expensive and time consuming, modifications can be made and implemented based on educational requirements.
7. *Cost-effective (to some extent):* Depending on what types of VR technologies are deployed, VR can be a cost-effective alternative to traditional methods of anatomy education for certain health-related disciplines under specific class settings. Anatomy materials can be easily accessed through certain types of VR setups, replacing the potentially expensive and hard-to-obtain human specimens.
8. *More accessible*: Stand-alone VR headsets and cardboard VR goggles, accessible to students outside the classroom, serve as convenient tools for remote learning, particularly amid the COVID-19 pandemic. These devices also provide equal access to educational resources for students who may not have access to traditional methods of medical education.

Limitations

1. *Software limitations*: To use VR technology for medical education, specialized VR education software is required, which can be a barrier to access for some institutions and students. Very few powerful VR anatomy education software with the highly detailed digitalized human body is available on the market that meets the requirements for a tertial medical curriculum. VR software with fewer details and less information may not be etiquette for professional degrees. Low-fidelity VR software with poor UI may disrupt the learning process.
2. *Hardware limitations*: Appropriate VR hardware is required for different teaching scenarios and disciplines. VR Equipment that is not powerful enough to accommodate sophisticated VR software will limit the effectiveness of VR technology in many preclinical teaching settings.
3. *Lack of tactile feedback*: While VR provides a highly immersive experience, it lacks tactile feedback, impacting the learning experience. VR dissection cannot fully replicate the authentic experience of dissecting a real cadaver or performing a surgical procedure.

Thus, it is recommended to use VR technology as a complement rather than a replacement for cadaver-based education.

4. *Limited application*: VR is best suited for anatomy and surgical training of specific regions with intricate structures, small and fragile objects, complex spatial relationships, or limited visual field. It may be less effective than other modalities for learning the anatomy of less-complex regions or isolated organs.
5. *Cost*: An etiquette number of VR sets with VR software is required to conduct effective VR anatomy dissection for a large cohort of students. The hardware will also be outdated in a relatively short time span (2–3 years), and additional investment may be required. The cost of maintaining a team of technicians to run the VR lab will also be substantial. It will be a considerable financial burden for the institutions.
6. *Learning curve*: There is a steep learning curve associated with operating VR systems, which may impact the effectiveness of the technology in medical education. Students and instructors may need additional training and support to use VR in the classroom effectively. Compared with traditional teaching methods, more teachers/instructors will be required to give timely feedback in the VR dissection lab.
7. *Cybersickness and hygiene*: Cybersickness may occur for a small percentage of the students, seriously affecting the learning experience. If the VR sets are to be shared, the hygiene of the equipment needs to be maintained, especially during the COVID-19 pandemic.
8. *Medical humanity and professionalism*: VR technology cannot provide the rich meanings of life behind the donated human body, an essential component of medical humanity and professionalism training in the medical curricula.

As VR technology continues to evolve, more compact and powerful headsets will be developed, and the price will likely decrease. VR technology will be much more accessible for institutions as well as students. More choices of advanced VR education software will be available to cater to special requirements for different settings and disciplines of medical education. VR equipment with better tactile feedback and a smoother user experience will also be developed to further enhance the learning experience.

Instead of using reality only as background, true-AR anatomy/surgical software, which can provide live information of real specimens/cadavers/patients (Terminator/Iron man style), may also be available in the near future. This will revolutionize how students and doctors interact with reality and enable drastic advancement in medical education and clinical fields.

4.2.2 Virtual Dissection Tables

Virtual dissection tables are computer-based systems used to simulate the process of dissection of human or animal specimens for educational purposes. The technology is packaged in a dissection-table-styled setup (which can be tilled to different viewing angles) and designed to replace traditional dissection methods with digital dissection, providing a more cost-effective and hygienic alternative for anatomy education (Fig. 4.2). The virtual dissection tables utilize high-resolution sectional imaging of human cadavers to reconstruct realistic 3D digital models, which students can then manipulate, dissect, and explore in detail. Limited histology and physiology features have also been added to the features. Currently, several tech companies worldwide produce virtual dissection tables based on the cadaver source data of visible human projects in the USA, China, and Korea.

Virtual dissection tables are designed to provide students with an interactive learning experience that mimics real dissection activities. Students can zoom in and out of structures, rotate models in 360 degrees, and perform dissections with surgical tools. One of the key benefits of the virtual dissection tables is that digitized bodies are perfectly aligned with sectional images as the virtual bodies are reconstructed from images of the extremely finely sectioned human body. Students can view the sectional image on

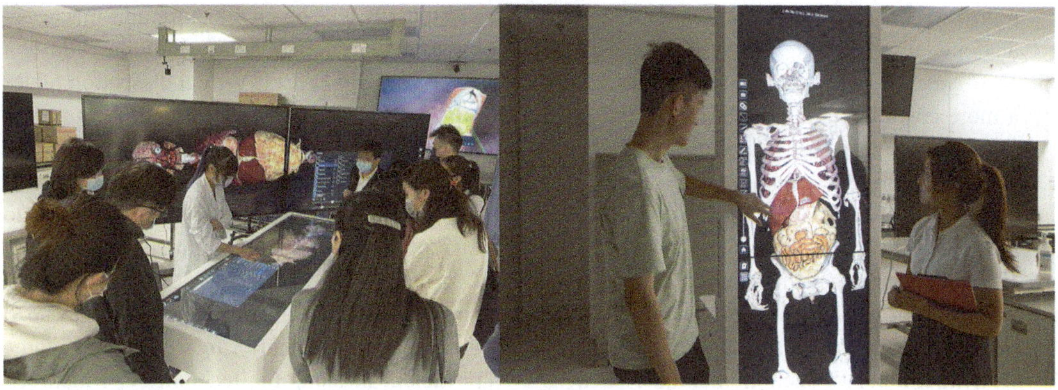

Fig. 4.2 Virtual dissection tables in action at different display angles

transverse, coronal, sagittal, or diagonal planes at any level of the body. These sectional images are also explicitly matched CT or MRI scans. These features make the virtual dissection tables exceptionally efficient for learning sectional anatomy and diagnostic radiology (Custer and Michael 2015; Bartoletti-Stella et al. 2021).

The latest models come with software that allows the creation of 3D models through CT/MRI DICOM (Digital Imaging and Communications in Medicine) files. This feature is excellent for educators to create their own virtual models for different anatomical variations and clinical pathologies. It also allows the users to access a massive database of pathological virtual models created by other users.

Another advantage of a virtual dissection table is that animal models are added as well for comparative anatomy and veterinary anatomy education. This eliminates the need for animal specimens, which can be a sensitive issue for some students and educators.

Some studies claimed that the virtual dissection tables generated the same or even better academic outcomes compared with traditional dissection in anatomy courses (Anand and Singel 2017; García Martín 2018). Although the debate on whether the virtual dissection tables can be a viable replacement for cadaver-based education for medical programs continues, the tables do provide students of many healthcare-related disciplines with a unique, engaging, and efficient way of learning anatomy (Amit Singh et al. 2018; Periya and Moro 2019; Alasmari 2021. The virtual dissection tables also give students more exposure to learning materials and allow students to repeat procedures as many times as needed, allowing for more repetition in learning anatomy.

Benefits

1. *Sectional anatomy and spatial relationships*: Virtual dissection tables allow students to explore different sectional planes and the spatial relationships between different anatomical structures in a unique way that is difficult to achieve with traditional methods. This is particularly important in diagnostic radiology, where identifying and locating structures accurately is crucial.
2. *Realistic representation*: Virtual dissection tables provide a highly realistic representation of the human body, with high-resolution images and accurate modelling of anatomical structures. This can be particularly useful in learning about sectional anatomy, where students need to understand the relationship between different sections of the body.
3. *Expendability for pathology database*: By creating digital models from DICOM files and accessing the data bank of custom-created models, the visual dissection tables provide unparalleled potential for educators to create clinically oriented anatomy learning activities.
4. *Repeatable procedures*: Virtual dissection tables allow students to repeat procedures as many times as needed, which can help

reinforce learning and promote mastery of anatomical structures.

5. *Cost-effective*: Virtual dissection tables can be a more cost-effective alternative to traditional methods, as the cost of acquiring cadaver/ human specimens, maintaining the facility and the technician team would likely be significantly higher.
6. *Reduce health and environmental impact*: Virtual dissection tables significantly reduce the health and environmental impact associated with traditional dissection methods, such as exposure to hazardous chemicals for the students and staff, potential biohazard contamination, and the disposal of hazardous chemicals.
7. *Ethical:* Virtual dissection tables eliminate the need for animal specimens, which can be a sensitive issue for some students and educators who may have moral or ethical concerns about the use of live animals in education.

Limitations

1. *Lack of tactile experience*: Virtual dissection tables do not provide the tactile experience of dissecting a real specimen, which may be necessary for medical students to develop skills for surgical procedures.
2. *Limited exposure to anatomical variation*: Virtual dissection tables may not offer the same exposure to anatomical variation, as usually multiple cadavers are provided for a large cohort of students during traditional dissection sessions. Although the latest software updates may alleviate this issue, as new digital bodies are being created and added to the software.
3. *Cost and access*: Virtual dissection tables can be expensive to acquire and maintain. Each dissection table may cost from tens of thousands of dollars to hundreds of thousands of dollars, depending on the brands. The hardware will also be outdated in a relatively short time span (2–3 years), and additional investment may be required. For a large cohort of students, multiple virtual dissection tables are required to conduct meaningful practical sessions. It will be challenging and inaccessible to many educational institutions.
4. *Technical difficulties*: Virtual dissection tables rely on the software and computing power of the PC incorporated, which may be subject to long loading time, occasionally malfunction, and other technical difficulties, leading to disruptions in the learning process.
5. *Learning curve*: As more and more functions and information are provided, students need time to adapt to the user interface and become proficient in using the technology, which can be a learning curve that detracts from their ability to focus on the anatomy itself.
6. *Quality of DICOM-file-created models*: The resolution and quality of the models generated by DICOM files may vary, depending on the quality of the original images.
7. *Medical humanity and professionalism*: As described previously, digital technologies cannot provide the rich meanings of life behind the donated human body, an essential component of medical humanity and professionalism training in the medical curricula.

In summary, virtual dissection tables provide a powerful tool for teaching anatomy in an engaging, interactive, and ethical way. Although there are limitations and the effectiveness of learning anatomy in different settings requires further investigation, the virtual dissection tables offer many benefits making them increasingly popular for anatomy education in academic and medical settings. Better user and learning experiences are expected with the technology development, as more virtual bodies and higher-quality pathology models will be incorporated into faster computers with higher-definition displays.

4.2.3 Digital Anatomy Apps

In recent years, digital anatomy apps have become popular in medical education due to their unmatchable accessibility and convenience. As smart personal mobile devices, such as smartphones, tablet computers, and notebook computers, become universal and essential for university students in most regions of the world, mobile learning apps (in our case, digital anatomy apps) play an increasingly crucial role in their everyday learning activities. These apps provide a platform for students to study human anatomy at their own pace, anywhere and anytime, without the need for physical textbooks, specimens, models, or teaching labs (Lewis et al. 2014; Golenhofen et al. 2020). During the COVID-19 pandemic, digital anatomy apps became a critical modality educator relied on for online/remote teaching (Iwanaga et al. 2021a, b).

Many digital anatomy apps are available on the market, providing the most cost-effective solution for anatomy education. Medical institutions may acquire the institutional license for their medical students to access the apps freely. These apps usually provide highly detailed 3D body models, allowing students to freely explore and interact with anatomical structures. Rich anatomy knowledge and clinical relevance would likely be added, correlating with different body regions. Some apps also contain simple animation to demonstrate movements of the joints, muscles, or organs (Motsinger 2020). They may also offer video tutorials and quizzes, which can help students to reinforce the concepts and knowledge learned. Histology and/or physiology knowledge may also be integrated into the apps. Augmented reality features may also be available, allowing students to integrate the digital human body into a familiar natural environment (Kucuk et al. 2016). Depending on the apps, the UI of some may be more intuitive than others. They may also incorporate a learning management system (LMS) for educators to design and manage learning sessions and assessments. These features make the digital mobile apps truly assessable, incredibly convenient for self-directed learning/revision, and provide possible solutions for educators conducting demonstrations and remote teaching (Mansouri et al. 2020; Harmon et al. 2022).

As the display on the mobile devices is 2D in nature, the anatomy apps cannot provide an immersive true-3D experience. The relatively small size of the display will also limit the visualization of complex structures. The smoothness of the user experience is also affected by the design of the UI and the specifications of the mobile devices. Like other digital technologies, the apps cannot provide the tactile and authentic experience of handling and dissecting a real cadaver. Unlike the virtual dissection table and some high-quality VR software, the digital body models in the apps are not reconstructed from sectional images of real human bodies. The quality and accuracy of the content can vary across different apps, which may lead to inconsistencies in learning outcomes. Although digital anatomy apps are not likely to be used as a replacement for cadaver-based education, the mobile nature of the technology and the features of these apps make them a powerful tool for both educators and students in many learning scenarios, especially in remote teaching and self-directed learning.

Benefits

1. *Accessibility*: Digital anatomy apps can be accessed from anywhere, making them a convenient tool for remote learning. Students can study human anatomy at their own pace, anywhere and anytime.
2. *Interactive learning experience*: Digital anatomy apps provide an interactive learning experience, allowing students to explore 3D anatomical structures freely, such as rotation, magnification, and isolation of different structures and tissues. This improves students' understanding and retention of knowledge.
3. *Features and resources*: Digital anatomy apps offer a range of features such as quizzes, animations, and video tutorials, which can help students to reinforce their understanding of the material.
4. *Cost-effectiveness*: Digital anatomy apps provide the most cost-effective complement to traditional anatomy education, compared with VR and virtual dissection tables. They provide the most accessible anatomy learning resources for students and institutions.
5. *Personalization*: Digital anatomy apps often allow for personalized learning experiences, as students can focus on areas where they

need more practice or spend more time on complex concepts.

Limitations

1. *Lack of tactile experience*: Like other digital anatomy technologies, the apps lack the tactile experience of handling and dissecting cadavers, affecting the depth of understanding of real human body structures.
2. *Quality and accuracy*: The quality and accuracy of the content can vary across different apps.
3. *Learning effectiveness and experience*: The UI may affect how students interact with the App. The small display size will cause difficulties in the visualization of the structures. The 2D display may also affect the understanding of the spatial relationships of the structures. Too detailed and fragmented information may be overwhelming for novice users to navigate.
4. *Technical limitations*: Some digital anatomy apps may require more powerful devices to run smoothly. Some apps are web based, and slow Internet speed will affect the user experience. Additionally, not all apps may be compatible with all devices.
5. *Limited scope*: While digital anatomy apps provide a valuable resource for studying human anatomy, they may not be as effective as VR or virtual dissection tables for active learning and clinical integration.

Digital anatomy apps are now widely used among medical schools worldwide to complement traditional anatomy education methods. They provide a convenient and interactive tool for medical students to supplement their learning of human anatomy (and histology/physiology, depending on the apps). However, they cannot be used as a replacement for traditional methods of anatomy education but in conjunction with other resources.

4.2.4 3D Printed Models

3D printing, as an emerging technology, has been gradually incorporated into anatomy and medical education in many institutions across the world. Its role in teaching human anatomy has been evaluated by several studies (Garcia et al. 2018; Ye et al. 2020; Yuen 2020). 3D printing technology has brought new possibilities in the way how medical students learn anatomy. With 3D printing, educators and clinicians can create physical models of anatomical structures that are accurate and detailed, allowing students to manipulate and study the structures in a way that is not possible with two-dimensional images or diagrams (Preece et al. 2013). Highly detailed and customized 3D printed models may also be used for visualizing complex regions, spatial relationships, pathologies, and anatomical variations that are difficult to observe through dissection alone. 3D printed models are also useful for preparing, developing, and practicing surgical procedures for surgeons. In addition, 3D printing technology has also been tested out in biochemistry and molecular biology education to provide students with better visualization of complex compound structures.

In anatomy education, 3D printed models are first created and utilized as an alternative teaching resource to counter the challenges caused by the shortage in the supply of human wet specimens and cadavers (AbouHashem et al. 2015). The process of creating a 3D printed model involves selecting the anatomical area of interest, creating the 3D geometry with 3D scans, optimizing the file for printing, and selecting the appropriate 3D printer and materials. 3D printed replicas of different parts of the human body have been produced for teaching purposes, such as bones, heart, liver, limbs, and torso. These replicas of specimens provide students with a hands-on learning experience, in the institute and regions where human specimens are difficult to acquire.

As 3D printing technology rapidly evolves, customized 3D printed models with the 3D geometry created from CT/MRI DICOM files have been gradually adopted in anatomy education as well as clinical/surgical education (Mcmenamin et al. 2014; Fredieu et al. 2015). These models can be designed to highlight certain regions, structures (nerves, blood vessels, lymphatics, etc.) (Fig. 4.3), anatomical variations, and

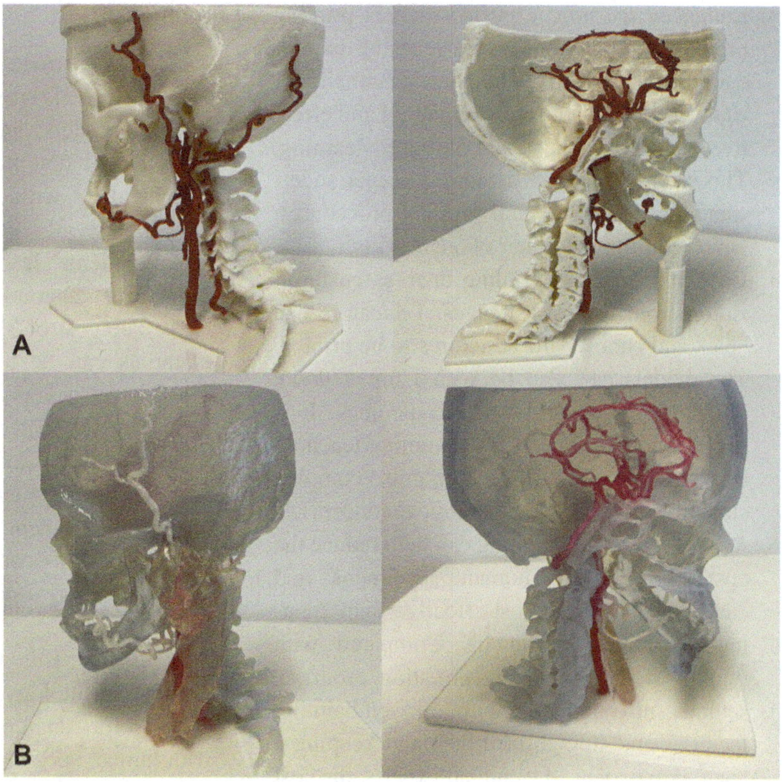

Fig. 4.3 High-fidelity customized models printed from CT images, highlighting head and brain circulation. (**a**) Model printed single materials. (**b**) Model printed with multiple materials

pathologies based on real patients. The models can also be scaled up or down based on the requirements, for displaying small yet complex structures or reducing the cost. These features give an edge to 3D printed models over traditional well-made plaster/plastic/silicone/rubber anatomical models in anatomy education (Jones et al. 2020). Teaching sessions and pedagogy can be tailor-made based on the features of the models that have been specifically designed by educators (Smith et al. 2018; Backhouse et al. 2019). Patient-specific 3D printed models with anatomical fidelity created from imaging datasets have the potential to significantly improve the knowledge and skills of newly graduated surgeons (Li et al. 2017). Furthermore, 3D printing technology is now increasingly used in medical applications including surgical planning, surgical guides, patient and trainee education, and implant fabrication (Ganapathy et al. 2022).

Benefits

1. *Improved accessibility*: 3D printing technology relieves, to an extent, the difficulty of acquiring human specimens and cadavers. It provides the students in the regions or institutions with limited access to human specimens with hands-on experiences in learning anatomy. It can also be utilized as additional learning material to increase the exposure to anatomical structures for the students in a dissection-based curriculum.
2. *Improved visualization*: 3D printed models provide a more intuitive and interactive way than 2D images and traditional anatomy models for students and healthcare professionals to visualize and understand complex anatomical structures. This can help improve learning outcomes and facilitate communication between healthcare professionals and patients.

3. *Enhanced learning experiences*: 3D printing provides a hands-on, tactile learning experience simulating real human specimens or cadavers, and can be more engaging and effective than traditional textbook-based learning. This can help students develop a deeper understanding of anatomy and surgical procedures, which can improve their clinical skills.
4. *Customization*: 3D printing enables customization from MRI/CT images. Educators can "pick and choose" the structures to be printed. The models can also be scaled up or down based on particular learning outcomes. It is particularly convenient for designing teaching activities that allow students to grasp complex structures and difficult anatomical concepts.
5. *Reduced costs*: 3D printing can help reduce the cost of acquiring human specimens and cadavers. It is more cost-effective than creating plastinated specimens. Compared with running a stainable body donation program, setting up and maintaining embalming facilities and storage facilities, and keeping a team of experienced technical experts, the cost of using 3D printing as alternative learning material is much more realistic and manageable for medical education institutions in many regions of the world.
6. *Increased fidelity*: 3D printing technology is capable to create high-fidelity models. This can be particularly useful in clinical settings, where accuracy is critical for diagnosis, treatment planning, and surgical outcomes. 3D printing enables the creation of patient-specific models, surgical guides, and implants, improving precision and reducing the risk of complications. This can be particularly useful in complex cases where traditional imaging techniques may not provide enough detail. 3D printed models can help train junior doctors as well as assist surgeons in planning complex surgeries by providing a detailed and accurate representation of patient-specific anatomy.

Limitations

1. *Fidelity dependent*: To achieve the perceived benefits of 3D printing technology, high-fidelity models are required. Low-fidelity models will not provide additional benefits, compared with traditional models. Low-quality 3D printed models may even affect students' learning outcomes and result in incorrect anatomical concepts.
2. *Cost*: The cost of 3D printing equipment and materials can be high, which may make it difficult for some institutions to adopt this technology. Especially for high-fidelity models, multiple materials may be used. The cost will be substantially high for large volume/size specimens. Additionally, the cost of creating patient-specific models can be expensive, which may limit their use in some clinical settings.
3. *Authenticity*: This limitation is not for 3D printing technology, but all alternative technology for anatomy education, compared with human specimens and cadavers. Limited by technology and materials, 3D printed models cannot provide the authentic dissection experience. It is difficult for 3D printed models to recreate small and delicate structures such as fasciae, connective tissue compartments, minor vessels, and small nerves. Despite the features provided by 3D printing technology, currently, it still cannot replicate the complexity and texture of the real human body.
4. *Medical humanity and professionalism*: Digital technologies cannot provide the rich meanings of life behind the donated human body, which is an essential component of medical humanity and professionalism training in the medical curricula.
5. *Regulatory challenges*: The use of 3D printing in clinical practice raises regulatory challenges related to quality control, safety, and liability. For example, there may be concerns about the quality and safety of 3D printed implants and surgical guides, which could result in

regulatory hurdles for healthcare professionals who want to use these devices.
6. *Limited materials*: 3D printing is currently limited to certain materials, which may not be suitable for all applications. For example, 3D printed models may not accurately represent the mechanical properties of certain tissues, which could limit their use in dissection/surgical simulations.
7. *Time consuming*: The process of creating 3D models can be time consuming, which may limit its use in some clinical settings.

As 3D printing technologies continue to evolve, the fidelity and quality of the models will improve, the cost will reduce, and more materials will be used for simulating authentic human tissue. This technology is already making a significant impact in the field of medical education, and it is expected to play a more substantial role in anatomy and clinical education, complementing cadaver-based education.

4.2.5 Virtual Microscopy

In the past decade, virtual microscopy has become widely adopted in histology education in universities worldwide, especially in the USA and the UK. This technology allows students to view high-resolution digital images of tissue sections, usually on a web-based platform, mimicking the experience of operating a real microscope (Pantanowitz et al. 2012). With virtual microscopy, students can freely navigate detailed-scanned real tissue sections, zoom in and out through low magnification to high magnification, annotate, and highlight different tissues, capture live images, and access related histological and pathological information. With the simple setup of desktop or laptop computers and internet access, virtual microscopy can be deployed in teaching labs as a replacement for optical microscopes and tissue specimens. It also allows students to browse the specimens online through mobile devices, enabling them to explore and analyze histological and pathological structures in an interactive and engaging way, anywhere and at any time, with the convenience at their fingertips. These features turned out to be especially valuable during the COVID-19 pandemic (Amer and Nemenqani 2020).

Many students consider traditional histology classes "boring," "irrelevant," and "too abstract." Students are easily confused, lost, and exhausted while looking through a real microscope. Rigid course designs may emphasize only identification and memorization rather than application. Lack of interaction and feedback may further decrease students' motivation, leading to poor learning experience and efficiency. Combined with task-oriented active learning strategies, an interactive online platform, collaborative and relaxed learning environment, virtual microscopy offers an excellent solution to modernize and enhance traditional histology education. During in-person or remote histology practical sessions, teachers may also provide timely feedback and synchronized live demonstrations through the administrator's unit of the virtual microscope. Many studies worldwide have demonstrated that virtual microscopy was as effective, and in many cases more effective, for achieving learning outcomes for histology and histopathology courses, significantly improving students' learning experience (Heidger et al. 2002; Dee 2009; Triola and Holloway 2011; Mione et al. 2013, 2016; Felszeghy et al. 2017). The fact that virtual microscopy is so cost-effective and efficient in increasing motivation, enhancing the learning experience, promoting interaction and peer collaboration, and improving assessment results, it has been adopted as a complete replacement for conventional microscopy by medical, dental, veterinary, nursing, and biomedical sciences programs in many universities in the UK, the USA, and worldwide (Harris et al. 2001; Kumar et al. 2006; Mills et al. 2007; Rosas et al. 2012; Maity et al. 2023).

The traditional microscope-based histology classes offer hands-on microscopy experience and experiential learning opportunities, allowing students to practice operating under a microscope with stereo vision, observing the same tissue specimen with staining and preparation variances. Despite all the advantages virtual microscopy offers, many institutions, including some of the top medical schools, choose to keep the traditional microscope-based histology teaching in their core curriculum. For them, virtual miscopy is used as an additional resource instead of a replacement.

Benefits

1. *Accessibility*: Virtual microscopy allows students to access high-quality digital images of tissue sections from anywhere with an Internet connection, eliminating the need for physical specimens.
2. *Cost-effective*: Virtual microscopy reduces the need for expensive microscopes and storage facilities for tissue specimens and equipment. It eliminates the cost of maintaining a technical team servicing the equipment and renewing the tissue specimens.
3. *Flexibility*: Virtual microscopy allows for more flexible teaching and learning, as students can access the images remotely and study them at their own pace. This is essential for online histology classes.
4. *Consistency*: Virtual microscopy offers consistent image and staining quality, whereas tissue specimens may vary in quality due to differences in preparation, storage, and decay.
5. *Learning experience*: Virtual microscopy improved the motivation and learning experience, promoting feedback and collaboration.
6. *Learning effectiveness*: It has been proven that virtual microscopy is significantly more effective than static histology images. Compared with conventional microscopy, it offers the same, if not better, quality of histology education and helps students to achieve better learning outcomes in histology and histopathology courses.

Limitations

1. *Lack of hands-on experience*: Virtual microscopy does not provide the same hands-on experience in operating a microscope and handling tissue specimens. Yet these skills can be essential for students who will be working with tissues in a clinical setting or performing procedures under operating microscopy in the future.
2. *Restriction on technology*: Virtual microscopy relies on the online platform, software, and Internet quality. Technical issues may interfere with access to images, potentially disrupting teaching and learning.
3. *Limited field of view*: Virtual microscopy may have a limited field of view compared to conventional microscopy, which may omit the relationships between certain adjacent structures. This may limit students' ability to form a comprehensive picture of the relevant tissue sections or regions.
4. *Tissue and staining variations*: Virtual microscopy provides almost no tissue and staining variability. This leads to limited exposure to morphological varieties of real human tissue for the students.

Virtual microscopy has proven to be a valuable and cost-effective tool in histology education, offering numerous benefits over conventional microscopy. It will potentially replace conventional microscopy as the core teaching modality in most universities and institutions worldwide.

4.2.6 Tablet Computers and 3D Scanning Technology

Tablet computers are becoming increasingly popular among university students and are convenient educational devices. They provide students with a flexible and accessible way to review and revise their learning (Patel and Burke-Gaffney 2018). This technology can be utilized in classroom settings as a pedagogical tool to supplement and enhance traditional preclinical education, such as cadaver dissection/prosection sessions and histology practical sessions (Fig. 4.4). Educators can use tablet computers simply as a didactic platform for displaying the instruction videos to provide students with a more in-depth

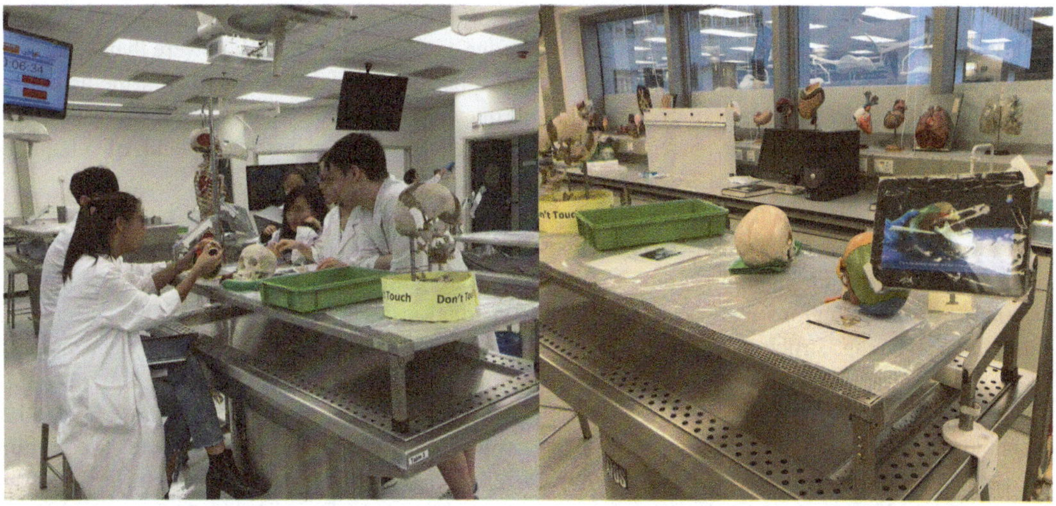

Fig. 4.4 Tablet computers deployed in prosection anatomy practical session

look at specific anatomical or histological structures or to show real-life clinical scenarios that illustrate the practical application of the knowledge (Scibora et al. 2018). It can also be used as an interactive platform for in-class task checklists, quizzes, and worksheets during practical sessions. Students can use the tablet computers assigned to each dissection table, demonstration station, or study group for support and self-guided active learning. AI-chat-bot may also be loaded onto these tablets and provide simple feedback and guidance to the students during the practical sessions. The tablet computer with video recording functions can also be the platform for learning activities that require deep learning, creativity, and peer collaboration (Yang et al. 2019). Educators can ask the students to create instruction videos or report interesting discoveries using the tablet computers provided through the video recording functions and share the videos on an online learning platform for peer support and evaluation. A tablet computer is a versatile tool that can be effectively integrated into many preclinical and clinical teaching settings, enhancing traditional education (Perez et al. 2013; Baumgart et al. 2017). It is best used in conjunction with other teaching modalities to ensure that students receive a well-rounded learning experience, and quality-designed materials and teaching activities are required. The cost of a tablet computer nowadays is relatively inexpensive. Equipping the dissection or histology lab with tablet computers will provide many options and possibilities for designing teaching sessions that provide better guidance, support, and learning experience.

3D scanning technology is being used to generate digitalized human specimens in many medical institutions and hospitals. The digital models will then be printed using 3D printers or incorporated into other systems, such as virtual reality software or digital anatomy apps (Thomas et al. 2016; Iwanaga et al. 2021a, b; Lotfi et al. 2022). Detailed and realistic digital models enable the applications for virtual dissection, VR dissection, customized 3D printing, and surgical planning, and enhance the learning experience for students and healthcare professionals. The quality of the 3D scanner and the digital model generated may vary, and poorly rendered digital models will affect the effectiveness of the downstream applications. Different types of 3D scanners can be used for anatomy and clinical education, including smartphones with 3D scanning apps, hand-held laser or colored-light scanners, and 360 high-definition camera arrays. The cost of a high-quality 3D scanner, editing software, and computer workstation can be high. The scanning process is time consuming, and scanning skill is required for a high-quality scan. Utilizing the scanning data for downstream applications in-house is technology demanding, which can be a barrier to adoption, and a technician with specialized skill sets is likely required. Furthermore, there are limitations in the current 3D scanning technology. For example, consumer 3D scanners, even with high-end models, cannot capture all the details and textures of the human body. It is particularly challenging for 3D scanners to capture regions with great depth and multiple overlaying structures. There are also likely to be restrictions on the size of the scanned specimens. With the development of 3D scanning technology, creating digital models with higher fidelity and better resolution at a reduced cost will be possible, and it is expected to be more accessible by medical schools worldwide for teaching and learning purposes.

4.3 Examples of the Application of Technology in Preclinical Curricula

4.3.1 Large Class Practical

4.3.1.1 Example of Integrating VR Dissection into Large Class Gross Anatomy Education

The VR dissection session was designed in 2020 for Bachelor of Medicine and Bachelor of Surgery (MBBS) 2nd-year undergraduate curriculum, gastrointestinal system block. The class size was around 300 students. We previously identified that the VR dissection sessions were most effective when delivered before cadaveric dissections (Cheung and Yang 2021). This arrangement helps the students visualize the complex anatomical structures and spatial relationships, consolidate difficult concepts, and prepare for the cadaveric dissection sessions. We integrated VR dissections sessions into each system block, and the ratio of VR dissection to cadaveric dissection is 1:3.

For the 2-h VR dissection session, two VR-enriched tasks (VRETs) were written by experienced anatomy teachers. These tasks were active learning-based and clinical-oriented, designed to promote deep learning and knowledge application. The VRETs were then validated and performed by other teachers in the anatomy education division to ensure that these tasks were written with clear instruction and of adequate difficulty (Supplement 1).

All VRETs were designed to be completed by at least two participants working in groups, one operating the VR device and the other giving instructions. Thus, the VR activities were team-based learning in nature, promoting peer interaction, and peer learning.

To comply with the university's COVID-19 restrictions, the 300 medical students were split into two large groups. Each VR dissection session hosted around 150 students. According to their PBL grouping, these students were further divided into 16 teams. As they have been working with the same PBL group for over a year, they should already have good synergy working together. We have tested random grouping, and the outcomes were not ideal. Each team was given four sets of VR devices, out of which one set was connected to a large-screen display for team demonstration. This setup (Fig. 4.5) allowed students to work in a familiar PBL group discussion environment, allowing them to perform the tasks as individual units as well as discuss the problems in PBL tutorial style. Five to six anatomy teachers and six postgraduate teaching assistants were on site throughout the session, providing instant feedback and academic support. Four technicians were standing by to provide technical support.

Students were given 40 min for each task. After each task, the leading teacher gave live quizzes and a debriefing session to consolidate the core anatomy concepts. No additional time was given to practice VR operation before the tasks, as the students had already attended the VR dissection session in the previous system blocks. Note that, due to the steep learning curve of VR technology, sufficient practice time should be given to first-time users before formal VR dissection sessions.

Fig. 4.5 VR dissection session on going

4.3.1.2 Example of Utilizing VR Biochemistry Platform in DNA and Protein Structure Practical

Four experienced teachers and one postgraduate student in our school have designed and created a VR biochemistry learning platform from 2020 to 2021. The platform was deployed in molecular biochemistry courses for Biomedical Sciences and Biological Sciences students. Immersed in the VR platform, students can observe, manipulate, interact, disassemble, and reassemble DNA and protein molecules. The VR platform allows students to visualize the 3D structures of complex molecules and perform self-assessment quizzes for knowledge consolidation.

In this example, Biological Sciences students used the VR biochemistry platform to explore DNA and protein structures. One hundred twenty students attended the practical. They were divided into 30 groups, and 30 sets of VR devices were provided. Four students shared each VR device. Two teaching staff were present, together with four postgraduate teaching assistants. Four technicians were standing by for technical support.

A pre-class test was given to gauge students' baseline knowledge, followed by a short briefing session from the leading teacher. We then gave the students detailed instructions on operating the VR system. The students were allowed sufficient time to practice VR operating skills before starting the one-and-half-hour VR learning activities. After exploring the DNA and protein structures and completing various tasks using the VR platform, a post-class test was given, followed by a debriefing session.

4.3.1.3 Example of Using Technology-Enhanced Stations as Part of a Traditional Prosection Demonstration Practical for 2nd Year Pharmacy Students

It is not always necessary to deploy a large number of VR equipment or virtual dissection tables to enable technology-enhanced preclinical education. One or two sets of the equipment can be effectively integrated into wet-specimen-based or plastinated model-based prosection practical sessions. To avoid cognitive overload to the students and streamline the logistics, it is recommended not to put two or more technologies with steep learning curves into one practical session.

We designed the practical session, Introduction to Human Anatomy, for Pharmacy second year students. The class size was 70. Two teachers with two postgraduate teaching assistants led the practical session. For the 180-min session, eight task-oriented demonstration stations were set, and students spent around 20 min for each station. One virtual dissection table and one computer with a large touch panel display were deployed as two technology-enhanced stations. The rest stations were conventional wet specimen prosection stations. The Virtual dissection table station was designed for the exploration of the major organs in the thoracic cavity. As we do not provide whole-body dissection to non-MBBS curriculum, the virtual dissection table is a viable alternative for other health professional students to study the concepts of body cavities, the surface anatomy of vital organs, and the spatial relationships of related anatomical structures. After observing the specimens of lungs and heart, students used the virtual dissection table station to explore the intact thoracic region, putting everything into perspective and consolidating the knowledge. The computer-based station was designed as an interactive debriefing platform, with self-test quizzes, drawing exercises, a summary of core knowledge, and additional information for clinical relevance.

4.3.2 Small Group Tutorial

4.3.2.1 Example of Utilizing Virtual Dissection Table for 3rd Year MBBS Advanced Anatomy Research and Education Program

In our MBBS curriculum, students will finish the preclinical course in the first 2 years. The 3rd year of the curriculum is called the enrichment year. Students will have the freedom to choose the learning activities for the entire year. These activities ranged from intercalation degrees in local or international institutions, research or clinical attachments locally or overseas, NGO service duties, elective education programs, etc. After meeting specific criteria and completing the enrichment year, they will proceed with the clinical years (starting from the 4th year).

There are six MBBS students enrolled in the Advanced Anatomy Research and Education Programme (one semester or whole year) for their enrichment year. In this program, they are required to conduct a clinical anatomy literature study on the topic of their choosing, complete an educational theory and pedagogy certificate course designed for near-peer teachers, enroll in the advanced clinical dissection course, carry out at least 60 h of near-pear anatomy teaching, attach to an anatomy-related research project, and submit a 6000 words dissertation.

In this example, we designed a 2-h tutorial session on the theme of clinical anatomy of complex pelvic fracture and trauma as part of the advanced clinical dissection course. One teacher will lead the tutorial, and one virtual dissection table will be used. The six students have finished 6 h of pelvic region dissection in their second year. Before coming to the tutorial, they should have studied the literature on the related topic and begun to work on a detailed dissection of the trunk.

For the first 20 min, the six students will play a quiz competition on pelvic structures, using the virtual dissection table's multiplayer (up to eight participants) game function. This activity allows students to refresh and reinforce their baseline

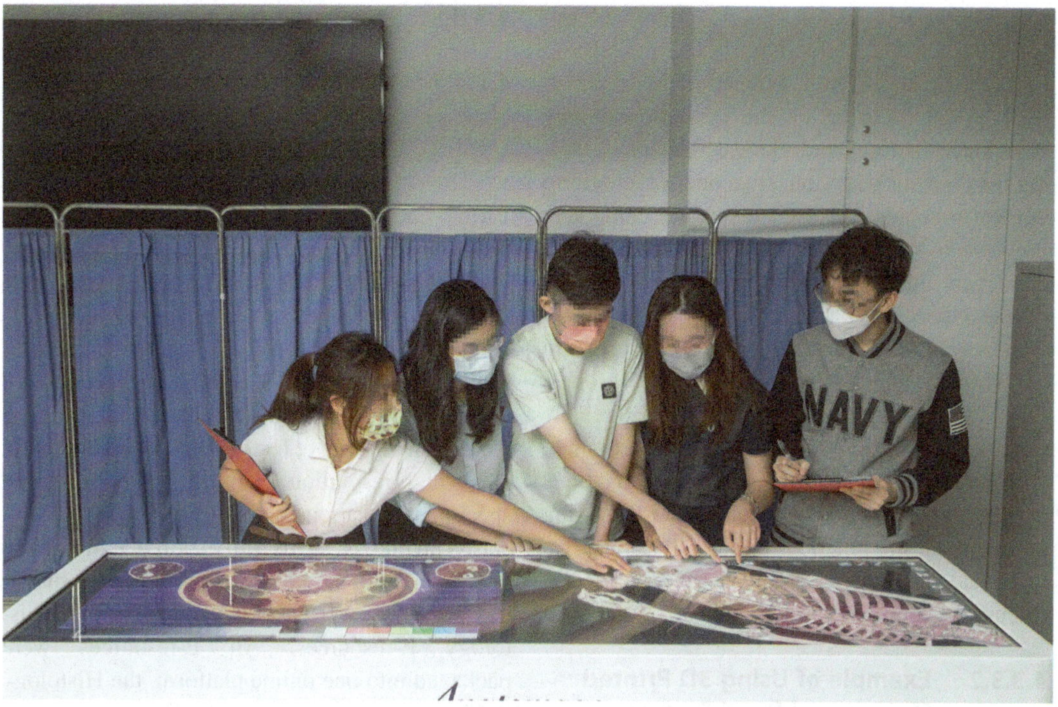

Fig. 4.6 Correlating sectional anatomy images with the structures

anatomy knowledge of the region. In the following 40 min, students will use the virtual dissection tools to analyze bones, muscles, fascia compartments, vascular supplies, nervous innervations, and lymphatic drainages of the pelvic and abdominal region, correlating these structures with the sectional anatomy images at each plane (Fig. 4.6).

With the latest software, the virtual dissection table has access to the bank of pathological 3D models rendered from DICOM files. In the following 30 min, the students will conduct a guided exploration of several clinical cases of complex pelvic fracture and trauma using the 3D models in the pathological model bank. In the last 30 min, the six students will discuss the potential signs, symptoms, and complications based on the clinical anatomy knowledge they have learned, facilitated by the teacher.

After the tutorial, the students will return to the dissection lab the following day and apply their knowledge to continue the clinical dissection study.

4.3.3 Peer Teaching and Assessment

4.3.3.1 Example of Utilizing the Dissection Peer-Support System (DPSS, Comprises Video Recording Tablet Computers and Online Learning Platform) in 1st Year MBBS Gross Anatomy Dissection Sessions

The DPSS comprises tablet computers mounted on each dissection table through specially designed mounting arms and an online learning platform (iclass, a real-time interactive teaching and learning platform on cloud, designed by the Department of Electrical and Electronic Engineering, the University of Hong Kong).

For this dissection session, we provided 34 cadavers for 300 MBBS students. Nine students worked in one group, sharing one cadaver and one set of DPSS. We split the class into two repeated sessions to comply with the university's COVID-19 restrictions. In each dissection session, there were 150 students in 17 groups. Five anatomy teachers, two surgeons, and six near-peer anatomy teachers provided on-site guidance and instruction.

Through DPSS, we assigned video tasks to each group, focusing on the dissection of key anatomical features and clinical relevance. During the dissection session, students in each group worked together and used the tablet computers provided to record 1 or 2 high-resolution 1-min dissection videos, according to the requirements of the video tasks (Figs. 4.7 and 4.8). These videos were then uploaded to the DPSS platform for viewing and evaluation by all 34 groups. Students used these peer-created videos to review essential dissection procedures and compare anatomical variations and pathological features of all 34 cadavers. Students also used the rating and commenting functions on the DPSS platform to evaluate the videos' quality and usefulness and discuss the relevant anatomy topics.

4.3.3.2 Example of Using 3D Printed Models and DPSS System to Promote Peer Learning in 2nd Year Dental Anatomy Practical

Customized 3D printed models are useful for teaching complex anatomy subjects. We utilized 3D printed models highlighting the blood supply of head and brain regions for one demonstration station for the dental 2nd-year anatomy practical, The Blood and Nervous Supply of Head and Neck. The 60 students in this class were grouped into six teams. The session was guided by two anatomy teachers and two postgraduate teaching assistants.

To promote higher cognitive learning and peer learning, the DPSS was also utilized in this station. The groups were asked to use the 3D printed models to study the complex structures, complete the questions and shoot a 90-s demonstration video. The videos were then displayed on the tablet computer mounted on the station for other groups to review. The groups were learning and teaching each other through this learning session, and the anatomy knowledge was reinforced.

4.3.4 Self-Directed Learning

4.3.4.1 Example of Histology E-Classroom Enables Effective Self-Directed Learning and Remote Teaching

Virtual microscopy has been adopted as a primary histology teaching modality worldwide. In our school, we utilize the Aperio system (by Leica Biosystems) with 150 scanned tissue specimens. We have adopted a hybrid approach, utilizing virtual microscopy to complement conventional microscope-based histology education. We have created an interactive histology E-atlas containing 200 high-resolution photos of the tissue specimens at various magnifications up to 1000× across all body systems. Fifty short demonstration video clips for the key structures of each histology specimen were also made to enrich the histology E-resources. All E-resources were packaged into one online platform, the Histology E-Classroom (Figs. 4.9 and 4.10).

The E-classroom is provided to all students for their self-directed learning and self-assessments. They can access the virtual microscope, histology E-atlas, demonstration videos, quizzes, and discussion forums on their computers or mobile devices anytime, anywhere. This system also serves as a reference resource for our junior teachers and fresh graduates. During the COVID-19 pandemic, Histology E-classroom became the core teaching modality enabling online histology practical learning for students.

4.3.4.2 Example of Technology-Enabled Open Learning Space Provides Accessible Digital Anatomy Resources and an Active Learning Environment for Students' Self-Directed Learning

One main challenge students face in learning anatomy is the need for more exposure to anatomy specimens and other learning materials after class. Although digital anatomy apps may help

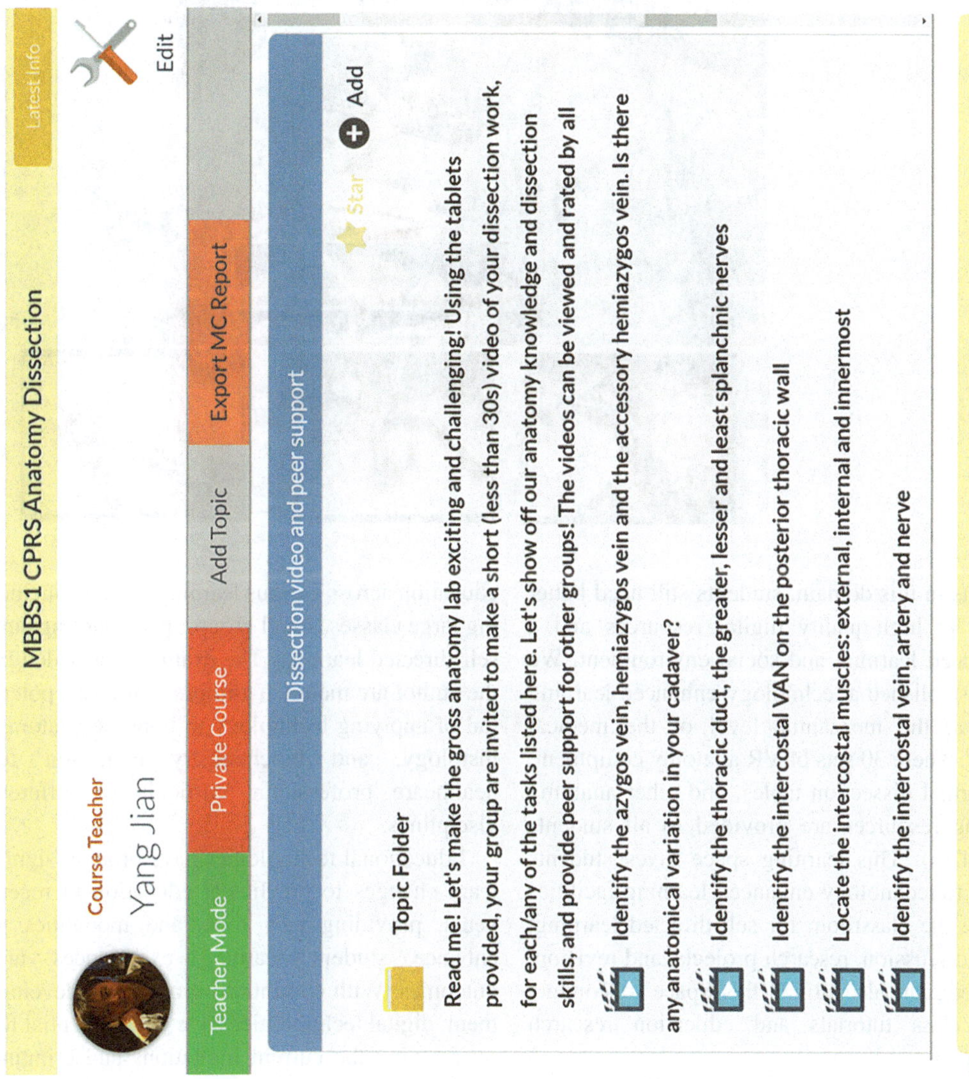

Fig. 4.7 The DPSS online platform and sample video tasks

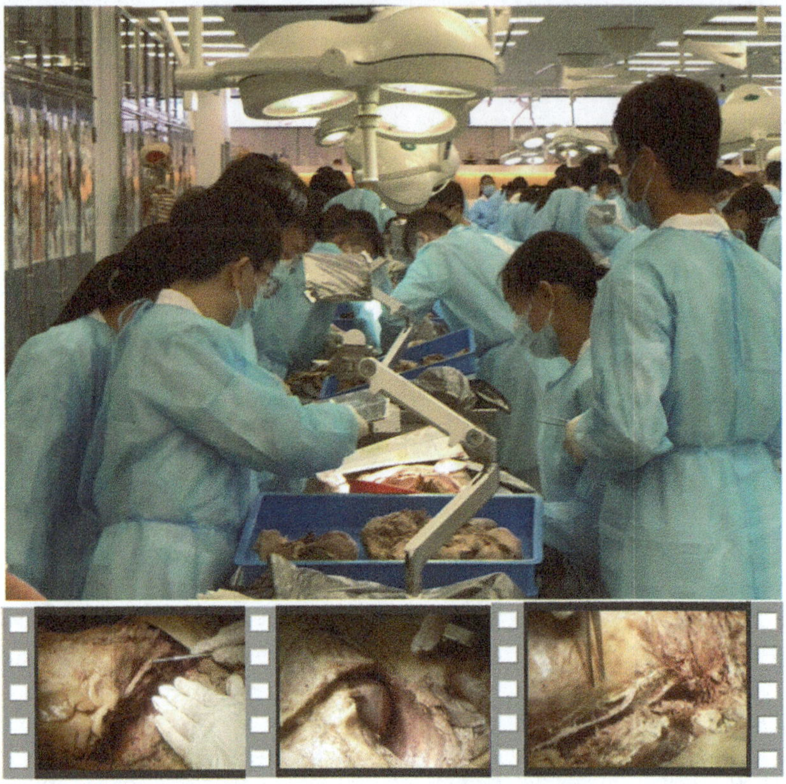

Fig. 4.8 Students using the DPSS system during dissection

students in this domain, students still need better access to high-quality digital resources and a designated learning and social environment. We have established a technology-enhanced learning space at the mezzanine level of the medical library, where 30 sets of VR anatomy equipment, two virtual dissection tables, and other anatomy learning resources are provided to all students (Fig. 4.11). This learning space gives students access to technology-enhanced learning facilities outside the classroom for self-directed learning, group discussion, research projects, and revision. Teachers can also utilize this space to conduct small class tutorials and education research projects.

4.4 Conclusion

Implementing technology in a pedagogically sound approach can enhance premedical education across various learning settings, including large classes, small groups, peer learning, and self-directed learning. The examples provided by the author are merely a glimpse of the vast potential of applying technology to improve anatomy, histology, and biochemistry education for healthcare professional curricula of different disciplines.

Educational technologies have brought significant changes to preclinical education in recent years, providing new tools and modalities to enhance student learning experiences and outcomes. With continuous growth and development, digital technologies have great potential for overcoming the current limitations and bringing further improvements and innovations to preclinical education. However, it is important to note that technology integration for preclinical education should align with pedagogical principles and be used in a student-centered and clinical-oriented approach.

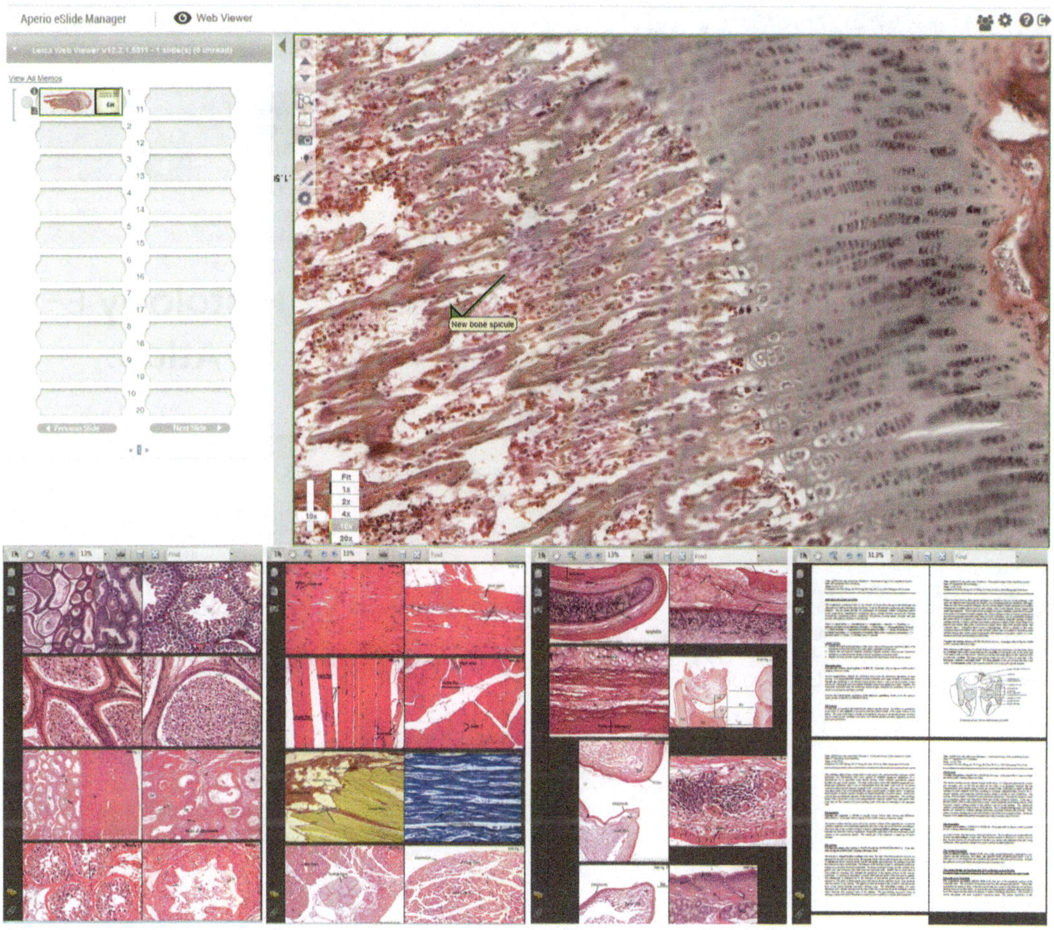

Fig. 4.9 Virtual microscopy platform and histology E-atlas

Figure Credits: All pictures are credited to LKS faculty of Medicine, The University of Hong Kong.

Apps, programs, and equipment used by author in the examples

VR equipment: HTC, VIVE Pro 2

VR Software: Digihuman, EDCH VR Anatomy

Virtual dissection table: Anatomage, Anatomage Table V8

Digital anatomy app: 3D4Medical, Complete Anatomy

3D printer: Stratasys J750 Digital Anatomy

3D Scanner: Shining 3D Einscan Pro HD

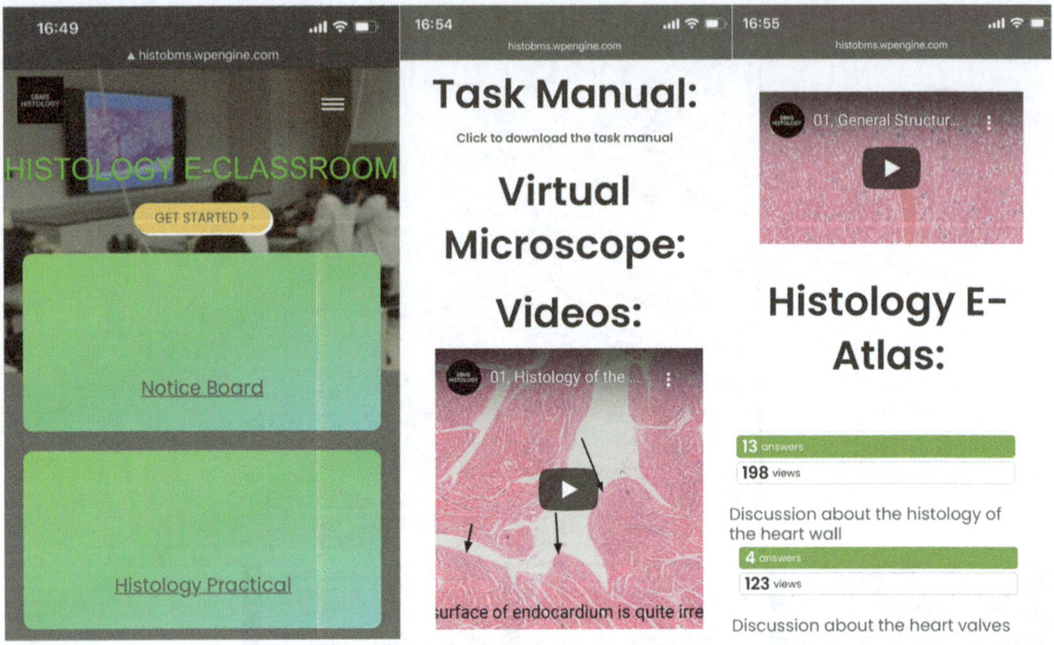

Fig. 4.10 The histology E-classroom on mobile devices

Fig. 4.11 The Technology-enriched learning space, TechMezz

References

AbouHashem Y, Dayal M, Savanah S et al (2015) The application of 3D printing in anatomy education. Med Educ Online 20:1. https://doi.org/10.3402/meo.v20.29847

Ahmad K, Khaleeq T, Hanif U, Ahmad N et al (2020) Addressing the failures of undergraduate anatomy education: dissecting the issue and innovating a solution. Ann Med Surg (Lond) 61:81–84. https://doi.org/10.1016/j.amsu.2020.12.024

Alasmari WA (2021) Medical students' feedback of applying the virtual dissection table (Anatomage) in learning anatomy: a cross-sectional descriptive study. Adv Med Educ Pract 12:1303–1307. https://doi.org/10.2147/amep.S324520

Amer MG, Nemenqani DM (2020) Successful use of virtual microscopy in the assessment of practical histology during pandemic COVID-19: a descriptive study. J Microsc Ultrastruct 8(4):156–161. https://doi.org/10.4103/JMAU.JMAU_67_20

Amit Singh B, Surya Kumari N, Rani VS (2018) A study on student perception of virtual dissection table

(Anatomage) at GSL Medical College. Rajahmundry Academia Anatomica Int 4(2). https://doi.org/10.21276/aanat.2018.4.2.8

Anand MK, Singel T (2017) A comparative study of learning with "anatomage" virtual dissection table versus traditional dissection method in neuroanatomy. Indian J Clin Anat Physiol 4(2):177–180

Anderson P, Ma M, Poyade M (2014) A haptic-based virtual reality head and neck model for dental education. https://doi.org/10.1007/978-3-642-54816-1_3

Backhouse S, Taylor D, Armitage JA (2019) Is this mine to keep? Three-dimensional printing enables active, personalized learning in anatomy. Anat Sci Educ 12(5):518–528. https://doi.org/10.1002/ase.1840

Bartoletti-Stella A, Gatta V, Mariani GA et al (2021) Three-dimensional virtual anatomy as a new approach for medical student's learning. Int J Environ Res Public Health 18(24):13247. https://doi.org/10.3390/ijerph182413247

Baumgart DC, Wende I, Grittner U (2017) Tablet computer enhanced training improves internal medicine exam performance. PLoS One 12(4):e0172827. https://doi.org/10.1371/journal.pone.0172827

Bork F, Stratmann L, Enssle S et al (2019) The benefits of an augmented reality magic mirror system for integrated radiology teaching in gross anatomy. Anat Sci Educ 12(6):585–598. https://doi.org/10.1002/ase.1864

Chen S, Zhu J, Cheng C et al (2020) Can virtual reality improve traditional anatomy education programmes? A mixed-methods study on the use of a 3D skull model. BMC Med Educ 20(1):1–10

Cheung CC, Yang J (2021) Large-scale integration of virtual reality-enriched tasks (VRETs) in gross anatomy curriculum during the pandemic. FASEB J 35 (S1). https://doi.org/10.1096/fasebj.2021.35.S1.02295

Cheung CC, Cecot TS, Tipoe GL et al (2020) Using VR-enriched tasks (VRETs) to promote active learning in pre-clinical gross anatomy education. FASEB J 34 (S1):1. https://doi.org/10.1096/fasebj.2020.34.s1.07575

Cheung CC, Bridges SM, Tipoe GL (2021a) Why is anatomy difficult to learn? The implications for undergraduate medical curricula. Anat Sci Educ 14(6):752–763. https://doi.org/10.1002/ase.2071

Cheung CC, Yang J, Tipoe GL (2021b) Multifactorial learning challenges in anatomy: what is the niche in current research? Anat Sci Educ 14(6):856–858. https://doi.org/10.1002/ase.2125

Chiou RJ, Tsai PF, Han DY (2017) Effects of a "silent mentor" initiation ceremony and dissection on medical students' humanity and learning. BMC Res Notes 10(1):483. https://doi.org/10.1186/s13104-017-2809-0. PMID: 28915916; PMCID: PMC5602928

Chytas D, Johnson EO, Piagkou M et al (2020) The role of augmented reality in anatomical education: an overview. Ann Anat 229:151463–151463. https://doi.org/10.1016/j.aanat.2020.151463

Codd AM, Choudhury B (2011) Virtual reality anatomy: is it comparable with traditional methods in the teaching of human forearm musculoskeletal anatomy? Anat Sci Educ 4(3):119–125

Cui D, Wilson TD, Rockhold RW et al (2017) Evaluation of the effectiveness of 3D vascular stereoscopic models in anatomy instruction for first year medical students. Anat Sci Educ 10(1):34–45. https://doi.org/10.1002/ase.1626

Custer T, Michael K (2015) The utilization of the anatomage virtual dissection table in the education of imaging science students. J Tomogr Simul 1:102. https://doi.org/10.4172/jts.1000102

Dee FR (2009) Virtual microscopy in pathology education. Hum Pathol 40(8):1112–1121. https://doi.org/10.1016/j.humpath.2009.04.010

Dobson D, Pearl R, Orsay C et al (2003) Virtual reality - new method of teaching anorectal and pelvic floor anatomy. Dis Colon Rectum 46:349–352. https://doi.org/10.1097/01.DCR.0000054639.29160.9E

Douglas-Jones R (2017) 'Silent mentors': donation, education, and bodies in Taiwan. Med Anthropol Theory 4:69. https://doi.org/10.17157/mat.4.4.454

Drake RL, McBride JM, Pawlina W (2014) An update on the status of anatomical sciences education in United States medical schools. Anat Sci Educ 7(4):321–325

Ekstrand C, Jamal A, Nguyen R et al (2018) Immersive and interactive virtual reality to improve learning and retention of neuroanatomy in medical students: a randomized controlled study. Can Med Assoc Open Access J 6(1):E103–E109

Ellington DR, Shum PC, Dennis EA et al (2019) Female pelvic floor immersive simulation: a randomized trial to test the effectiveness of a virtual reality anatomic model on resident knowledge of female pelvic anatomy. J Minim Invasive Gynecol 26(5):897–901. https://doi.org/10.1016/j.jmig.2018.09.003

Erolin C, Reid L, McDougall S (2019) Using virtual reality to complement and enhance anatomy education. J Vis Commun Med 42(3):93–101. https://doi.org/10.1080/17453054.2019.1597626

Farey JE, Bui DT, Townsend D et al (2018) Predictors of confidence in anatomy knowledge for work as a junior doctor: a national survey of Australian medical students. BMC Med Educ 18:174. https://doi.org/10.1186/s12909-018-1280-5

Felszeghy SZ, Pasonen-Seppänen SM, Koskela et al (2017) Student-focused virtual histology education: do new scenarios and digital technology matter? MedEdPublish 6:10.15694/mep.2017.000154

Fredieu J, Kerbo J, Herron M et al (2015) Anatomical models: a digital revolution. Med Sci Educator 25: 183–194. https://doi.org/10.1007/s40670-015-0115-9

Fruhstorfer BH, Palmer J, Brydges S, Abrahams PH (2011) The use of plastinated prosections for teaching anatomy: the view of medical students on the value of this learning resource. Clin Anat 24:246–252

Ganapathy A, Chen D, Elumalai A et al (2022) Guide for starting or optimizing a 3D printing clinical service.

Methods 206:41–52. https://doi.org/10.1016/j.ymeth.2022.08.003
García Martín J (2018) Possibilities for the use of anatomage (the anatomical real body-size table) for teaching and learning anatomy with the students. Biomed J Sci Tech Res 4. https://doi.org/10.26717/BJSTR.2018.04.0001094
Garcia J, Yang Z, Mongrain R et al (2018) 3D printing materials and their use in medical education: a review of current technology and trends for the future. BMJ Simul Technol Enhanc Learn 4(1):27–40. https://doi.org/10.1136/bmjstel-2017-000234
Ghosh SK (2017) Cadaveric dissection as an educational tool for anatomical sciences in the 21st century. Anat Sci Educ 10(3):286–299. https://doi.org/10.1002/ase.1649
Golenhofen N, Heindl F, Grab-Kroll C et al (2020) The use of a mobile learning tool by medical students in undergraduate anatomy and its effects on assessment outcomes. Anat Sci Educ 13(1):8–18. https://doi.org/10.1002/ase.1878
Hajj I, Dany M, Forbes W et al (2015) Perceptions of human cadaver dissection by medical students: a highly valued experience. Ital J Anat Embryol 120(3):162–171
Harmon DJ, Burgoon JM, Kalmar EL (2022) Development and assessment of an integrated anatomy mobile app. Clin Anat 35(5):686–696. https://doi.org/10.1002/ca.23895
Harris T, Leaven T, Heidger P et al (2001) Comparison of a virtual microscope laboratory to a regular microscope laboratory for teaching histology. Anat Rec 265(1):10–14. https://doi.org/10.1002/ar.1036
Heidger PM Jr, Dee F, Consoer D et al (2002) Integrated approach to teaching and testing in histology with real and virtual imaging. Anat Rec 269(2):107–112. https://doi.org/10.1002/ar.10078
Hennessy CM, Kirkpatrick E, Smith CF et al (2016) Social media and anatomy education: using twitter to enhance the student learning experience in anatomy. Anat Sci Educ 9(6):505–515. https://doi.org/10.1002/ase.1610
Iwanaga J, Loukas M, Dumont AS et al (2021a) A review of anatomy education during and after the COVID-19 pandemic: revisiting traditional and modern methods to achieve future innovation. Clin Anat 34(1):108–114
Iwanaga J, Terada S, Kim H-J et al (2021b) Easy three-dimensional scanning technology for anatomy education using a free cellphone app. Clin Anat 34(6):910–918. https://doi.org/10.1002/ca.23753
Izhakoff N, Abram J, Gomez F et al (2022) Utilization of virtual 3D anatomy platforms with prosected deceased donations in the teaching of female pelvic floor anatomy. FASEB J 36(S1). https://doi.org/10.1096/fasebj.2022.36.S1.R5786
Javaid MA, Chakraborty S, Cryan JF et al (2018) Understanding neurophobia: reasons behind impaired understanding and learning of neuroanatomy in cross-disciplinary healthcare students. Anat Sci Educ 11(1):81–93. https://doi.org/10.1002/ase.1711
Jones JFX, McCaul C, Gorman L et al (2020) 3D printing in anatomy. In: Chan LK, Pawlina W (eds) Teaching anatomy: a practical guide. Springer International Publishing, pp 349–357. https://doi.org/10.1007/978-3-030-43283-6_35
Kucuk S, Kapakin S, Göktaş Y (2016) Learning anatomy via mobile augmented reality: effects on achievement and cognitive load. Anat Sci Educ 9:411. https://doi.org/10.1002/ase.1603
Kumar RK, Freeman B, Velan GM et al (2006) Integrating histology and histopathology teaching in practical classes using virtual slides. Anat Rec B New Anat 289(4):128–133. https://doi.org/10.1002/ar.b.20105
Lewis TL, Burnett B, Tunstall RG et al (2014) Complementing anatomy education using three-dimensional anatomy mobile software applications on tablet computers. Clin Anat 27(3):313–320. https://doi.org/10.1002/ca.22256
Li K, Kui C, Lee E et al (2017) The role of 3D printing in anatomy education and surgical training: a narrative review. https://doi.org/10.15694/mep.2017.000092
Little WB, Dezdrobitu C, Conan A et al (2021) Is augmented reality the new way for teaching and learning veterinary cardiac anatomy? Med Sci Educ 31(2):723–732. https://doi.org/10.1007/s40670-021-01260-8
Lotfi F, Atouf A, Atlasi MA et al (2022) Creating three-dimensional anatomy models: effects on different aspects of learning anatomy, 27 October 2022, PREPRINT (Version 1) available at Research Square. https://doi.org/10.21203/rs.3.rs-2198122/v1.
Maity S, Nauhria S, Nayak N et al (2023) Virtual versus light microscopy usage among students: a systematic review and meta-analytic evidence in medical education. Diagnostics (Basel) 13(3):558. https://doi.org/10.3390/diagnostics13030558
Mansouri M, Bigdeli S, Dehnad A et al (2020) Exploring the features of mobile phone application of anatomy in basic medical sciences: a qualitative study. BMC Med Educ 20:231. https://doi.org/10.1186/s12909-020-0214
Maresky HS, Oikonomou A, Ali I et al (2019) Virtual reality and cardiac anatomy: exploring immersive three-dimensional cardiac imaging, a pilot study in undergraduate medical anatomy education. Clin Anat 32(2):238–243. https://doi.org/10.1002/ca.23292
McBride JM, Drake RL (2015) Student perceptions of an interprofessional educational experience: the importance of goal articulation. Anat Sci Educ 8:381–385
McBride JM, Drake RL (2018) National survey on anatomical sciences in medical education. Anat Sci Educ 11(1):7–14. https://doi.org/10.1002/ase.1760
Mcmenamin P, Quayle M, McHenry C et al (2014) The production of anatomical teaching resources using three-dimensional (3D) printing technology. Anat Sci Educ 7:479. https://doi.org/10.1002/ase.1475
Mills PC, Bradley AP, Woodall PF et al (2007) Teaching histology to first-year veterinary science students using virtual microscopy and traditional microscopy: a

comparison of student responses. J Vet Med Educ 34(2):177–182. https://doi.org/10.3138/jvme.34.2.177
Mione S, Valcke M, Cornelissen M (2013) Evaluation of virtual microscopy in medical histology teaching. Anat Sci Educ 6(5):307–315. https://doi.org/10.1002/ase.1353
Mione S, Valcke M, Cornelissen M (2016) Remote histology learning from static versus dynamic microscopic images. Anat Sci Educ 9(3):222–230. https://doi.org/10.1002/ase.1572
Mitchell R, Batty L (2009) Undergraduate perspectives on the teaching and learning of anatomy. ANZ J Surg 79:118–121. https://doi.org/10.1111/j.1445-2197.2008.04826.x
Moro C, Štromberga Z, Raikos A et al (2017) The effectiveness of virtual and augmented reality in health sciences and medical anatomy. Anat Sci Educ 10(6):549–559. https://doi.org/10.1002/ase.1696
Moro C, Birt J, Stromberga Z et al (2021) Virtual and augmented reality enhancements to medical and science student physiology and anatomy test performance: a systematic review and meta-analysis. Anat Sci Educ 14(3):368–376. https://doi.org/10.1002/ase.2049
Motsinger SK (2020) Complete anatomy. J Med Libr Assoc 108(1):155–157. https://doi.org/10.5195/jmla.2019.853
Nakai K, Terada S, Takahara A et al (2022) Anatomy education for medical students in a virtual reality workspace: a pilot study. Clin Anat 35(1):40–44
Nguyen N, Wilson TD (2009) A head in virtual reality: development of a dynamic head and neck model. Anat Sci Educ 2(6):294–301. https://doi.org/10.1002/ase.115
Onigbinde OA, Ajagbe AO, Oyeniran OI et al (2021) Post-COVID-19 pandemic: Standard operating procedures for gross anatomy laboratory in the new standard. Morphologie 105(350):196–203. https://doi.org/10.1016/j.morpho.2020.10.004
Owolabi J, Bekele A (2021) Implementation of innovative educational technologies in teaching of anatomy and basic medical sciences during the COVID-19 pandemic in a developing country: the COVID-19 silver lining? Adv Med Educ Pract 12:619–625. https://doi.org/10.2147/amep.S295239
Pantanowitz L, Szymas J, Yagi Y et al (2012) Whole slide imaging for educational purposes. J Pathol Inform 3:46. https://doi.org/10.4103/2153-3539.104908
Parikh M, Rasmussen M, Brubaker L et al (2004) Three dimensional virtual reality model of the normal female pelvic floor. Ann Biomed Eng 32:292–296. https://doi.org/10.1023/B:ABME.0000012749.79488.d6
Patel S, Burke-Gaffney A (2018) The value of mobile tablet computers (iPads) in the undergraduate medical curriculum. Adv Med Educ Pract 10(9):567–570. https://doi.org/10.2147/AMEP.S163623
Patra A, Asghar A, Chaudhary P et al (2022) Integration of innovative educational technologies in anatomy teaching: new normal in anatomy education. Surg Radiol Anat 44(1):25–32. https://doi.org/10.1007/s00276-021-02868-6
Perez BA, von Isenburg MA, Yu M et al (2013) Exploring the impact of tablet computers on medical training at an academic medical center. J Med Libr Assoc 101(2):154–157. https://doi.org/10.3163/1536-5050.101.2.012
Periya SN, Moro C (2019) Applied learning of anatomy and physiology: virtual dissection tables within medical and health sciences education. Bangkok Med J 15(1):121–121
Preece D, Williams SB, Lam R et al (2013) "Let's get physical": advantages of a physical model over 3D computer models and textbooks in learning imaging anatomy. Anat Sci Educ 6(4):216–224. https://doi.org/10.1002/ase.1345
Rosas C, Rubí R, Donoso M et al (2012) Dental students' evaluations of an interactive histology software. J Dent Educ 76(11):1491–1496
Santos VA, Barreira MP, Saad KR (2022) Technological resources for teaching and learning about human anatomy in the medical course: systematic review of literature. Anat Sci Educ 15(2):403–419. https://doi.org/10.1002/ase.2142
Saw A (2018) A new approach to body donation for medical education: the silent mentor programme. Malays Orthop J 12(2):68–72. https://doi.org/10.5704/MOJ.1807.015
Scibora LM, Mead TP, Larson C (2018) The influence of iPads on course performance and student perceptions of learning in human anatomy. J Teaching Learn Technol 7(1):108–124. https://doi.org/10.14434/jotlt.v7i1.23973
Singh R, Shane Tubbs R, Gupta K et al (2015) Is the decline of human anatomy hazardous to medical education/profession?—a review. Surg Radiol Anat 37(10):1257–1265
Smith DM, Oliker A, Carter CR et al (2007) A virtual reality atlas of craniofacial anatomy. Plast Reconstr Surg 120(6):1641–1646. https://doi.org/10.1097/01.prs.0000282452.22620.25
Smith CF, Martinez-Álvarez C, McHanwell S (2014) The context of learning anatomy: does it make a difference? J Anat 224(3):270–278
Smith CF, Tollemache N, Covill D et al (2018) Take away body parts! An investigation into the use of 3D-printed anatomical models in undergraduate anatomy education. Anat Sci Educ 11(1):44–53. https://doi.org/10.1002/ase.1718
Stepan K, Zeiger J, Hanchuk S et al (2017) Immersive virtual reality as a teaching tool for neuroanatomy. Int Forum Allergy Rhinol 7(10):1006–1013. https://doi.org/10.1002/alr.21986
Stirling AC, Moro C (2020) The use of virtual, augmented and mixed reality in anatomy education. In: Chan LK, Pawlina W (eds) Teaching anatomy. Springer, Cham. https://doi.org/10.1007/978-3-030-43283-6_36
Taylor L, Dyer T, Al-Azzawi M et al (2022) Extended reality anatomy undergraduate teaching: a literature

review on an alternative method of learning. Ann Anatomy – Anat Anz 239:151817. https://doi.org/10.1016/j.aanat.2021.151817

Thomas DB, Hiscox JD, Dixon BJ et al (2016) 3D scanning and printing skeletal tissues for anatomy education. J Anat 229(3):473–481. https://doi.org/10.1111/joa.12484

Trelease RB (2016) From chalkboard, slides, and paper to e-learning: how computing technologies have transformed anatomical sciences education. Anat Sci Educ 9(6):583–602. https://doi.org/10.1002/ase.1620

Triola MM, Holloway WJ (2011) Enhanced virtual microscopy for collaborative education. BMC Med Educ 26(11):4. https://doi.org/10.1186/1472-6920-11-4

Turney BW (2007) Anatomy in a modern medical curriculum. Ann R Coll Surg Engl 89(2):104–107. https://doi.org/10.1308/003588407x168244

Yang J, Chan LK, Fok WWT et al (2019) Application of dissection peer-support system in preclinical gross anatomy education. Exp Biol 2019 Meeting Abstracts. FASEB J 33(S1). https://doi.org/10.1096/fasebj.2019.33.1_supplement.605.4

Ye Z, Dun A, Jiang H et al (2020) The role of 3D printed models in the teaching of human anatomy: a systematic review and meta-analysis. BMC Med Educ 20:335. https://doi.org/10.1186/s12909-020-02242-x

Yuen J (2020) What is the role of 3D printing in undergraduate anatomy education? A scoping review of current literature and recommendations. Med Sci Educ 30(3):1321–1329. https://doi.org/10.1007/s40670-020-00990-5

Zargaran A, Turki MA, Bhaskar J et al (2020) The role of technology in anatomy teaching: striking the right balance. Adv Med Educ Pract 11:259

Zhang Q, Deng J, Yan C et al (2020) Who is willing to donate their bodies in China? Perceptions, attitudes and influencing factors among citizens of Changsha. Ann Anat 229:151483. https://doi.org/10.1016/j.aanat.2020.151483

Integration of Gross Anatomy, Histology, and Pathology in a Pre-matriculation Curriculum: A Triple-Discipline Approach

5

Gongchao Yang, William Daley, and Dongmei Cui

Abstract

In recent years, many schools have cut back on their teaching and laboratory hours, and courses in the anatomical sciences are more integrated into the horizontal and longitudinal curriculums. Traditionally, teaching in anatomical science classes consists of lectures and laboratory sessions. Usually, gross anatomy and histology are the two main courses in pre-matriculation programs. The summer pre-matriculation program at the University of Mississippi Medical Center (UMMC) was designed to better prepare students for success in their first year of medicine and dentistry. The course provided these students with an opportunity to study gross anatomy and histology ahead of time and develop their learning skills for the coming academic year. Historically, gross anatomy and histology courses have been taught separately with a different emphasis. We have designed a new approach to implement gross anatomy, histology, and pathology—all three disciplines—in a single lecture in the organ section of the histology course. This triple-discipline (triple lecture) approach allows three professors in anatomy, histology, and pathology to work together in a two-and-a-half-hour review lecture. This approach allows students to develop their critical thinking skills and better prepares them to transfer basic anatomic knowledge into their future clinical practice. Students indicated this was their first experience with the triple-discipline approach, and they remarked that it helped them integrate anatomy and histology as well as demonstrate the clinical importance of the structures and organs. This approach can be used at different stages of medical education and helps faculty and students better integrate various disciplines and is more suitable in the modern curriculum. A similar approach was used in our fourth year course in medical histology.

Keywords

Histology · Anatomy · Pathology · Integration · Triple lecture · Teaching

G. Yang · D. Cui (✉)
Department of Advanced Biomedical Education, University of Mississippi Medical Center, Jackson, MS, USA
e-mail: gyang@umc.edu; dcui@umc.edu

W. Daley
Department of Pathology, University of Mississippi Medical Center, Jackson, MS, USA
e-mail: wdaley@umc.edu

5.1 Introduction

Medical schools in the United States offer several basic anatomical courses, and course titles vary according to the preference of each institution. In general, basic science courses in the anatomical

D. Cui et al. (eds.), *Biomedical Visualisation*, Advances in Experimental Medicine and Biology 1431,
https://doi.org/10.1007/978-3-031-36727-4_5

sciences include medical general anatomy, medical developmental anatomy, medical histology (microscopic anatomy) and medical cell biology, and medical neurobiology (medical neuroanatomy). These courses are usually taught in pre-clinical years by basic science faculty who have expertise in the anatomical sciences and their own research background. These courses are very important to students as they are the basis for their future clinical practice. Introducing and integrating clinical relevance in early basic science courses will guide students in their studies by allowing them to focus on clinical-oriented issues, thus enhancing their problem-solving skills. For many traditional basic science anatomy courses, there is very little clinical relevance in the content. The lack of communication between basic science teachers and clinicians has created a tremendous gap between basic science teaching and clinical practice. Studies have shown that students had very positive attitudes toward the relevance of anatomical materials to clinical medicine, especially final-year medical students (Moxham et al. 2016, 2017, 2022).

Many schools have launched pre-matriculation programs to bring underprepared students up to speed before the official school year begins (Crump and Fricker 2015; Heck et al. 2017). Medical schools endeavor to admit and graduate an academically qualified student body that is broad in its cultural and ethnic composition as well as rich in diverse life experiences (Tucker 2008). In order to ensure the success of students coming from non-traditional academic backgrounds, most medical schools in the United States offer a summer or pre-matriculation program to help prepare these students for the rigors of medical education (Shields 1994; Strayhorn 1999; Schneid et al. 2018). Health professional educators share the common goal of creating a learning environment that allows students with broad prerequisite knowledge to successfully progress through the curriculum (Schneid et al. 2018). Most pre-matriculation programs provide a large quantity of content in a short time. A number of factors can increase the risk of academic difficulty for some students, such as longer intervals between graduations from college, students from rural areas, students with relatively lower grade point averages (GPAs), and students having difficulty with standardized tests (Huff and Fang 1999; Tucker 2008; Kleshinski et al. 2009; Andriole and Jeffe 2010; Dunleavy and Kroopnick 2013; Schneid et al. 2018). Recognition of at-risk medical students who may require special counseling or tutoring as soon as possible is important for these students' success (Segal et al. 1999; Tucker 2008). The pre-matriculation program at our institution offers basic science courses to first-year medical and dental students to enhance the preparedness of these students before their formal first-year course begins. Most students who participate in the pre-matriculation program are from rural areas in Mississippi. These students have been accepted into the schools of medicine and dentistry at our school and will begin their first year of study in the fall semester. The pre-matriculation program is offered during the summer, and its courses include human gross anatomy, histology (microscopic anatomy), and biochemistry. The program involves an intensive 6-week training of these three courses which allow participants to increase their competitive base through early exposure to some of their professional studies. The program also offers training in study skills and test strategies. Both gross anatomy and histology classes consist of lectures and lab sessions. In order to ensure students, understand anatomical knowledge, engage in deep learning of the material, and receive an enhanced integration of clinical information into their anatomy and histology (microscopic anatomy) knowledge, we integrated gross anatomy, histology, and clinically orientated pathology, incorporating clinical cases in a one-lecture format in the histology course. The goal of the triple-discipline lecture approach was to benefit students' learning of clinical concepts by enhancing their understanding of the basic medical sciences and their clinical relevance in the early stages of their pre-clinical studies.

5.2 Course Design

Within the pre-matriculation program at UMMC, the gross anatomy course involves the study of anatomical regions, organs, and other structures—all visible to the naked eye. The gross anatomy course during the most recent summer term consisted of 16 lectures, 10 lab sessions, and 2 block exams. The lab sessions entailed cadaveric dissection and pre-lab discussions of two blocks' worth of content, including the regions of the back and upper extremities and of the head and neck. The histology course, on the other hand, involves the study of organ systems through the exploration of small-scale anatomical organs, tissues, cells, cellular ultrastructural features, and other structures—all requiring the aid of microscopy to be viewed. The histology course during the most recent summer term was composed of 13 group lectures, 8 laboratory sessions, 1 triple-discipline review lecture, 1 lecture review session, 2 laboratory review sessions, 2 block exams, and 2 post-exam review sessions on lecture and laboratory materials. Included at the beginning of the course was an introductory lecture which included the course organization, course materials, course contents, and course schedule as well as a discussion of online access to course materials, laboratory information, and a general overview of the four basic tissues, tissue preparation, and exam information. The course consisted of two blocks: a basic tissue block and an organ system block; two examinations covered the materials of each block. Post-exam reviews reviewed the exam materials from both lecture and laboratory, and analyzed and explained correct and incorrect answers.

5.2.1 Basic Tissue Block

This block covered the epithelium and exocrine glands, connective tissue, muscle tissue, and nerve tissue during the first 2 weeks. Each topic included learning objectives, tissue classification details, the function and location of each tissue type, and the specific cells and structures related to each type of tissue. Full image view of light micrographic images, illustrations, and transmission electron microscopy (TEM) and scanning electron microscopy (SEM) images were selected for each tissue type. There were six lectures, including an introductory lecture and lectures for all basic tissues: (1) introduction to histology and microscopy; (2) epithelium; (3) connective tissue; (4) cartilage and bone; (5) muscle tissue; and (6) nervous tissue. Each lecture lasted about 45 min, and each lecture focused mainly on the histological structure and function of basic tissues, the classification of basic tissues, and some examples of clinical correlations pertaining to basic tissues. In addition, there were three lab sessions: (1) lab session one: Meet microscopy; epithelium tissue and glands; (2) lab session two: connective tissue; cartilage and bone; (3) lab session three: muscle tissue and nervous tissue. Laboratory sessions cover all of the content in the basic tissue block. Each lab session involved a computer, histological slides, and microscopy connected with a camera and monitor that took about 90 minutes to complete. These three lab sessions covered content on epithelial tissue, types of glands; connective tissue, fibers, cells, and types of connective tissue; muscle tissue and associated structures and cells; nervous tissue, peripheral receptors, sensory and motor ganglia, peripheral nerve and spinal cord, brainstem, and cerebellar and cerebral cortex, respectively. In this block, students were able to identify each tissue type and subtype and describe the main functions and locations of each tissue type. They were able to recognize the particular cells and structures associated with the tissue types and subtypes and provide examples of each type and subtype of tissue.

5.2.2 Organ System Block

This organ system block contained nine lectures, including triple-discipline lectures (trip lectures), four laboratory sessions, and one laboratory review session. The content areas included the circulatory system, lymphoid organ system, oral cavity, digestive tract, digestive glands, respiratory system, endocrine system, male reproductive system, and female reproductive system. Each topic included learning objectives, an overview of organs in that system, the unique structures and cells associated with each organ, the histological features of each organ, the functions and locations of each organ, and the relevance of clinical correlations. Each lecture was about 45 minutes long, and each lecture included objectives, outline, and highlighted important topics of individual organ system illustrations, light microscopic images, TEM and/or SEM images, and flow charts. Students were also free to ask questions during the lectures. The four lab sessions lasted about 90 min each, and we covered the content in each laboratory session are: (1) Lab session one: circulatory system and lymphoid organ system; (2) Lab session two: oral cavity, digestive tract, and glands; (3) Lab session three: respiratory system and endocrine system; (4) Lab session four: male and female reproductive system, respectively. In these lab sessions, students focused primarily on identifying cells, structures, and organs, and on describing their functions and locations in the body. In this organ block, students were able to describe the structural organization and function of each organ. In addition, students were able to identify each organ and the cells, structures, and histological features associated with each organ and to describe the locations and functions of these cells, structures, and features.

5.2.3 Triple-Discipline Lecture (Triple Lecture) Format

Although relevant clinical correlations were included in most of the course lectures, the triple-discipline lecture was the highlight and summary review of the course. The triple lecture was implemented in the organ system block, when students had learned the content from the cardiovascular system, lymphoid system, digestive tract, digestive glands, respiratory system, endocrine system, and male and female reproductive systems. At that time, students had a basic knowledge of tissues and organs in the histology course and had completed most aspects of the gross anatomy content associated with organs in an anatomy course. This collective prior knowledge provided ideal timing to integrate anatomy, histology, and pathology in a triple-discipline lecture format that allowed students to review histological content while considering its gross anatomical and clinical contexts. Discussions of the clinical cases aided students in comparing and contrasting the pathological details of organs and the details of their normal microscopic and macroscopic anatomy. The triple-discipline, or triple, lecture lasted approximately two and a half hours and included a brief break approximately halfway through the triple-discipline lecture session. During this triple lecture, the anatomy professor focused on the anatomical orientation, the position and location of the organs, and their blood supply. The histology professor reviewed the cells, tissues, and features of the organs, as well as their functions and locations. The pathology professor, who is also a clinician and pathologist, then discussed the clinical correlation of each organ and structure and their components that the anatomy and histology professors had reviewed. In this session, a new clinical case was provided and discussed in the classroom. For example, first, the anatomy professor reviewed the anatomy of the liver, its blood supply, its location in the human body, its different lobes, and its relationship to the gallbladder. Second, the histology professor reviewed the microscopic structures of the liver, such as the portal triad (bile ductule, hepatic portal vein, and hepatic artery), the lobules of the liver (classic lobules, portal lobules, and hepatic acini), and hepatocytes. In this specific example, three zones of histological hepatic acinus were examined, and the students were asked the following questions: (1) In the case of a lack of blood supply, insufficient oxygen, and nutrients, to the liver which zone will be affected first? (2) In a case in which a patient is exposed to blood-borne toxins, which area of the liver is most susceptible to injury? Once the students had responded to the questions, the pathologist explains the pathological change in the liver. He also gave an example of a case involving an individual with an

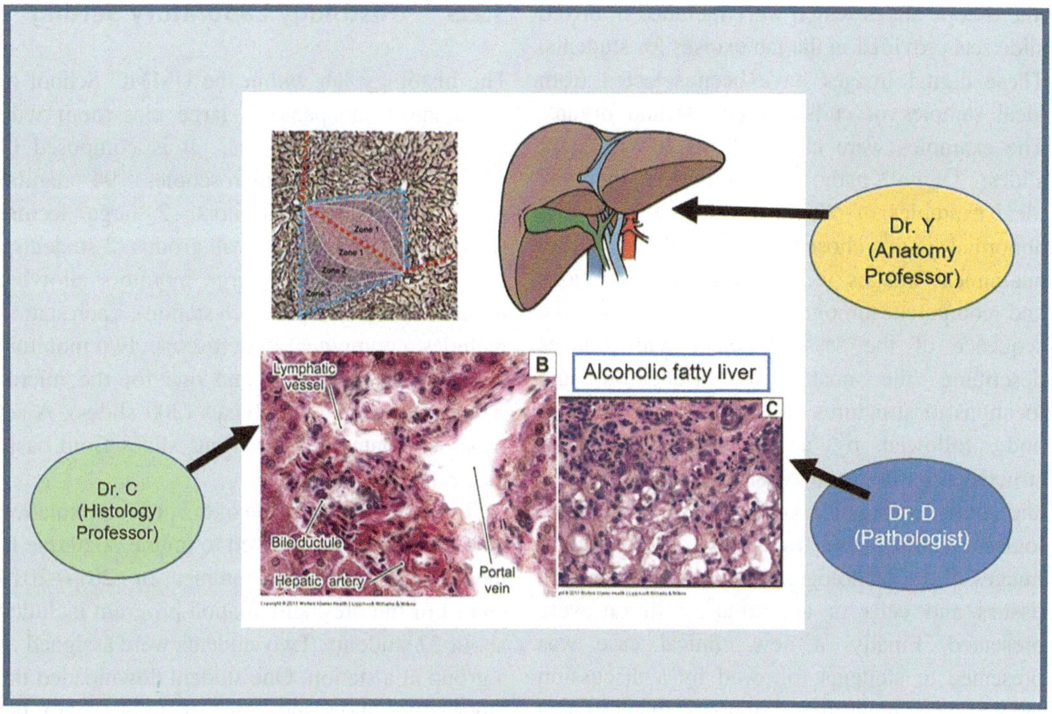

Fig. 5.1 An example of the PowerPoint slide used in the triple-discipline lecture. The slide contained an image of liver anatomy which was reviewed by anatomy professor, histological images reviewed by the histology professor, and a pathological image reviewed by the pathologist (Photo credit: reproduced with permission from Wolters Kluwer/Lippincott Williams & Wilkins. Cui et al. 2011 *Atlas of Histology with Functional & Clinical Correlations*)

alcoholic fatty liver which results from alcoholic fatty liver disease, or hepatic steatosis, an asymptomatic and reversible liver condition associated with alcohol consumption in which lipids accumulate in the hepatocytes (Fig. 5.1).

The professors also provided a different clinical case related to the liver so that students could diagnose the disease. There were five potential answers for students to consider, and only a single correct answer was accepted. Once students had provided their thoughts and answers, the pathologist discussed why they were correct or incorrect. The following items were discussed in this clinical case: diagnosis, pathological changes (normal vs. abnormal), signs and symptoms associated with the condition of the patient, and treatment options. For instance, one clinical case of hepatocellular carcinoma was presented to the students for discussion. The associated question inquiring about the identity of the discussed pathology included five answer choices: (A) Cirrhotic nodule, (B) Hemangioma, (C) Hepatocellular carcinoma, (D) Liver adenoma, and (E) Metastatic colon cancer. Most students chose hepatocellular carcinoma, but some students chose other options. The pathologist explained why hepatocellular carcinoma was the correct answer and discussed the diagnostic pathological changes, signs, and symptoms related to the patient's condition and treatment.

5.2.4 Digital Image Integration and Sequence of Triple-Discipline Lecture

Most digital histology images used in the lecture and the lab guide were obtained from glass microscope slides which were included in boxed slide sets provided in the lab excises for students. These digital images have been selected from ideal samples of cells, structures, and organs. The examples were chosen from 300 different slides. Digital pathological images included ideal examples of clinically relevant cases and abnormal tissues chosen by a pathologist. Gross anatomical images were based on illustrations and computed tomography (CT) images. The sequence of the session began with images describing the anatomical orientation and locations of structures and organs in the human body followed by a microscopic view of structures, tissues, and cells in normal condition and about a discussion of their associated histological features and functions. Subsequently, images of the pathological views of the relevant tissues and cells in abnormal condition were presented. Finally, a new clinical case was presented to students followed by a discussion of the case regarding diagnosis, pathological changes, signs, symptoms, and treatment. The sequence of the triple-discipline lecture is described in Fig. 5.2.

5.2.5 Histology Laboratory Setting

The histology lab within the UMMC School of Medicine is a spacious, large classroom with around 100 movable seats. It is composed of 47 computers, 49 microscopies, 94 regular monitors, 4 large monitors, 2 large lecture screens, 47 stations for small groups (2 students), and 4 stations with 4 large monitors allowing flexible use (3 students each station). Each station includes a computer, a microscope, two monitors (one for the computer and one for the microscope), and three slide boxes (300 slides). A set of slides contains 300 various slides from basic tissues, cells, and organs.

The laboratory session of the pre-matriculation histology class was moved to online 2020 due to the COVID-19. The summer of 2017–2019 cohort of the pre-matriculation program included about 22 students. Two students were assigned as a group at a station. One student downloaded the laboratory guide and other materials onto the computer screen, and the other student operated the microscope to project slides onto the monitor connected to the microscope. Slides could be

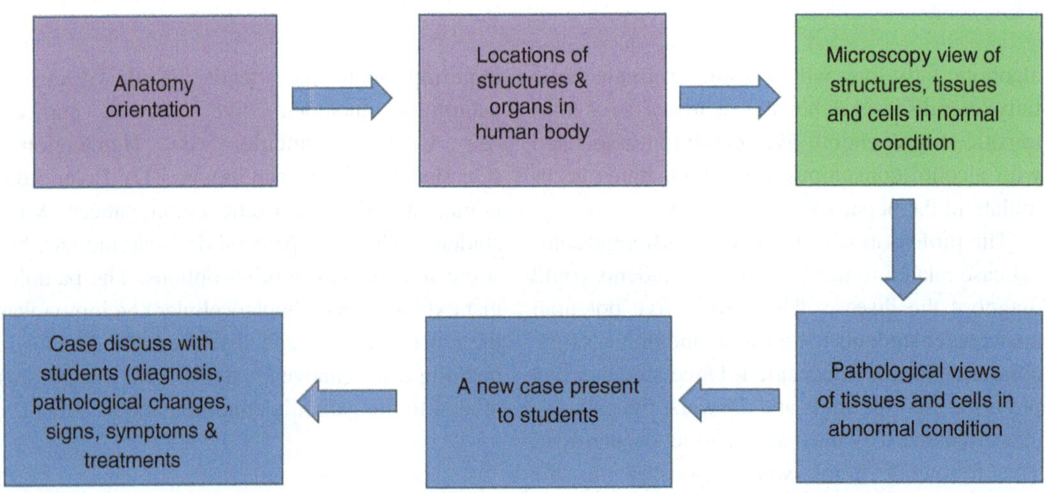

Fig. 5.2 The sequence of the triple-discipline lecture. The sequence of this lecture is shown in this flow chart. The anatomical orientation and locations of structures and organs were presented first by the anatomy professor. Microscopic views of structures, tissues, and cells in normal condition were reviewed second by the histology professor. Thirdly, pathological views of tissues and cells in abnormal condition were shared by a pathologist. Finally, a new case of a patient in the form of a clinical-vignette style question was presented to students, and the pathological changes, signs, symptoms, diagnosis, and treatment options were discussed with the students

Fig. 5.3 Overview of the laboratory setting in the UMMC Histology/Computer Lab. Each station within the histology lab is composed of one microscope and two monitors (one connected to the computer and another connect to the microscope and camera). Each station also includes a blue cup and a red cup to notify faculty when assistance is not needed and when it is needed, respectively (At each station, the red cup is underneath or inside the blue cup)

easily moved upward or downward, moved left or right, and magnified or reduced. Students may have one or more screen views of the pictures. If they found an ideal example of a specimen, structure, or cell, they were able to photograph it, study it later, or share it with the class. During lab sessions, the histology professor utilized the master microscopy station in the lab to show good examples of cells and structures and to demonstrate the navigation techniques and tips for how to locate them. This master microscopy station includes one computer, one microscope, and one monitor, whose display can be presented on the students' monitors at their stations, and it is located in the front of the classroom where the histology professor administered laboratory quizzes to the students at the end of each laboratory and review session before exams. In addition to the instructor's capability of presenting the view of slides displayed on the master monitor onto the monitor at each student's workstation, instructors could also display images of their microscope slide views as presented on the large screens in the front and screens along the walls of the lab. During the lab sessions, blue cups were used as a signal that no assistance from faculty was required by students at their stations. In other words, when students were working on their own and did not need help locating or viewing cells, tissues, or structures using the microscope, they left the blue cups in their designated locations at their respective stations. However, if any student had questions or needed assistance, they would uncover the red cup beneath the blue cup and place it on top of the blue cup at their respective stations. This protocol notified faculty when students needed help without students' having to keep their hands raised until a faculty member arrived at their respective stations. This method was particularly useful when multiple groups of students had questions at the same time as it allowed the students to continue working efficiently while the faculty approached them systematically one by one. Figure 5.3 depicts this described histology laboratory setting.

5.3 Online Access

Students had access to course material for the duration of the 6 weeks of the pre-matriculation program. Canvas was used within the course as the online learning management platform which allowed professors to make announcements, upload materials to modules, issue assignments to students, administer quizzes to students, and store students' grades which could be automatically communicated to students once grades were posted. Canvas also provided an avenue for communication between students and faculty and had the capability of serving as an online teaching and learning platform in conjunction with BigBlueButton. This capability is highly useful in the event of another pandemic in the future or in the case of a need for physical distance learning offered to students in remote locations. The following materials were made available to students in their Canvas modules: course contents and a course schedule, reading suggestions, lecture PowerPoints, a laboratory introduction, histology laboratory session materials (laboratory guides), laboratory review materials, course references (specimen preparation, tissue staining techniques, glossary terms for describing cells, and terminology for organs), and histology materials for each block.

5.4 Course Feedback from Students

For the most part, faculty received positive feedback from students on several aspects of the course. The students also provided faculty with helpful feedback to enhance the quality of the course for future students and educational improvement.

5.4.1 Students' Feedback on the Traditional and Triple-Discipline Approaches

Students indicated that this course was their first experience with the triple-discipline approach, and remarked that it helped integrate anatomy and histology and demonstrate the clinical importance of relevant structures. The course feedback from students was positive. Eight four percent of students preferred the integrated triple-discipline lecture over the traditional lectures, 5% preferred traditional lectures over the triple lecture, and 11% preferred both lectures (Fig. 5.4). All

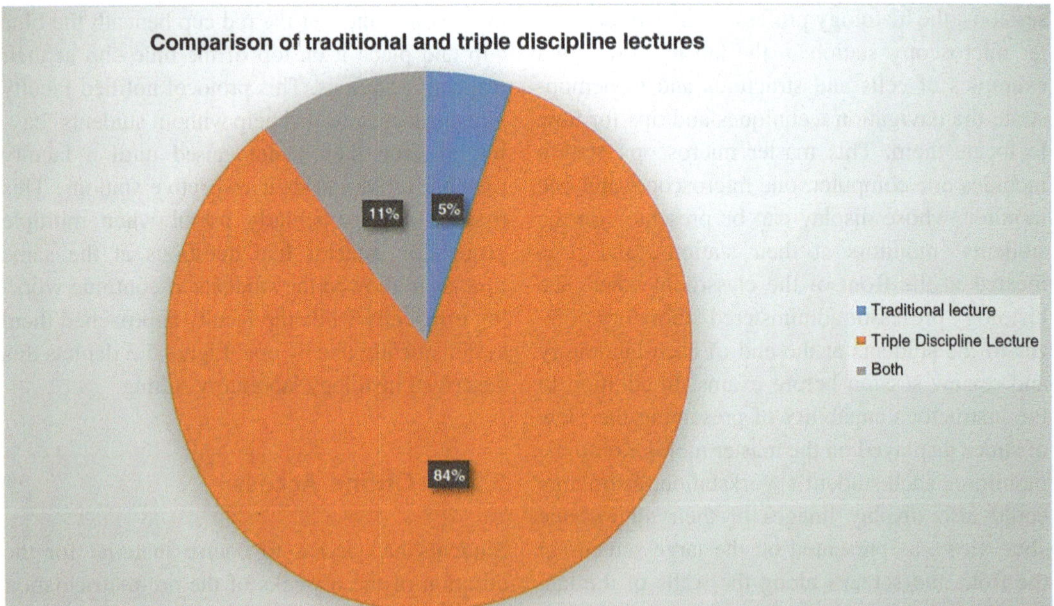

Fig. 5.4 Students' perceptions of traditional and triple-discipline lectures. This pie chart shows students' feedback regarding the comparison of traditional and triple-discipline lectures

students either strongly agreed or agreed that the triple lecture was helpful to learn the important relationships between anatomy, histology, and pathology/clinical correlations. All students also agreed that the triple-discipline approach helped them to think about and explain the human body from gross anatomical, histological, and pathological perspectives. Students offered comments, such as "I like the triple lecture idea. It allowed me to make more connections in my gross anatomy class." "I like the integration of disciplines and very well-organized course." "I enjoyed the triple lecture because it keeps the class very engaged." "I enjoy the triple lecture and it was very helpful and interesting!" "I did enjoy the triple lecture thought I took more away than regular PowerPoints." A student also suggested incorporating an additional triple lecture during the first 3 weeks (after the basic tissue block).

5.4.2 Students' Course Feedback on the Laboratory Setting

Overall, students enjoyed studying in the lab. They indicated that they liked to spend more time in the lab than in a lecture session and that they liked to have a mandatory lab review before testing. All students either strongly agreed or agreed that a laboratory setting that allows the examination of slides with a microscope and the monitor is very useful for learning histology (Fig. 5.5). Students have also commented, "I really liked the triple lecture and also enjoyed the lab portion of the class." "I liked how we were given lab documents that had pictures of the exact slides we viewed in the lab. I was able to use the labelled slides from the document to study the slides in the lab." "I think more lab time would be beneficial so all the slides for each section can be viewed and studied."

The students also recommended that faculty needed to include more clinical correlations and

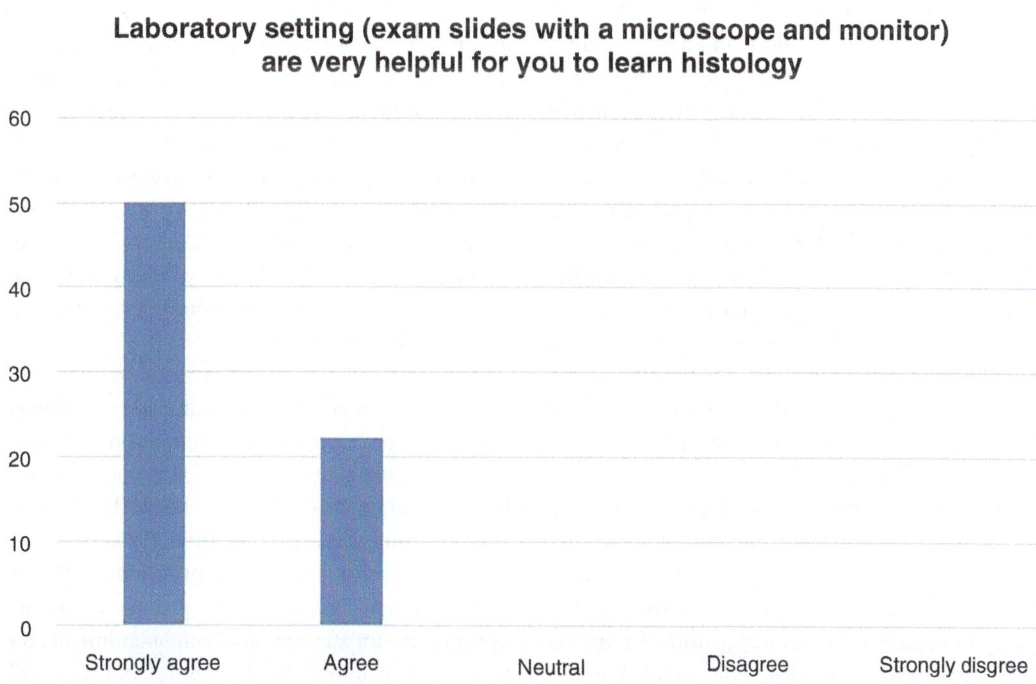

Fig. 5.5 Students' views on traditional microscopy in the histology lab. This column chart depicts on a 5-point Likert scale students' feedback regarding the use of traditional microscopy with a monitor to examine slides to study histology in the laboratory setting

identifications of abnormalities on the microscope slides. In other words, they would have appreciated seeing more clinical applications. Overall, they indicated that the course was very well structured and beneficial to learning histology; the teaching method was very beneficial, and each topic was explained in detail.

5.5 Discussion

The pre-matriculation program varies by school. The program is intended to provide academic enrichment and support services to disadvantaged students and students from various backgrounds. This type of program has the potential to improve students' performance during their first year of medical school (Schneid et al. 2018). First-year medical and dental courses are challenging for students, and the pre-matriculation program can help reduce students' stress and increase students' confidence early in the school year. The strategy and method of designing courses and programs are extremely important, especially when the duration of the course is short, and a large amount of content needs to be covered. The triple-discipline lecture format is a novel approach to teaching methodology. The UMMC pre-matriculation program initially implemented this new triple-discipline lecture approach in the histology course which has helped students learn the important relationships between anatomy, histology, and pathology/clinical correlations. Students indicated that this approach was more effective than regular lectures. To our knowledge, there is insufficient evidence that this triple-discipline lecture approach has been implemented in the pre-matriculation programs at other institutions. This approach allows for the integration of anatomy, histology, and pathology knowledge into a single course. It also helps students to think and explain the three-dimensional (3D) orientation of structures within the human body and their relationships to one another as well as their relationship across these three disciplines. Students also indicated that other clinical correlates should be added to the course.

Since many medical schools have reduced lecture hours and laboratory hours (Cottam 1999; Drake 1998, 2014; Drake et al. 2002, 2009, 2014; Verhoeven et al. 2002; Moxham and Plaisant 2007; Yeung et al. 2012), and in many cases, they have moved their physical conventional microscopy histology laboratory sessions completely to virtual microscopy online or virtual slides are available for students' self-study (Cotter 2001; Boutonnat et al. 2006; Campbell et al. 2010; Bloodgood 2012; Barbeau et al. 2013; Caruso 2021). In these situations, students have lost the opportunity to study as a group in the physical conventional microscopy laboratory. To some degree, the educational culture has lost connection with how histological tissue samples are prepared and handled, how slides are viewed under a microscope, and how students can work together to discuss observations from microscope slides and relevant histological concepts in a lab setting that fosters guided inquiry and critical thinking. At UMMC, students still have the opportunity to work as a small group within the histology lab and examine slides with the use of an updated light microscope, camera, computer, and monitors under the guidance of instructors in the lab. Students are allowed to take pictures and share them with the class. Overall, students in the most recent pre-matriculation cohort loved the teamwork aspect engendered by the laboratory environment and exuded excitement when they found ideal examples. Students enjoyed the lab setting and liked working in the lab, even asking for more time in the lab.

Pre-matriculation programs have traditionally covered selected topics that have allowed students to develop a strong foundation for underlying principles and then to better apply higher-level, problem-solving skills when the material was encountered a second time (Schneid et al. 2018). The pre-matriculation program at UMMC is no different, but it has also provided a unique opportunity for students studying anatomical content to do so in a triple-discipline approach with in-class lectures and in-person laboratory sessions, which combine microscopy and computer displays. The participating students studied ahead of time and reached the goal of gaining preparation for their upcoming first year of medical and dental studies.

5.6 Conclusions

Our methods of course arrangement, laboratory setting operation, and the triple-discipline approach have benefitted students' learning in the pre-matriculation program. The students liked the organization of the course and enjoyed both the triple lecture and learning in the technologically enhanced traditional microscopy laboratory environment. Limitations of the course included the fact that too much material was covered in the 6-week course, so reducing course content or increasing course hours might be considered in the future. The current trend is towards a greater increasing clinical correlation in medical education. Clinical information is important for medical and dental students during both their pre-clinical and clinical years of study because the clinical relevance helps improve the retention of content for dental (Persky et al. 2017) and medical students (Malau-Aduli et al. 2019). Therefore, the amount of clinical correlation content should be increased gradually from year to year. Integration of clinical contents in the pre-matriculation program has provided a good starting point for this gradual increase in the emphases of clinical applications in medical education. The triple discipline approach, which integrates gross anatomy, histology, and pathology in pre-matriculation curriculum is a novel and achievable teaching method that can be implemented in premed, medical, and health-related school curricula in the future.

References

Andriole DA, Jeffe DB (2010) Prematriculation variables associated with suboptimal outcome for the 1994-1999 cohort of US medical school matriculants. JAMA 304(11):1212–1219

Barbeau ML, Johnson M, Gibson et al (2013) The development and assessment of an online microscopic anatomy laboratory course. Anat Sci Educ 6:246–256

Bloodgood RA (2012) Active learning: a small group histology laboratory exercise in a whole class setting utilizing virtual slides and peer education. Anat Sci Educ 5:367–373

Boutonnat J, Paulin C, Faure C et al (2006) A pilot study in two French medical schools for teaching histology using virtual microscopy. Morphologie 90:21–25

Campbell G, Demetriou LA, Arnett TR (2010) Virtual histology in the classroom and beyond. Med Educ 44:1124–1125

Caruso MC (2021) Virtual microscopy and other technologies for teaching histology during Covid-19. Anat Sci Edu 14:19–21

Cottam WW (1999) Adequacy of medical school gross anatomy education as perceived by certain postgraduate residency programs and anatomy course directors. Clin Anat 12:55–65

Cotter JR (2001) Laboratory instruction in histology at the University at Buffalo: recent replacement of microscope exercises with computer applications. Anat Rec 265:212–221

Crump WJ, Fricker S (2015) A medical school prematriculation program for rural students: staying connected with place, cultivating a special connection with people. Teach Learn Med 27(4):422–430. https://doi.org/10.1080/10401334.2015.1077709

Drake RL (1998) Anatomy education in a changing medical curriculum. Anat Rec 253:28–31

Drake RL (2014) A retrospective and prospective look at medical education in the United States: trends shaping anatomical sciences education. J Anat 224:256–260

Drake RL, Lowrie DJ Jr, Prewitt CM (2002) Survey of gross anatomy, microscopic anatomy, neuroscience, and embryology courses in medical school curricula in the United States. Anat Rec 269:118–122

Drake RL, McBride JM, Lachman N (2009) Medical education in the anatomical sciences: the winds of change continue to blow. Anat Sci Educ 2:253–259. https://doi.org/10.1002/ase.117

Drake RL, McBride JM, Pawlina W (2014) An update on the status of anatomical sciences education in United States medical schools. Anat Sci 7:321–325. https://doi.org/10.1002/ase.1468

Dunleavy DM, Kroopnick MH (2013) The predictive validity of the MCAT exam in relation to academic performance through medical school: a national cohort study of 2001-2004 matriculants. Acad Med 88(5): 666–671

Heck AJ, Gibbons L et al (2017) A survey of the design of pre-matriculation courses at US medical schools. Med Sci Educ 27:229–236

Huff KL, Fang D (1999) When are students most at risk of encountering academic difficulty? A study of the 1992 matriculants to U.S. medical schools. Acad Med 74(4): 454–460

Kleshinski J, Khuder et al (2009) Impast of preadmission variables on USMLE step 1 and step 2 performance. Adv Health Sci Educ Theory Pract 14(1):69–78

Malau-Aduli BS, Alele FO, Heggarty P et al (2019) Perceived clinical relevance and retention of basic sciences across the medical education curriculum. Adv Physiol Educ 43:292–299. https://doi.org/10.1152/advan.00012.2019
Moxham BJ, Plaisant O (2007) Perception of medical students towards the clinical relevance of anatomy. Clin Anat 20:560–564
Moxham BJ, Emmanouil-Nikoloussi E et al (2016) The attitudes of medical students in Europe toward the clinical importance of embryology. Clin Anat 29(2): 144–150. https://doi.org/10.1002/ca.22667
Moxham BJ, Emmanouil-Nikoloussi E et al (2017) The attitudes of medical students in Europe toward the clinical importance of histology. Clin Anat 30(5): 635–643. https://doi.org/10.1002/ca.22889
Moxham BJ, Brenner E et al (2022) The attitudes of European medical students towards the clinical importance of neuroanatomy. Ann Anat 239:151832. https://doi.org/10.1016/j.aanat.2021.151832
Persky AM, Wells MA, Sanders KA et al (2017) Improving dental students' long-term retention of pharmacy knowledge with "Medication Minutes". J Dent Educ 81:1077–1084. https://doi.org/10.21815/JDE.017.062
Schneid SD, Apperson A et al (2018) A summer prematriculation program to help students succeed in medical school. Adv in Health Sci Educ 23:499–511. https://doi.org/10.1007/s10459-017-9808-8
Segal SS, Giordani B et al (1999) The academic support program at the University of Michigan School of medicine. Acad Med 74:383–385
Shields PH (1994) A survey and analysis of student academic support programs in medical schools focus: underrepresented minority students. J Natl Med Assoc 86:373–377
Strayhorn G (1999) Preadmissions programs and enrollment of underrepresented minority students before and during successful challenges to affirmative action. J Natl Med Assoc 91:350–356
Tucker (2008) Performance in a prematriculation gross anatomy course as a predictor of performance in medical school. Anat Sci Ed 2008(1):224–227
Verhoeven BH, Verwijnen GM, Scherpbier AJ et al (2002) Growth of medical knowledge. Med Educ 36:711–717
Yeung JC, Fung K, Wilson TD (2012) Prospective evaluation of a web-based three-dimensional cranial nerve simulation. J Otolaryngol Head Neck Surg 41:426–436

Methods for Assessing Students' Learning of Histology: A Chronology Over 50 Years!

6

Geoffrey T. Meyer

Abstract

Over the past 50 years technology has made teaching and learning more interesting and has been a useful tool to motivate students to learn.

Students today can learn using computers, iPads and mobile devices to access increasing numbers of educational websites and many eLearning apps. E-learning is now an integral component of most teaching delivery. There have been many innovations utilising both technology-enhanced and interactive learning strategies to revolutionise histology teaching. Now learning resources can be successfully delivered via the Internet, so students can complete all learning outcomes away from the traditional histology classroom environment (i.e. study histology online).

The rapid developments in technologies and computer-based learning opportunities coincided with the appearance of Teaching and Learning Centres in Universities promoting teaching practices and supporting more improved learning strategies.

This chapter describes how assessments in histology have become an important tool for student learning. A chronological documentation of various assessment opportunities enabled by the increasing use of technology will be described. In particular, assessment packages that engage students and return immediate feedback make histology learning efficient and significantly improve students' performance on examinations.

Assessment strategies described here may be useful for "early career" histology teachers engaging eLearning and/or new teaching departments at new universities embarking on presenting curricula in newly established medical, dental and biomedical/health science programmes.

This chapter is also an historical account of at least some assessment practices in histology over the past 50 years. Perhaps similar practices were used to assess student learning in other closely related disciplines such as anatomy including neuroanatomy.

Keywords

Histology images · Online learning · Virtual microscopy · Microanatomy · Histology quizzes · Histology atlas

6.1 Introduction

In general, it may seem that today's students have it easier since teaching pedagogy has taken enormous advances coupled with technologies now focussed on engaging students and enhancing learning. But academics teach what they research

G. T. Meyer (✉)
School of Human Sciences, The University of Western Australia, Perth, WA, Australia
e-mail: geoff.meyer@uwa.edu.au

D. Cui et al. (eds.), *Biomedical Visualisation*, Advances in Experimental Medicine and Biology 1431,
https://doi.org/10.1007/978-3-031-36727-4_6

and what is constantly being discovered, so students today must learn a lot more knowledge than students did 50 years ago. Certainly, technology has made teaching and learning more interesting and has been a useful tool to motivate students to learn.

Unlike students today, apart from attending traditional formal classes such as lectures and laboratory practicals or tutorials, the bulk of learning 50 years ago involved using a textbook and spending hours in the university libraries turning pages of what seemed an infinite number of reference books and scientific journals. A student 50 years ago lived in a world without the Internet and the extensive numbers of technologies available for learning today. Students today enjoy educational strategies that have embraced the use of the Internet using computers, iPad and mobile devices to access increasing numbers of learning websites and subscribe to a host of eLearning apps. E-learning is now an integral component of most teaching delivery with emerging technologies for use in education occurring on an almost daily basis.

The COVID pandemic has highlighted the essential role of digital technology in education. The more affluent universities have been able to cope and support academics needing to move to the online delivery of course content. But at the same time, the COVID pandemic also highlighted inequalities in university education worldwide with some universities not being able (financially) to implement eLearning strategies with academics lacking the infrastructure and expertise/advice to present content in an eLearning mode. This has triggered opportunities for academics in one university to share learning content with other academics in many other universities and for their students to access content that is presented in a modern pedagogical environment.

In a previous article (Meyer 2023) innovations utilising both technology-enhanced and interactive learning strategies to revolutionise histology teaching were described. With the successful delivery of learning resources via the Internet, students can complete all learning outcomes away from the traditional classroom environment (i.e. studying histology online). These innovations address and help solve cognitive challenges that students experience in histology learning. But also, students worldwide can access these resources as well as their histology teachers to complement their delivery of teaching content.

6.1.1 Where We Were Teaching Histology 50 Years Ago and Where We Are Now: What Has Changed?

When I first started the journey as an academic at the beginning of the technological era academic colleagues and many university administrations were resistant to change, and there were no significant allocations of university budgets to support teaching and learning. Professional development opportunities in teaching and learning did not exist. The general teaching practice across all disciplines (until the early 1990s) was to present the curriculum and at the end of the teaching period test students' learning with a formal examination. Mostly lecturers practised the same pedagogies as their lecturers did before them. The advent of Teaching and Learning Centres in Universities began the transformation of promoting teaching practices and supporting more improved learning strategies, and this was accompanied by rapid developments in technologies and computer-based learning opportunities. With regards to teaching histology, new innovative technologies allowed learners to visualise histology anywhere, at any time. These are outlined in Meyer (2023), and Chapman et al. (2020) describe flexible self-directed learning through social media, live streaming and virtual reality.

6.1.2 Innovations in Teaching and Learning Histology Has Revolutionised the Uses of Assessment

The purpose of this chapter is to extend the descriptions of innovative histology learning platforms presented in the previous article (Meyer 2023) and describe how assessments in histology have become an important tool for student learning. This chapter provides a chronological documentation of various assessment opportunities enabled by the increasing use of technology. Emphasis is on the use of technologies to create assessment packages to engage students and those that provide immediate feedback which makes histology learning efficient and can significantly improve students' performance on examinations.

What prompted this chapter to be written was a recent casual meeting with students who were amazed that when I studied histology (1973), assessments were very formal, and limited technologies prevented the use of assessments as an important learning strategy as they are used today. This is not a chapter describing assessment practices within universities or a critical review of those practices, but my own personal experiences developing these assessment opportunities as I embraced each developing technology over at least the past three decades. I am sure similar assessment practices are done in many histology teaching courses throughout universities worldwide, but the descriptions presented here may be useful for "early career" histology teachers engaging eLearning and/or new teaching departments at new universities embarking on presenting curricula in newly established medical, dental and biomedical/health science programmes. It also serves as a historical account of at least some assessment practices in histology over the past 50 years and may be like other closely related disciplines such as anatomy including neuroanatomy.

The first section of this chapter will briefly give an account of how technology has developed over the past 50 years and has provided better learning tools for students and assessment strategies for histology teachers. Then, an account of various assessment opportunities using these technologies will be described—in particular, the concept of assessment practices for student learning.

6.2 A Histology Examination in 1973: 50 Years Ago

As a student, I completed a histology course in 1973—50 years ago at the University of Western Australia (UWA). The histology course was over three terms of 9 weeks each, and the course enrolment was about 60 students. Today's enrollment in a similar histology course, with all the changes within the university department's teaching mission, is over 270 students which necessitated a major change in teaching practice over at least two decades.

My histology course in 1973 was presented as 1 lecture per week and a 3-h laboratory practical session each week. The curriculum covered an entire histology course namely, "Basic Tissues" and "Human Organ Systems".

The typical practical histology class involved students viewing histological sections of specific tissues and organs for each topic presented in the curriculum. They viewed the sections using a microscope and were guided by a printed laboratory practical manual. All formal learning using resources such as histology slides and microscopes needed to be completed within the 3-h timeframe because the histology classroom was a multipurpose room, and students could not enter the room to study histology slides outside their allocated 3-h timeslot each week. So, any review or revision was only available if a student could fit it into subsequent laboratory sessions or on a very restricted timeframe during "study vac"—the week prior to examinations. Nowadays, with the use of technologies, students can review content anytime and anywhere.

6.2.1 In 1970s Continuous Assessment, Practice Quizzes etc., Were Very Limited

In the 1970s continuous assessments, practice quizzes etc., were not part of any histology course structure. The only assessment was an "end of term test" each term in "practical histology" and a final written examination at the end of the academic year. Apart from 1 or 2 "mystery slides" set up at the back of the histology classroom, each practical session "practice quizzes" with feedback either as a learning resource or as a preparation for formal assessments were not provided.

There was no provision for an end-of-term formal practical histology examination in the University examination calendar, so an "end of term practical test" was scheduled during the allocated practical laboratory time in the last week of each semester 1, 2 and 3. Each practical test consisted of "musical chair" rotations, each 1.5 min through 40 well-structured questions—mostly using a microscope to identify a tissue or organ. There were not many light microscope images of cells, apart from electron micrographs of a collection of cells of which only about five were included in the test. Because there were 40 questions in each examination, it meant that about five questions could be set on each of the eight topics covered each term. Each of these three practical histology tests contributed 10% towards the final mark.

The questions in the practical histology tests were usually only either "identify this tissue or organ" or "name the function/s of this tissue or organ". Questions for each "end of term test" only related to the content delivered in each of the three terms. The only feedback after each test was a copy of the test paper (with the correct answers included) pinned on the noticeboard in the practical histology laboratory classroom. Students could come to the noticeboard and check their responses because their test answer sheet was returned after the marking period—usually 2–4 days later.

A final 3-h written examination (worth the remaining 70%) was scheduled during the final examination period at the end of the third semester (mid-late October). The final examination examined the content presented over the three terms. It was divided into three sections with students required to answer any two of four questions asked in each of the three sections which tested content presented in Term 1 (Section 1), Term 2 (Section 2) and Term 3 (Section 3). The suggestion was to devote 30 min to each of the six questions within the 3-h exam time period. All questions were of equal value. Questions were essay or extended answer questions. An example of *two* of the *four* choices ONLY in each section is shown in Table 6.1.

The histology curriculum covered about 24 topics (i.e. 8 topics per term). A "one off" formal examination of this type could not test all topics; in fact, only 50% of the topics could be tested since each question was confined to 1—topic content only. It was a bit of a lottery as to what topic was to be examined. In such circumstances, students found this very frustrating as they had spent a lot of time completing the learning to find *that* specific learning was neither assessed nor able to contribute to their final scores or grade.

There was no opportunity for any feedback in this final examination. The student's final mark was provided in his/her "end of year" final results.

6.2.2 Summary

Histology teaching aids or resources were very limited in the 1970s. The only learning opportunities were attending a 45-min lecture, consulting a textbook and attending a 3-h laboratory practical session. At the end of each of the three terms, there was a well-structured practical test worth a total of 30% and a formal final examination worth 70% but rather limited in the amount of content it could test. In retrospect, perhaps the practical test should have been worth 70% and the formal final examination only 30%. Alternatively, a formal 3-h examination consisting of short answer questions and multiple-choice questions (MCQs) (rather than a written examination) could have tested a lot more

Table 6.1 1973 written histology examination questions (examples "in part" only)

Section	Question (2 of 4 shown only)
1 (Term 1 content)	1. Describe the various types of epithelia and how they are related to their functions. 2. Describe the processes involved in intramembranous and endochondral ossification.
2 (Term 2 content)	1. Compare and contrast the histological structure of the three types of muscle tissue. 2. Describe the histological appearances of all the types of glial cells in an H&E section and state their major function/s.
3 (Term 3 content)	1. Compare and contrast the mucosa of the stomach, duodenum and the colon. 2. Describe the classic liver lobule, the portal lobule and the liver acinus.

learning of the content. This was what evolved over the next five decades when expected learning outcomes were clearly defined and all learning outcomes were assessed.

6.3 Emerging Technologies Initiated New Assessment Strategies

The appropriate type of assessment ultimately depended on the course format and the time available to provide the learning but also what technology was available. Fifty years ago, technology did not play a major part in learning and assessments. In the coming paragraphs, advances in teaching practices over the past five decades as well as the opportunities for changes in assessment strategies (all as a result of emerging technologies) will be described.

6.4 Use of the Light Microscope

At least up until about the last decade, proper use of the microscope was an important component of the skills of learning histology. Certainly, back in the 1970s and up until about 2005, exam questions often tested the parts of a light microscope and physical principles such as resolving power etc. In the practical histology tests during that period, one question required the student to position a histological slide on the stage of a microscope and focus on a particular histological feature at 10× eyepiece and ×40 objective lens with correct illumination whilst supervised by a proctor (usually a graduate student). The histological feature was easy to locate (e.g. skin epidermis or the lumen of a large tubular structure such as the oesophagus) as it was an exercise in correct use of the microscope. There were always a small number of students who had trouble focussing because they had placed the slide on the stage upside down! Students were allocated marks for both correct use of the microscope but also for successfully locating the correct histological structure. Often students failed to successfully locate the histological structure in the time provided.

6.4.1 Lecture Formats (1973 Onwards)

Various new teaching aids were available over time to improve the visual presentation of knowledge to studentsduring "en-face" lectures. A short description of some of these aids are presented below (Sects. 6.4.2–6.4.4).

6.4.2 Blackboards, Whiteboards and Overhead Projectors

Lectures in a lecture theatre were originally presented using a blackboard or a whiteboard with very little or no images. In lecture theatres that did not have a blackboard or whiteboard, the lecturer used ink pens and transparent sheets on an overhead projector. A one-page handout of the lecture synopsis was usually provided to students. Textbooks were considered an essential resource for studying histology as they had images of cells, tissues and organs as well as more detailed descriptions of the structures presented in the lectures that were often only skimmed over due to time restraints of the 45-min lecture format. Interestingly, today most histology textbooks (Mescher 2013; Ovalle et al. 2013; Ross and Pawlina 2016; Lowe et al. 2018; Gartner and Lee 2022) contain references to many common pathologies and contain details of "Cell and Molecular Biology" probably support what was stated in the "Introduction" that "*students today must learn a lot more knowledge than students did 50 years ago*".

Concentrating on listening to what information the lecturer provided as well as taking adequate notes was a challenging exercise for most students. Clearly "audio" learners were at an advantage. There was no means for having the lecture repeated or recorded although audio recorders were often used by the more affluent students, but most lecturers did not allow these to be used. Lecturers in those early days needed to be very skilful at maintaining the students' focus and presenting a motivating learning environment. He/she could not rely significantly on visual aids etc. to supplement the learning exercise or help explain a complex histological structure.

6.4.3 Kodachrome Slides

During the late 1970s and early 1980s, kodachrome slides were introduced, and lecture theatres were fitted with projection facilities. The lecturer would load a carousel of projector slides that he/she could project when required during the lecture presentation. One downfall of this was that lectures often had to be prepared well in advance because the visual media facilities at the university needed adequate time to produce the projector slides. Common practise was to photograph sections of cells, tissues and organs at a variety of magnifications and have them made into slide sets to project during lectures. Eventually textbook publishers offered for an additional cost to academics slide sets of textbook images and other visual aids which were used and complemented the directive for students to refer to some important text sections for which time did not permit to mention in a formal lecture.

To provide more focussed directives, a pre-lab presentation was provided for about 15–20 min where kodachrome slides were projected to indicate what essential histological features students needed to identify in the histology sections in their class slide boxes.

The use of these kodachrome slides was limited due to the effort and expense of photographing histological features in the histological sections. Good photography prior to creating the kodachrome slide required an expensive photographic addition to a very good microscope (at that time) and the cost of creating slides by the University photographic unit or even a private photographic facility limited the scope to create kodachrome slides to use as a significant assessment resource. In my experiences as a histology student, kodachrome slides were never used as an assessment tool or to provide any practice quizzes—mostly because the curriculum was quite extensive, and the teaching hours did not provide a timeframe to provide such practice quizzes. Today, images of histological features can be easily taken as a "screenshot" and even annotations can be added which has opened many opportunities for assessment exercises.

6.4.4 PowerPoint Presentations

PowerPoint was launched in April 1987, originally for Macintosh computers only, until Microsoft acquired PowerPoint a few months later. In terms of an image-intensive discipline such as histology, PowerPoint revolutionised the delivery of histology content because it enabled the lecturer to do away with kodachrome slides and create his/her own presentations rather rely on a media department in the university—and most importantly be able to add or delete a slide or content right up to the time of the lecture presentation. But also, the PowerPoint lecture could be printed and copied as an extensive handout to which students could refer to during the lecture presentation. All these innovations enabled content to be presented in a clearer format for student learning.

Embedding an annotated "screenshot" of a histological structure into a PowerPoint presentation created many opportunities for more flexible assessment practices and the ability to use assessment packages as a component of the learning process. Initially, this was merely adding a few PowerPoint slides at the end of a lecture handout with some accompanying review questions as a practice quiz. But as explained in Sect. 6.19.2, they were used more commonly for "practical examination" formats.

6.5 Assessment: Written Examinations and Practical Examinations (Mid-1980s to Mid-1990s)

My academic career began in 1981 with an appointment in the School of Anatomy at UWA. After an initial 2–3 years teaching Human Biology subjects I became responsible for the teaching of "Histology" in the school. During the mid-1980s, the UWA academic year structure changed from three, 9-week terms to two, 13-week semesters. A change in the undergraduate degree programmes within the School of Anatomy resulted in two histology courses being offered to students studying Science or Biomedical Sciences over two semesters rather than during the earlier 3-term academic year. A histology course "Cells and Basic Tissues" was offered as a second-year undergraduate course in semester 1, and then an advanced histology course covering "Human Organ Systems" was offered in semester 2. A "pass" in the "Cells and Basic Tissues" course was a prerequisite for enrolling in the "Human Organ Systems" course.

Histology and anatomy were offered as separate study courses in a medical/dental curriculum and assessed separately. Many more hours of teaching had been allocated to these areas of study in both the medical and dental programmes. However, from about the year 2000 and certainly nowadays, histology and anatomy contents have been most commonly combined (and also reduced) given the substantial changes to the medical and dental curricula and their emphases to suit everchanging medical and dental issues within our modern society. Assessment was also a combined exercise.

The descriptions that follow relate to the development of assessment strategies in the histology courses "Cells and Basic Tissues" and "Human Organ Systems". Traditional histology lectures were presented using initially kodachrome slides but then using PowerPoint slides up to the present time.

Histology practical laboratory sessions in these two courses were 2 h with 4× rotations of 50 students (because there were only 50 microscopes) to cater to about 200 students. Each session usually included about five microscopes set up as a "practice quiz"—a similar format to what students could expect in their mid-semester and final histology practical examination.

These increased numbers of students during the late 1980s and mid-1990s made traditional examination procedures (e.g. requiring multiple practical class examinations to cater for up to 200 students) and marking the examination responses very onerous tasks particularly if only one academic staff member was the course coordinator. Then there were the added efforts needed to provide for students who required special considerations, e.g. longer exam time and/or needed to defer their examination. These practical examinations took a long time to set up—often requiring the exam to be set up the day prior to the scheduled examination. Some questions (e.g. a blood smear) needed to be set at 100 oil immersion to focus on perhaps a monocyte or an eosinophil, and a pointer embedded in the eyepiece was superimposed over the cell. Whilst clear instructions were for the student to only focus and not to move the slide (even with the slide mount taped) invariably the student knocked the microscope or managed to somehow shift the eyepiece so the next student did not know what cell was being questioned, and so a staff member was continually resetting the question. Assessment practices needed to be made easier to set up, and so mostly due to increased academic workloads, a paradigm shift began that encompassed emerging technologies and a change in what we needed to assess.

6.5.1 Before Computers and the Internet: Use of Power Point for Assessment

To make setting up "Practical histology" assessments more efficient, the first "developmental" phase was projecting examination questions in a PowerPoint format in the histology classroom or a lecture theatre so all students could complete the examination in just one attendance. That practice introduced some issues that were either resolved or ignored. Firstly, students could possibly copy answers since they sat near one another, but this was resolved by booking a much larger lecture theatre, so students sat at least one seat apart. Secondly, it was deemed that students no longer needed to demonstrate competency using the light microscope to scan a histological section to identity the tissue or organ. This was resolved as the requirement for competency with using the microscope was reduced particularly in the two histology units where the graduate destination of the student was in areas where skills in microscopy were not a major requirement. For students who entered their third-year level of a degree programme and/or research programme that employed microscopy as a major laboratory instrument, a third-year programme of study was available in "Advanced histology and histological techniques" in our School and students did learn the use of various sorts of microscopes, but the number of students was only 25–30. The ability to identify tissues and organs were still emphasised, so the PowerPoint evaluation involved projecting three images of several tissues or organs taken at low, medium and high magnification (Fig. 6.1) which included key characteristic features of a tissue or organ.

One other exercise that was introduced (which students enjoyed) was presenting images of tissues or organs (particularly taken at medium magnifications) and asking students to list characteristic features of the tissue or organ present in the image *and* a separate list of characteristic features that are *not* evident in the image. This was in preparation for a more challenging quiz (or formal examination) question which was to project images of a tissue or organ and ask what characteristic features enabled you to identify this tissue or organ. Students lost marks if they included a feature that was *not* evident in the image. For example, images of a kidney slice might have been shown but might have lacked glomeruli; the posterior pituitary (or neurohypophysis) might have shown axons and pituicytes but not have shown "Herring" bodies; cardiac muscle might not have shown intercalated discs.

6.5.2 Are You Over-Assessing?

The introduction of PowerPoint to present assessments coincided with a directive from the University Vice Chancellery that we over-assess students and that perhaps for a normal 39-h course (i.e. 1 lecture and a 2-h practical class per week for 13 weeks), assessment should be reduced to be not more than 4 h duration. It was not any major issue to omit the traditional formal written final examination and create two formal assessments, each of about 1.5-h duration, to be held midterm and again at the end of the course. About 60 questions would be projected to students in a large lecture theatre over the 90-min period. The midterm examination tested content delivered in the first 6 weeks and the end-of-semester examination tested the remaining content for the 6 weeks after the midterm only. The examinations were usually held during the study/non-teaching week (midterm examination) and the study week (for the end-of-semester examination) before the final examination period. This change in assessment schedule and practice

Low magnification	Medium magnification	High magnification
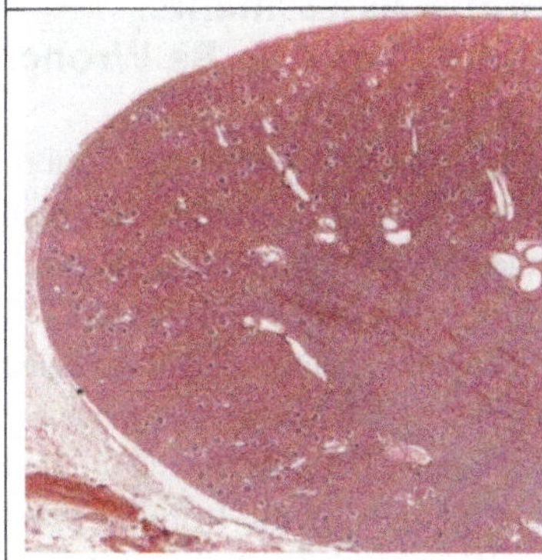	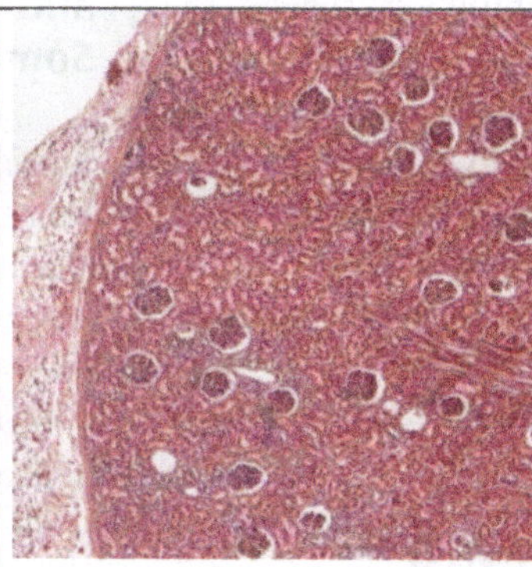	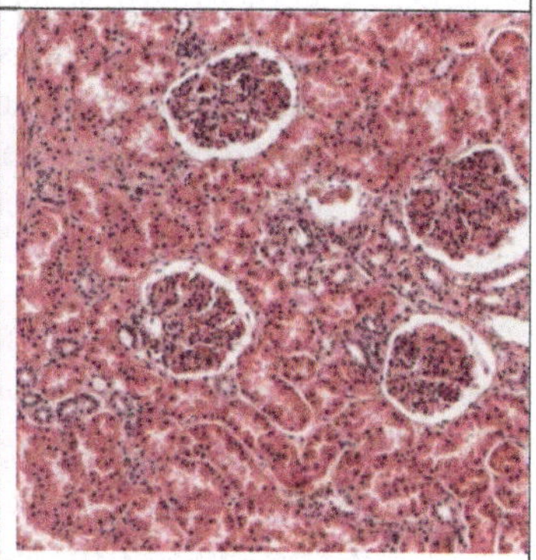

Fig. 6.1 Images of a kidney section taken at three magnification levels used in an "identification" examination question

changed the sort of questions that were asked. More theoretical items based on lecture content were combined in an equal blend with a more "practical histology" type items (Table 6.2).

Most questions required a one-word answer or just a short sentence. Examination papers were manually marked. During the late 1990s, I was fortunate enough to form a collaborative teaching arrangement with Professors John Campbell (now deceased, 2021) and Michael Hall, both at the University of California Los Angeles (UCLA). Both Professor Campbell and Professor Hall assisted in my histology teaching commitments at UWA and, in turn, I travelled to UCLA and helped with their teaching of histology to UCLA Dental Students. These two academics and friends helped write content for the interactive histology atlas contained in https://www.histology-online.com that was to be created over the next two decades. The three of us manually marked the examination papers—usually sitting around a table for about 2 h. Later (Sect. 6.15.1) we were replaced by computer-based automated marking systems!

Similar teaching practices were then put in place at UCLA for students studying histology in the dental curriculum until both Professor Campbell and Professor Hall retired and I no

Table 6.2 Combined practical examination item and theoretical item. Example only. Both items would not be asked in the same assessment

Question X (Practical content)	*Question X (Theoretical content)*
Name the organ whose epithelial lining is arrowed in the high-magnification image below. 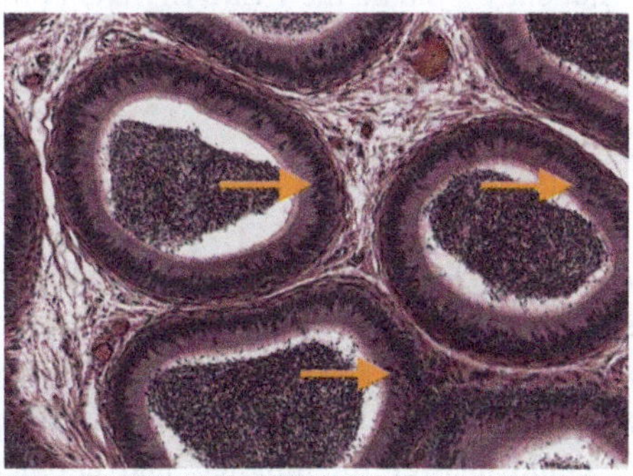	The highly coiled structure in the male reproductive tract lined by pseudostratified columnar epithelium containing irregular microvilli called stereocilia is called the ________________.
Answer Ductus epididymis/epididymis	*Answer* Ductus epididymis/epididymis

longer continued the collaborative teaching experiences at UCLA. However, Professor Hall and I did present histology classes to students studying through UCLA Extension from 2015 to the present time. These courses were offered completely online and are described in Meyer (2023).

6.6 Late 1990s–2000: "Histology Practical Assistant" and the Demise of the Histology Practical Laboratory Classroom

The physical layout of the histology classroom changed with the advent of computers. A renovation of the histology classroom in the mid-1990s increased the number of work stations (120) of which 60 stations had a computer, and also 60 new microscopes were purchased for the remaining 60 stations. The maximum capacity (120) was two students sharing a class slide box, and each had access to a microscope or a computer. Practical classes were reduced to only one (2-h) repeat each week.

Students used a very early version of "Meyer's Histology" (https://www.histology-online.com) called the "Histology Practical Assistant". It was a computer-based learning resource used to guide students when viewing the histological slides using the microscope. As we encouraged collaborative learning, more and more students worked in pairs or even in groups of three to complete the laboratory practical exercises.

Until the late 1990s, access to the computer-based resources was confined to the histology classroom *only* via the internal University IT networks. Each teaching unit had its own webpages housed on the School or University servers. As the web-based histology learning resources continued to be developed from the initial "Histology Practical Assistant", more often students spent most of their time sitting at the computer rather than using the microscope.

6.7 Collaborative Learning and Group Assessments: Sometimes You Can Be Wrong!

A colleague and I continued to promote collaborative learning as that seemed to be the pedagogic drive at that time. We encouraged students to work together on the internal web-based histology resources (using the computers) and assist each other in completing the learning exercises—even to the extent of reaching out to the student pairs adjacent and completing the learning exercise as a group. Our motivation for this was "When you get to work in your professional jobs you will need to collaborate so why not learn in a collaborative environment?" We were pleasantly surprised that formal feedback to us was that this collaboration was considered a very favourable class activity. Based on this favourable feedback, the following year we promoted this activity to the new student cohort. But then we took it too far! We decided to assess the students using a similar reason and motivation: "When you get to work in your professional jobs you will need to collaborate, so why not learn *and be assessed* in a collaborative environment?" Note the added three words in italics! So, during a 20-min period towards the end of the laboratory practical class, we proceeded to move amongst the students and gather them in groups of five and ask questions to each of them in turn (e.g.

To Student A—name ONE component of the portal triad.

To Student B—name ANOTHER component of the portal triad.

To Student C—name the REMAINING component of the portal triad.

To Student D—liver sinusoids converge into what structure?

To Student E—a tubular space between two hepatocytes is called a what?)

And we would keep going with more in-depth questions to a maximum of 10 questions (about two questions for each student).

If a student did not know the answer another student could answer for him/her, and another question was asked. We were very careful not to allow any one student to dominate and also not to be in any way critical of a student who could not answer correctly; the group assessment was intended to promote learning and make learning and assessment fun! We would informally allocate the group a mark out of 10 and as an informal competition, we would then post a list of the "team" scores. At the end of the teaching term, the "best" team won the box of chocolates. It did become quite competitive and again was viewed favourably by the students.

6.7.1 Then We Made the Big Mistake!

For the next cohort of students, we decided to introduce this as a formal part of the assessment—worth 10% of a student's final mark. Each student was to equally receive the mark (out of 10) for his/her team regardless of how each individual student contributed to the score. I probably do not need to list the reasons why this was not received well by the student cohort, and so in the end, we "reneged" on the idea. As it happened within the next year or so the practical histology classes were completed away from the histology classroom as learning resources moved to a Learning Management System (LMS) and a histology website (https://histology-online.com) (see Sect. 6.10).

6.8 Virtual Microscopy: The Ability to "Screenshot" Images of Tissues and Organs

By 2000, all learning resources housed in the UWA servers began to be available via the Internet. At the turn of the century, the emergence of virtual microscopy replaced the emphasis for students to use a light microscope to examine histological slides.

Use of virtual microscopy enabled histology teachers to screenshot images of tissues and organs with ease and incorporate these into PowerPoint presentations but to also create PowerPoint presentations purely related to a practice quiz by including images and having them annotated.

The introduction of a virtual microscopy database revolutionised how practical histology classes could be completed. No longer did the student need to attend an allocated practical class; he/she could complete the class objectives away from the classroom as long as they could access the Internet. This then presented an opportunity to also provide more extensive assessment packages for students to act as a learning tool. It was an onerous task to set up "practice quiz" questions each week in a formally attended laboratory practical class. Now each topic presented on their LMS included assessment packages called "Weekly Graded Quizzes" (see Sect. 6.14.2).

Concomitant with this change, students began purchasing their own laptops, iPads and other mobile devices so they could bring these into the histology classroom and use these to access the histology learning resources. This alleviated another issue and that was the limitation of the number of students able to attend a laboratory practical when it had transitioned to be solely computer based. The previous limitation of 60 students able to access a computer (as there were only 60 computers) was lifted, and a class was then about 120 students most using their own computers. This had several implications when it came to assessment.

The availability and preference for students to complete laboratory practical classes online (i.e. away from the histology classroom) provided an opportunity for them to attend their allocated laboratory practical session and ask questions about lecture content or laboratory exercises—particularly if they could not locate a histological feature on the digital section. They were encouraged to attend to briefly view the "real section" on a traditional histology slide and perhaps maintain skills in using a light microscope. However, over time, attendance diminished as the traditional slide was not an importance emphasis for the students, particularly with the advent of discussion forums etc., on the LMS and their realisation that the digital slide was no different from the traditional histology slide in their class slide box. With increasing numbers of enrolments correct use of the microscope was not assessed as it was too time consuming.

6.9 2000 (Onwards): Computer-Based Assessment

Students needed to complete a "Mid-semester" and "Final" examination, each worth 45% and 55%, respectively. The "Mid-semester" examination was held at about the mid mark of the semester schedule and held during their allocated laboratory practical class time to complete the examination. These examinations were computer based in the sense that questions were provided via a PowerPoint presentation of annotated histology images posted on their Histology webpage, but students completed their responses on an answer sheet and handed them in to be marked at the end of the examination. Again, my UCLA visiting colleagues and I manually marked all the papers.

6.10 Learning Management Systems

Learning Management Systems (LMS) were created in the late 1990s mostly because of eLearning but their use has been widespread today mostly as a result of the need to shift to online teaching and learning formats and an even greater need due to the COVID-19 pandemic. Each specific LMS was used to enable students to access all their required learning content. Lecture capturing or recording and their inclusions in a Learning Management System (LMS) allowed lectures with PowerPoint slides to be stored and reviewed anytime by students. Links to external websites and videos could be provided. Discussion forums and Announcements were an avenue through which the teacher communicated with the student cohort. But one of the major features of any LMS was the ability to create assessment packages with several different question formats. This was a major success story when teaching histology. Meyer (2000) described using an LMS to present histology courses (including formal assessments) completely online. With the advent of an LMS all the web-based teaching resources in our School were linked to the LMS, but by this stage, the "Histology Practical Assistant" was now available via the Internet at https://www.histology-online.com and was being used by a number of universities throughout Australia and several universities worldwide. In 1999, I was awarded an Australian Award for University Teaching in the category "Flexible Delivery/Learning" for the creation of this resource.

6.11 Histology Teaching, Learning and Assessment: An Online Format

PowerPoint, Virtual Microscopy, LMS, Video recordings, Web-based learning software and the Internet provided the technologies to expand the learning tools for students from the early 2000s. By 2010 much of the histology learning was transferred to a completely online format as described in a recent article (Meyer 2023). Emerging technologies enabled examinations to be completed by viewing the exam present on the LMS. Students were allocated a time to complete the exam and within hours after the exam was closed, results were released as the question formats had automatised marking functionalities, and even questions requiring a written explanation could be very quickly marked. But, the advent of these emerging technologies and computer-based learning opened many opportunities for very innovative quiz/examination formats to be created which also contributed to increased student engagement and better learning outcomes. Even the traditional lecture opened the opportunity to incorporate quizzes as the content was being presented live (Sect. 6.14.1).

6.12 Theory of Assessment Methods

The purpose of this chapter was not to provide details of various assessment practices as they are well documented elsewhere, but it is useful to summarise a few recommendations that have been adopted in my various assessment packages during over three decades of teaching histology when significant technologies (mentioned above) were available and described below.

6.12.1 MCQ and "Fill in the Blank" Question Formats

MCQs are the best option as they can have automatic grading, but "fill in the blank" questions or matching questions are also useful as the automatic grading can also be applied with the need sometimes to review responses that are not included in the marking key. MCQs are also a good choice to monitor both validity and reliability. One important exercise I perform after a review quiz or formal assessment is to look at firstly, student performance between successive class cohorts and secondly, the student responses on each question in the assessments. This gives me some indication of how valid the assessment is or, more specifically, how well written the question is, so that there is no unclear or ambiguous component likely to confuse the student and the reliability of the assessment.

MCQs can test a significant amount of content, but questions should focus on the learning objectives to be valid. For a course like histology, most of the MCQs simply require students to select the best option to recall basic knowledge rather than testing more higher-order skills.

Keeping the questions simple and easy to read is important when there is a time limit on the assessment, such as in the case of the midterm and final examinations; however, that is relaxed for review quizzes that were constructed where a time limit does not have an effect. So, to promote the MCQ as a learning resource as well as an assessment in the practice or preview quizzes often the opening statement reviews/repeats some knowledge before the actual stem of the question is provided (Sect. 6.15.2).

"Fill in the blank" questions also have the advantage of ensuring correct spelling of histological terms areevaluated. They also have the advantage of ensuring correct spelling of histological terms are evaluated. Marks could be deducted for repeated spelling errors in any one assessment.

6.12.2 Learning Objectives Define the Scope of Assessments

The opening page of all the topics in the histology curriculum [e.g. "Cardiovascular System" (Fig.6.2)] begins with a short description or summary of the system and then lists the expected "learning outcomes". I find it essential to list "Learning outcomes" when designing the curriculum content and ask myself the question—"-What do the students need to know?" Firstly, it enables me to assemble the learning content in a logical order as well as obtain the relevant images and write the appropriate text (learning content) to ensure that the learning objective can be completed by the students. Secondly, I then announce these learning outcomes in the student course guide, so it is very clear to the students what learning is required. Thirdly, it defines the scope of the assessments because I make sure that the questions I formulate for any examination are related directly to the learning objectives I have announced. Students comment that they do use the learning objectives as a guide in their preparations for formal examinations (Table 6.16: Item 4). As a "discussion forum", I have asked students to post questions under each learning objective on the LMS.

What is also important is to be mindful that adequate time is allocated to complete the learning and any formative assessments. The expectation would normally be about 8 h for each topic each week, and students agreed that was an adequate time (Table 6.16: Item 5).

The curriculum content for all the histology courses presented was the core histology curriculum for Medical/Dental Schools so the desired

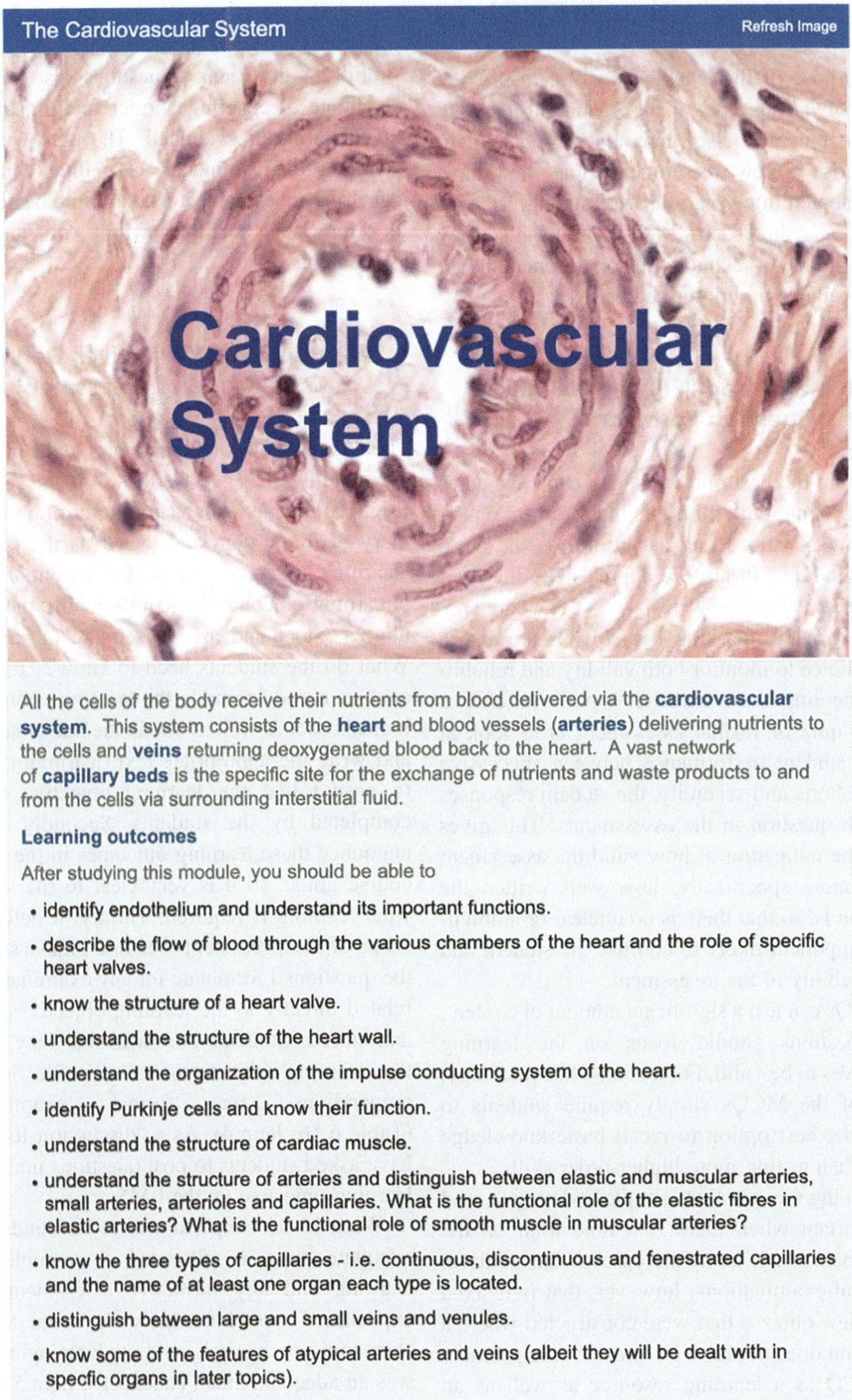

Fig. 6.2 Learning outcomes for "Cardiovascular System"

outcomes would be a thorough understanding of that curriculum content. Since "Histology" teaching is in a Pre-med degree or the early years of a medical/dental curriculum, there is no need for the outcomes to align with professional competencies or any requirement for a professional accreditation or license.

6.12.3 Types of Assessment for Histology: Reasons for Assessment

Much of histology education is presenting factual knowledge that requires students to identify and/or describe a structure or relate functions to microscopic structures. To assess higher-order cognitive processes, written responses like essays can be helpful but with increasing student numbers, it becomes an onerous task marking essay type questions for a student cohort of 350.

For the subject area "Histology", assessments not only provide a mechanism for feedback but also provide motivation for a medical or dental student's future learning of pathology.

As explained in Sect. 6.15.1, each topic should be preceded by very clear announcements of what the learning outcomes are and what knowledge students should be expected to attain. Each item of content on the resources then facilitates the student to learn the content to meet those learning outcomes. Review quizzes and formal assessments enable the students to determine whether they have achieved the expectations.

Quizzes or any other form of assessments are used not just to evaluate students but most importantly to also contribute to the learning process as they:

1. Indicate students' knowledge or mastery of the histology content.
2. Motivate students to learn and reinforce their learning of histology.
3. Provide accurate gradings/ranking of the student cohort.
4. Provide feedback to histology teachers.

6.12.4 Feedback

When feedback is provided, it enhances student learning, and it is a critical factor in student success (https://visible-learning.org). This is even more important when presenting online learning as you do not "feel the vibes" from students as you do in any enface course delivery mode. Coupled with a discussion forum or a portal to ask questions, feedback has enabled an insight into students' progress and their understanding of the content. It also has been another avenue for exchange between the student cohort and the instructor.

More frequent review quizzes with feedback keep students engaged and motivated (Table 6.16: Item 6).

6.12.5 Mastery Learning

Many platforms offer a variety of quiz formats. On completion of a quiz, there is an option for instant feedback to students, making histology learning efficient and rewarding, after having significantly improved students' grades/marks on examinations.

Online formative assessments are perceived as tools that promote self-directed learning, improve knowledge and tailor learning for individual learning needs and styles (Nagandla et al. 2018). Formative assessments are presented within the histology resources in different formats.

The interactive descriptions of histological structures and direct links to the virtual microscopy sections meet an essential requirement for delivering online course content in an interactive and engaging manner (Rueshle et al. 1999). Mastery learning (i.e. mastering each learning component before moving on to the next) (Rueshle et al. 1999) was provided by extensive, interactive review quizzes.

6.13 The Transition to a Completely Online Histology Course

Given all the technologies mentioned earlier, I have presented and supervised completely online histology courses to students attending the University of Western Australia (UWA) but also enrolled in histology courses offered by UCLA Extension (https://www.uclaextension.edu) since about 2009. The histology courses provided for students at UWA and UCLA Extension were identical in content and so were the assessments.

6.14 Assessment Practices Used Over the Last Two Decades When Teaching Histology Completely Online

6.14.1 Lecture Presentations and Quiz Functions

For video recordings, I used Camtasia—a software suite created and published by TechSmith, for creating and recording video tutorials and presentations via screencast, or via a direct recording plug-in to Microsoft PowerPoint. Meyer (2023) lists other video recording facilities available. Initially, the "Camtasia" video recordings were used to add short "Welcome" videos and videos outlining the course schedule etc., to students. The purpose of these short videos was for me to put a personal touch to the correspondences on the LMS. A series of review videos were created which included a series of quizzes whereby I would load a video showing a number of histological slides, and using a pointer, I would ask questions. I would not post answers to these on the LMS and challenged the students to post their answers on a student forum with a clear explanation as to why they chose that answer. I would monitor and add my contribution if the answer was not arrived at by student consensus.

Presentation of histology lectures became more creative and very professionally recorded using various software packages (e.g. Camtasia), but perhaps the most successful lecture capture facility now is available from "Lecturio.com".

All the various disciplines in a medical school curriculum, such as anatomy and histology, have been visualised on a specialised learning platform. "Lecturio.com" is one of the most popular lecture platforms available (https://lecturio.com) for delivering medical education. Lecturio's mission as stated "*is to provide students and faculty with the best learning tools, proven effective by learning science*". Lecturio has "*created a high-quality digital medical education resource, which is affordable, adaptive, and personalized*". Lecturio designed their platform "*with the needs of learners and faculty in mind, combined with the latest state-of-the-art learning technology and comprehensive monitoring and assessment features*".

All the features about the histology lectures provided by Lecturio.com have been previously described (Meyer 2023), but for the purpose of this chapter, once the full lecture is recorded they are then presented to users as "mini-lectures". So, the length of the lecture topic is not confined to a traditional 45-min lecture that would normally be delivered "en-face" in university lecture theatres. This presented the opportunity to cover the content of each lecture topic in much more detail. For example, the student has listened for 54 s and has 7 min 55 s remaining (Fig. 6.3a) to view the mini-lecture on "Spermatogenesis" on the Lecturio platform. These lectures have review questions embedded in the lecture presentation. For example, the mini-lecture on "Spermatogenesis" contains six questions (Fig. 6.3b).

At certain points during each mini-lecture, an MCQ quiz question would "pop-up" to test the knowledge gained by the user after viewing a specific explanation of content. But the viewer has the option of deferring access to quiz questions until the mini-lecture is completed by turning off the "pop-up" questions in the quiz settings (Fig. 6.3c) prior to listening to the mini-lecture. At the end of the lecture the student can then "Start the quiz" (Fig. 6.3d) and complete all

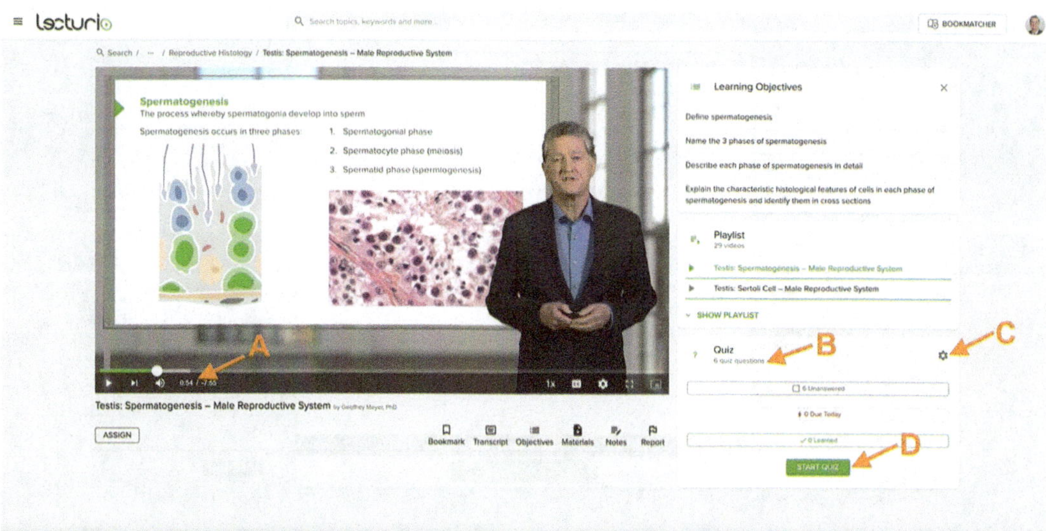

Fig. 6.3 Lecturio lecture platform—Histology lecture series

six questions. Whether students preferred to have the questions "pop-up" during the lecture or defer until the end was almost even numbers (Table 6.16: Items 7 and 8).

If the student prefers to have the questions "Pop-up" at the appropriate times in the mini-lecture, then when the question appears on the screen (Fig. 6.4), the student selects his/her answer (Fig. 6.4a) and then has the option of using the review buttons at the base of the question frame if he/she is unsure, or the student can submit his/her answer if she/he is happy with it by clicking on the "smiley icon" (Fig. 6.4b) and get immediate feedback (Fig. 6.5b). In this case, the student's selected choice (Fig. 6.4a) is "Correct".

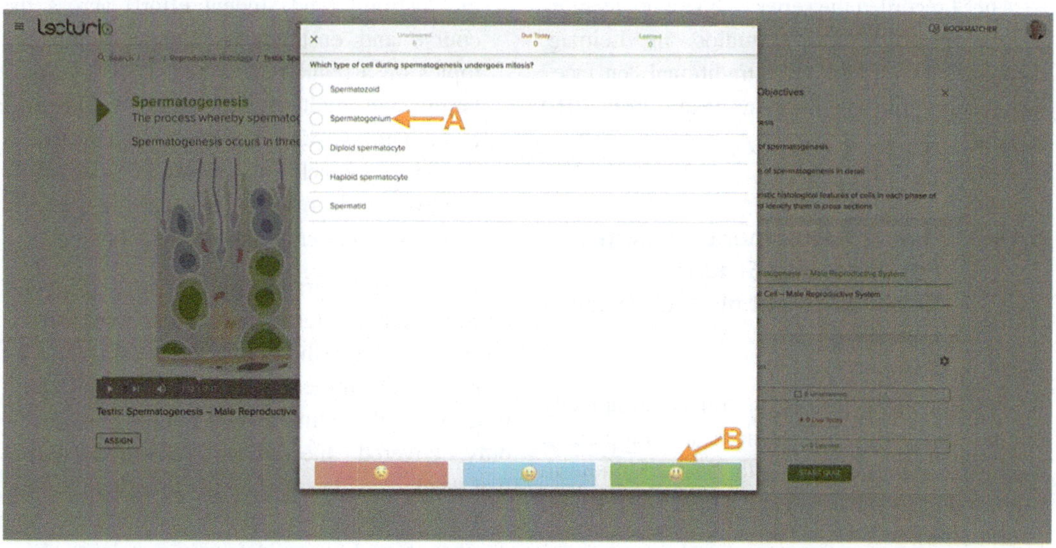

Fig. 6.4 Lecturio lecture platform—Histology lecture series "pop-up" quiz question attempt

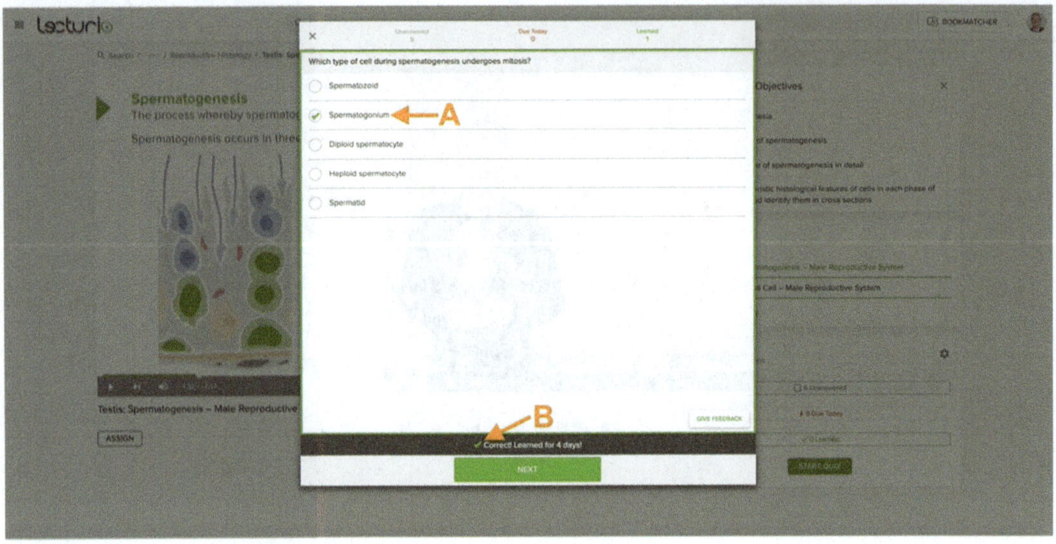

Fig. 6.5 Lecturio lecture platform—Histology lecture series "pop-up" quiz question feedback

One student comment demonstrated the favourability of this feature: *"Completing the quizzes engaged me in the content being presented"* (Table 6.16: Item 9).

A discussion forum is an important part of the learning platform for every mini-lecture. Viewers can post a question about the content—perhaps to clarify some issues they may have. A response by the lecturer is usually posted within 24 h.

After I recorded the series of histology lectures at the "Lecturio.com" studios in Leipzig, Germany, I no longer gave traditional "en face" lectures in a university lecture theatre from 2010 onward.

6.14.2 Use of Assessment Packages: For Assessing Student Knowledge but also for Student Learning

In introducing a histology course completely online, students were required to complete a weekly graded quiz" (worth 5%), a midterm examination (worth 25%) and a final examination (worth 30%).

The purpose of introducing weekly graded quizzes was:

1. To motivate students to achieve the "Learning outcomes". Thus, these quizzes were an important and very useful learning tool rather than simply an assessment. The knowledge gained by completing quizzes each week adequately prepared students for Midterm/Mid-semester and Final Examinations.
2. To require students to complete a short, graded quiz each week. This arrangement distributed the learning (and student effort) across the course and ensured the student completed topics on a regular basis and did not "cram" before the midterm/mid-semester and final examinations. My experience is that this is the best approach for students to successfully achieve all learning outcomes and perform well and confidently on examinations.

Each weekly graded quiz opened from 12:01 a.m. Saturday (at the end of the week) to 11:59 p.m. Friday (the following week) (i.e. for 7 days). Students could complete the quiz anytime during those times. Each weekly graded quiz only covered the topics presented during that week.

Each weekly quiz contained 20 questions randomly selected from a database containing about 40–50 questions. Each question was worth 2.5

points, and students had 18 min to complete the quiz. Students had two attempts at each quiz (but, of course, different questions may be selected on any second attempt) and the highest score was recorded. The expectation was that students would not have time to consult notes etc., when completing the questions in the allotted time. Even if a student scored 100% on his/her first attempt, he/she was encouraged to complete the second attempt and the opportunity to view some different questions.

Many of the UCLA Extension students were working full time and often had family commitments etc., so I did understand that on an occasion, it may be a difficult task to complete the rigid learning time schedule, and a student may not perform as well as he/she would like in a particular weekly graded quiz. So, a student's worst/lowest score on any quiz attempted was *not* included when calculating his/her final mark.

Students completed a Midterm Exam that only examined material presented in the first half of the course. The Final Examination only examined material presented in the last half of the course. Two assignments were worth 5% each (Sects. 6.19.2 and 6.19.3).

UWA students attended the histology classroom and used the computers in that room for their Midterm and Final examinations. For a UCLA Extension student, in order to accommodate people's work schedules, both the Midterm and the Final Examinations opened at 6:00 a.m. Saturday of the exam week and closed at 11:59 p.m. the following Monday.

Each examination contained 50 questions worth 5 points (Midterm Examination) or 7 points (Final Examination) each. Students had 55 min to complete each exam. This was plenty of time for students to give each question careful thought. Questions were very similar in content and format to those completed in the weekly graded quizzes.

6.14.3 Assessments and Academic Integrity: Use of ProctorU

The most common question I was asked about online assessments was "How do you deal with academic integrity when students complete an assessment online?" For UWA students, both the midterm examination and the final examination were supervised by myself and my UCLA colleagues. But for UCLA Extension students, supervision was by "ProctorU" (https://www.proctoru.com). ProctorU is a leading provider of remote proctoring and integrity safeguards for online testing and includes trained human test proctors. Before students register a time to complete each of these examinations, they have three documents available for them to read through—"ProctorU—How it Works for the Test Taker", "ProctorU—Information for Students" and "ProctorU—Process for Students". Students have expressed no issues with this proctoring system.

6.15 Question Formats Provided for Online Assessment: The Key to Mastery Learning

6.15.1 The Simplest MCQ or "fill in the blank" Question Format

When delivering the learning content in any histology curriculum, teaching emphasises that three general learning outcomes are to be achieved. The student cohort needs to be provided with these general learning outcomes:

1. Specific cells, tissues and organs are identified.
2. Identifiable characteristics of specific cells tissues and organs can be recognised.
3. Functions of a cell, tissue or organ can be explained.

A typical weekly graded quiz question, midterm examination question or final examination question would incorporate these three outcomes (Table 6.3). To test knowledge of (for example) the "Goblet cell", a student may be assessed using four separate items—or in combination. The format could be "short answer" or "fill in the blank".

Item 1 (Table 6.3) would be considered the easier of the four items—only a simple identification is required. Item 2 (Table 6.3) would be considered slightly more difficult (than Item 1)

Table 6.3 A selection of quiz/examination items

Assessment image	Item type
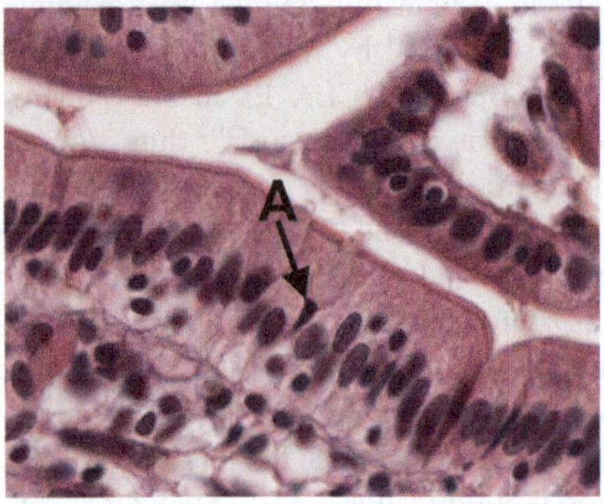	**Item 1:** Name the cell whose nucleus is labelled **A** in this medium magnification image. **Item 2:** Name the major locations of the cell whose nucleus is labelled **A** in this medium magnification image. **Item 3:** What is the secretory product of the cell whose nucleus is labelled **A** in this medium magnification image? **Item 4:** What is the function of the secretory product of the cell whose nucleus is labelled **A** in this medium magnification image?

because again the student needs to correctly identify the cell *but then* know where its specific location/s is/are in the human body. Item 3 (Table 6.3) would be considered more difficult again because the student needs to correctly identify the cell and then know its secretory product/s to provide the correct answer. Item 4 (Table 6.3) would be considered even more difficult because the student needs to correctly identify the cell and then know what it secretes and then the function of the secretory product to provide the correct answer.

To pose all four items in the one examination would penalise the students excessively if they could not correctly recognise the cell in the image, but perhaps also if only two items were asked.

My normal practice would be to include only one of the items in any single mid-semester or final examination. In a weekly review quiz, I would ask another of the choices. But, most importantly, I would direct the students to make sure they could answer any of the four types of items when studying and preparing for examinations. This was what I called "assessments as a learning practice" exercise. When PowerPoint questions were presented for assessments (Sect. 6.15.1) and I would project an image and ask students to identify a histological structure, I would casually inform them that I could have asked this or this…as I pointed at other structures. This was to reinforce my suggestions to create other questions when they reviewed their responses or during their learning and examination preparations.

All four items in Table 6.3 could be presented as four MCQs with four other alternatives (as incorrect answers) or an item format requiring the student to type in the correct answer (if the quiz was accessed online). Usually, the quiz formatting on the most common LMS used in universities today enables correct answers to be entered for automatic marking. However, this can only be successful (without checking by the assessor) if the most common correct alternative answers are entered. For example, alternative answers may be provided by students—some of which deserve partial marks (Table 6.4).

One issue is that students may enter different responses and even with minor spelling errors. Also, it is very important to provide "Instructions on how to fill in responses" so students get used to the process. For example:

1. Do not enter any letter in upper case (e.g. goblet NOT Goblet).
2. If an item prompts you to name a cell, do NOT add "cell" in your response (e.g. goblet NOT goblet cell).

Table 6.4 Answers for a quiz/examination item (Table 6.3) and possible responses and marking guide

Item	Correct answer	Alternative answers	Mark (out of 1)
1	Goblet	Goblet Unicellular gland Mucous Mucus-secreting Secretory	1 0.75 0.5 0.5 0.25
2	Respiratory and gastrointestinal tracts		1 1
3	Mucin/mucus	Mucous glycoprotein	1 0.75
4	To serve as a protective barrier	Protection	1

3. If an item prompts you to name a type of fibre, do NOT add "fibre" in your response (e.g. collagen NOT collagen fibre).

By the time students complete several review quizzes etc., they adjust to entering the terms to enhance the accuracy of the marking process.

Most LMSs had a facility to enter a "wildcard" whereby the key term (e.g. "goblet") was the only word it recognised in the marking key. This was most helpful for items requiring more than one or two words. There were always a few students who either misspelled terms or wrote their responses in a different way than was anticipated, so I would always open the quiz once it was completed. If the student had entered a response that had been marked incorrect but thought it would be accepted on review, then he/she could email me and ask me to look at that particular item and his/her response.

Formative feedback certainly engaged student learning, but it also provided insightsinto student errors and difficulties in learning. Performances in assessments and discussion forums indicated students found difficulty with understanding certain components of the "Eye" and the "Ear". Without more extensive descriptions of the anatomy of these two organs, which would overload the course, I decided to make these two topics optional in the curriculum (and not assessed) but emphasised the importance of at least understanding the structure of the eyelid, cornea, lens and retina and the crista ampullaris maculae of the utricle, and organ of Corti. Surprisingly, most students did study these histological components.

6.15.2 Using more Complex MCQs or "Fill in blank" Questions to Reinforce Learning—Formative Assessments

During a histology course, I employed formative assessments and summative assessments. The weekly graded (formative assessment) worth 2.5%–5% was to provide feedback to students about their learning progress and the level of knowledge they have attained. Multiple-choice question (MCQ), matching and short answer or "fill in blank" type items were used and written in such a way so as to also reinforce learning and confirm learning outcomes are being addressed. As described later in this chapter, summative assessments are constructed differently—but again also to reinforce learning.

For example,

Question: The apical surface of many epithelial cells contains surface specialisations such as microvilli, stereocilia and cilia. Microvilli increase the area of the absorptive cell surface in epithelium lining the

A. Oesophagus
B. Stomach
C. **Duodenum**
D. Colon
E. Rectum

Most of the stem of this item (above) is informative. The item could have omitted the first sentence "The apical surface of many epithelial cells contains surface specializations such as microvilli, stereocilia and cilia", but that loses the effect of reinforcing knowledge already attained. It is important to not have severe time restrictions on the completion of the formative assessment, so the students have time to read the question and benefit from the reinforced learning. Eighteen minutes was enough time for students to complete a weekly graded quiz with extended text in the question stem (Table 6.16: Item 5).

Most importantly, if students are familiar with the "assessments as a learning practice" concept I explained then in their learning and especially their review of such an item they should substitute "microvilli" for "stereocilia" and "cilia" and determine the correct response. This practice enables students to be confident answering perhaps more "testy" items (see Sect. 6.15.3) in the "summative assessments". Students commented that the extended review content in the stem of items in the formative assessments reinforced their learning (Table 6.16: Item 10).

6.15.3 Using More Complex MCQs or "Fill in blank" Questions for Summative Assessments

Multiple-choice question (MCQ), matching and short answer type items are also used for "summative assessments", such as midterm and final examinations. The questions are perhaps more "testing" and mainly designed/written to determine the level of learning that has been achieved and indicate a student is ready to progress to further study in a related area (e.g. pathology). So informative text in the stem that may be used in a formative assessment (see Sect. 6.15.2) is deleted (e.g. "*The apical surface of many epithelial cells contains surface specializations such as microvilli, stereocilia and cilia*"). A more complex question could be asked.

For example,

Question: Match the structural modification with the appropriate region of the human body given (Table 6.5).

Answer: Table 6.6.

Table 6.5 Structural modification on epithelial cell surfaces and regions of the human body

Structural modification	Region
Microvilli	Uterine tube
Cilia	Duodenum
Stereocilia	Trachea
	Proximal renal tubule
	Ductus epididymis

Table 6.6 Structural modification on epithelial cell surfaces matched to correct regions of the human body

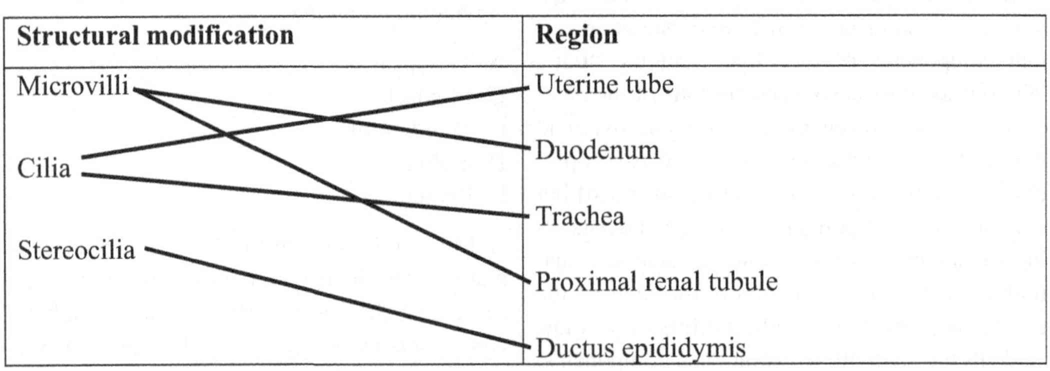

6.15.4 Assessment Question Design: Do You Choose an MCQ or an "Image based fill in answer"?

Think about what you wish to test (e.g. If you intended to ask a question about identifying regions of the adrenal gland and knowledge about their secretory products, how would you begin?).

Firstly, the question should match the statement in the "Learning outcomes" provided to the students. Secondly, it is so important to have listed the learning outcomes clearly. Thirdly, do it properly the first time with a definite plan based on the learning outcomes listed. Set several questions covering the learning outcome and store them in a question bank or database to recall when needed in examinations year after year.

6.15.5 Assessment of the Histology and Functions of the Adrenal Gland: Example Only

The learning outcomes listed for the adrenal gland might be:

Be able to

1. Identify the adrenal gland from other human organs.
2. Identify the adrenal cortex and the adrenal medulla.
3. Identify zones of the adrenal cortex (i.e. zona glomerulosa, zona fasciculata and zona reticularis).
4. Identify the specific types of cells in the adrenal medulla.
5. Understand unique features of the blood supply to the adrenal cortex and the adrenal medulla and how they may affect secretions from the adrenal gland.
6. Name the general class of hormones secreted by each zone of the adrenal cortex and the adrenal medulla.
7. Name the general functions of the secretory products of each zone of the adrenal cortex and the adrenal medulla.

In order of complexity/difficulty, the items can be simply testing each of the seven learning outcomes listed above. The adrenal gland is of course really two separate glands—an adrenal cortex and adrenal medulla but they are considered as a single unit for the purpose of the descriptions below.

Note: As much as possible, use images now that it is easy to create screenshots etc. Histology is a very visual discipline about the learning of details of microscopic structures so assessment should reflect that emphasis

The examples below indicate various item formats to assess learning outcomes 1, 2, 3 and 6 and 7 above. Assessing learning outcomes 4 and 5 could be similar in format.

In the item in Table 6.7, the automated marking key would have a number of alternatives

Table 6.7 Level 1 Item (Testing adrenal gland learning outcome 1) "Fill in the blank" format

Assessment image	Item and answer
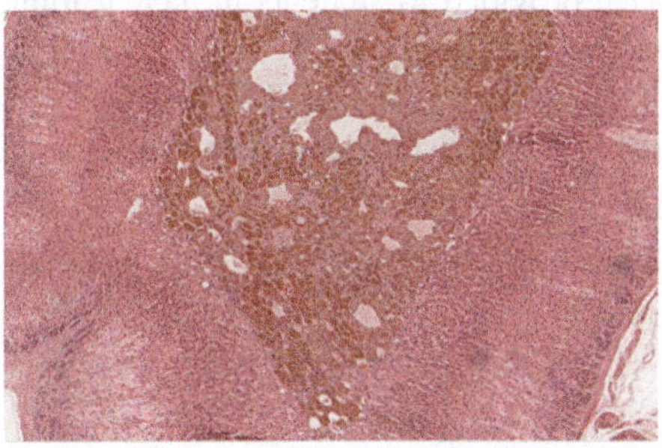	**Item:** Name the organ imaged here at low magnification. ______________________. **Answer:** Adrenal Adrenal gland Adrenal cortex and medulla Adrenal cortex and adrenal medulla Adrenal medulla and cortex Adrenal medulla and adrenal cortex Suprarenal gland

Table 6.8 Level 1 Item (Testing adrenal gland learning outcome 1) "MCQ" format

Assessment image	Item and answer
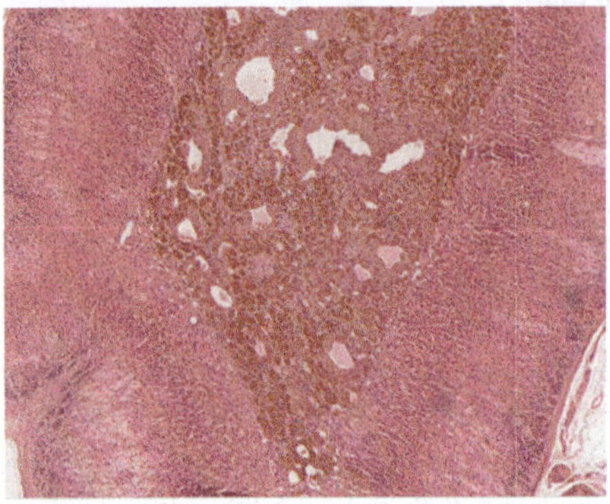	**Item:** Name the organ imaged here at low magnification. A. Spleen B. Thymus **C. Adrenal gland** D. Liver E. Pancreas **Answer:** C Adrenal gland

Table 6.9 Level 1 Item (Testing adrenal gland learning outcome 2) "Fill in the blank" format

Assessment image	Item and answer
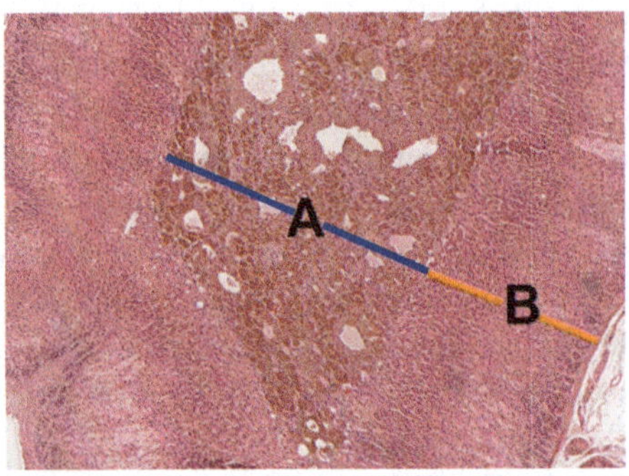	**Item:** The region of this gland labelled "A" is the ________________ and the region of this gland labelled "B" is the ____________________. **Answer:** A = Adrenal medulla B = Adrenal cortex

entered as acceptable answers. Some LMS systems enable you to set only key terms as the important component of the answer.

In the question in Table 6.8, the automated marking key would have "C" entered as the acceptable answer.

In Table 6.9, alternative items could be:

1. Name the region of the adrenal gland labelled A and the region of the adrenal gland labelled B in this medium magnification image.or
2. Name the region of the gland labelled A in this medium magnification image.

The first alternative item stem already lets the student know it is the adrenal gland, but the latter question stem needs the student to realise that it is the adrenal gland before arriving at the correct region or zone.

There was always a cohort of students that answered version 1 correctly, but if asked version 2, they often faulted.

Advice provided to students in their learning involved encouragement to confirm the organ under scrutiny, even if it was obvious, by looking for the characteristic/identifying histological features.

Table 6.10 Level 1 Item (Testing adrenal gland learning outcome 3) "Fill in blank" format

Assessment image	Item and answer
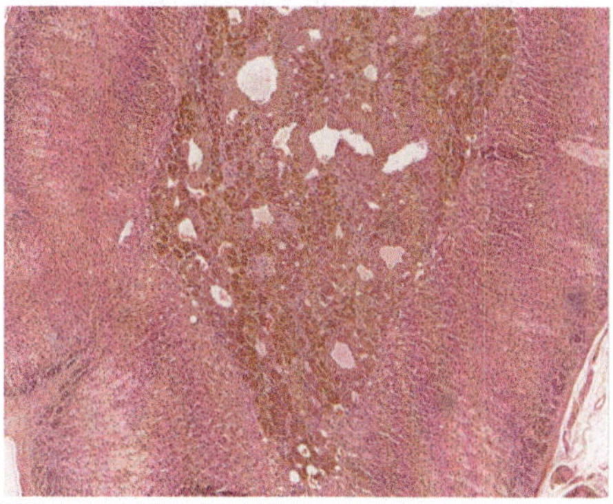	The three zones of the adrenal cortex extending from the outer capsule to the border of the adrenal medulla are the zona ____________________, zona ____________________, and the zona ____________________. **Answer:** Glomerulosa, fasciculata, reticularis

The automated marking key would have a number of alternatives entered as acceptable answers. Of course, in a similar repeated item, "A" and "B" could be switched (relabelled).

Level 1 item (Testing adrenal gland learning outcome 3) "Fill in blank" format.

The three zones of the adrenal cortex extending from the outer capsule to the border of the adrenal medulla are the zona ____________________, zona ____________________,

and the zona ____________________.

The automated marking key would have had the terms "glomerulosa", "fasciculata" and "reticularis" entered in the correct order.

This item could be substituted for an image-based item (Table 6.10) not necessarily labelled but to prompt the student in the right direction.

Again, the automated marking key would have had the terms "glomerulosa", "fasciculata" and "reticularis" entered in that correct order.

6.16 Use of a Single Image Duplicated and Annotated Differently with Each Duplicate Can Provide a Lot More Opportunity for More Items

6.16.1 Level 1 Item (Testing Adrenal Gland Learning Outcome 3) MCQ Format

Like a normal MCQ which tests quite a lot of information particularly the functional aspects of a cell, tissue or organ, a lot of knowledge can be assessed using the same image. Usually, each image would contain several important histological components. Setting an MCQ for testing the learning outcome 3 presents an ideal means for using an assessment item as a learning tool. Rather than labelling several histological features on the one image, label only one but repeat the majority of the stem of the item several times to include the other items (e.g. regions of the adrenal gland). The item could be in the form of an MCQ or short answer as it is easy to incorporate the answer into an automated marking system (Table 6.11).

The items (Table 6.11) are repeated (and test more knowledge—or motivate more learning) by substituting "A" for another region and changing the correct answer in the automated marking key. This means there are potentially four different items possible that could be posed.

Table 6.11 Level 1 Item (Testing adrenal gland learning outcome 3) "Fill in blank" format

Assessment image	MCQ question and answer	"Fill in the blank" question and answer
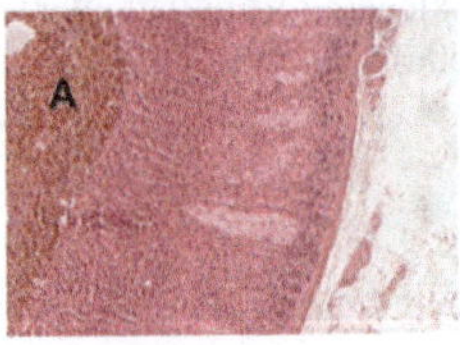	The region labelled **A** is the A. Adrenal cortex B. Zona glomerulosa C. Zona fasciculata D. Zona reticularis E. Adrenal medulla **Answer:** Adrenal medulla	The region labelled **A** is the __________ **Answer:** Adrenal medulla
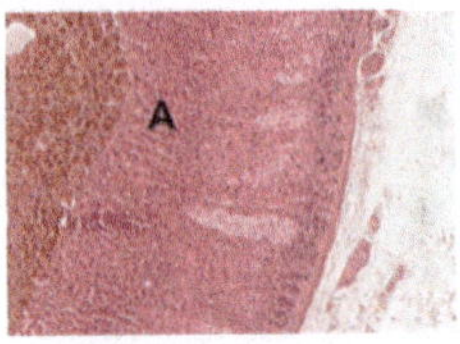	The region labelled **A** is the A. Adrenal cortex B. Zona glomerulosa C. Zona fasciculata D. Zona reticularis E. Adrenal medulla **Answer:** Zona reticularis	The region labelled **A** is the __________ **Answer:** Zona reticularis
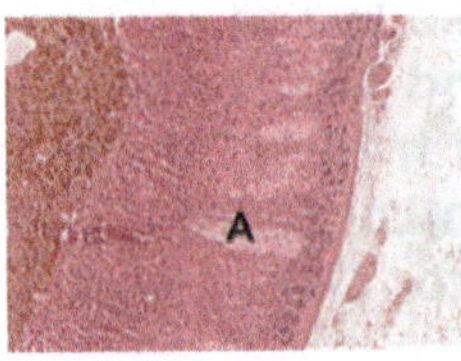	The region labelled **A** is the A. Adrenal cortex B. Zona glomerulosa C. Zona fasciculata D. Zona reticularis E. Adrenal medulla **Answer:** Zona fasciculata	The region labelled **A** is the __________ **Answer:** Zona fasciculata
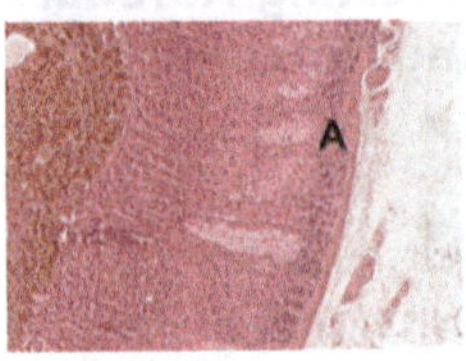	The region labelled **A** is the A. Adrenal cortex B. Zona glomerulosa C. Zona fasciculata D. Zona reticularis E. Adrenal medulla **Answer:** Zona glomerulosa	The region labelled **A** is the __________ **Answer:** Zona glomerulosa

Table 6.12 Multi-labelled images are best reserved for "summative assessments"

Assessment image	Item and answer
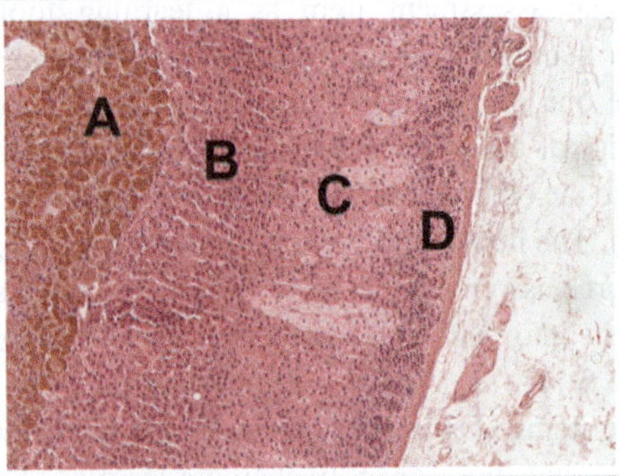	Which **One** of the following alternatives is **Correct?** A. A = Adrenal cortex, B = Zona glomerulosa, C = Zona fasciculata, D = Zona reticularis B. A = Adrenal medulla, B = Zona glomerulosa, C = Zona fasciculata, D = Zona reticularis C. A = Adrenal cortex, B = Zona glomerulosa, C = Zona reticularis, D = Zona fasciculata D. A = Adrenal medulla, B = Zona reticularis, C = Zona fasciculata, D = Zona glomerulosa E. A = Adrenal cortex, B = Zona reticularis, C = Zona fasciculata, D = Zona glomerulosa **Answer: D**

**Note:* Normally I would select ONLY one of the questions in Table 6.11 but recommend students in their review to construct the other alternative questions—an example of the *"Assessments as a learning practice" concept.*

Table 6.13 Level 2 Item (Testing adrenal gland learning outcome 6) MCQ and "Fill in blank" format

Assessment image	MCQ question and answer	"Fill in the blank" question and answer
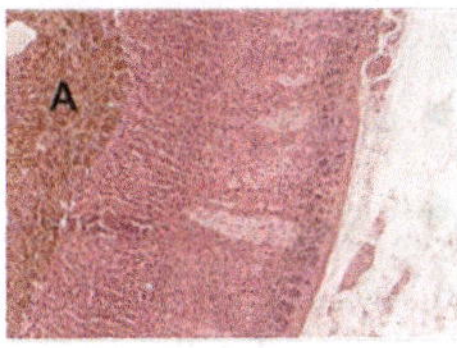	The general class of secretory products synthesised by the region **A** are A. Mineralocorticoids B. Sex hormones C. Glucocorticoids D. Catecholamines E. Peptides **Answer:** Catecholamines	The general class of secretory products synthesised by the region **A** are ________________ **Answer:** Catecholamines
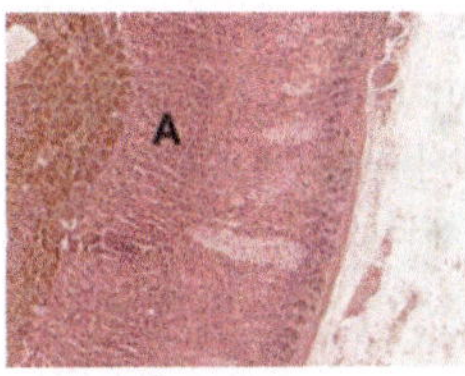	The general class of secretory products synthesised by the region **A** are: A. Mineralocorticoids B. Sex hormones C. Glucocorticoids D. Catecholamines E. Peptides **Answer:** Sex hormones	The general class of secretory products synthesised by the region **A** are ________________ **Answer:** Sex hormones
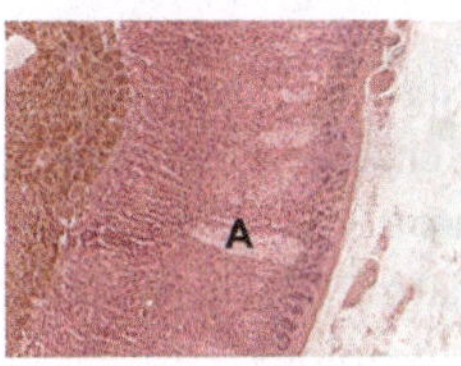	The general class of secretory products synthesised by the region **A** are: A. Mineralocorticoids B. Sex hormones C. Glucocorticoids D. Catecholamines E. Peptides **Answer:** Glucocorticoids	The general class of secretory products synthesised by the region **A** are ________________ **Answer:** Glucocorticoids
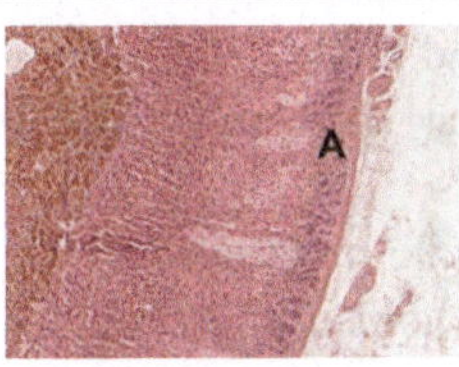	The general class of secretory products synthesised by the region **A** are: A. Mineralocorticoids B. Sex hormones C. Glucocorticoids D. Catecholamines E. Peptides **Answer:** Mineralocorticoids	The general class of secretory products synthesised by the region **A** are ________________ **Answer:** Mineralocorticoids

Why not use an item with the following format (Table 6.12)?

Such a question (Table 6.12) may be okay for a summative assessment, but for "Mastery learning", items depicted in Table 6.11 are preferred if students then create alternative items during their revisions and preparation for examinations.

There are several other items related to histological characteristics of the adrenal cortex and medulla that could be posed to test the learning outcomes 4 and 5 above, but they need not be outlined here.

6.16.2 Level 2 Items Relate Structure and Function and May Even Be Allocated More Marks

Apart from simply a series of identification items on histological structures/features of the adrenal gland, there are item types that will test knowledge about the function of each of the regions listed above. Then it becomes a little more difficult to enter correct choices into an automated marking key unless the item clearly requests or asks for just the general functions [e.g. secretion of steroid hormones (for the adrenal cortex and individual zones) or secretion of catecholamines (from the adrenal medulla)]. An item of more specific functions of each zone might be designed as either a "fill in blank" item or an MCQ (Table 6.13).

The item is easily repeated and tests more knowledge by substituting different regions and changing the correct answer in the automated marking key to include "Mineralocorticoids (aldosterone)", "sex hormones (androgens and estrogens)" and "catecholamines (adrenaline and noradrenaline)" for the zona glomerulosa, zona reticularis and the adrenal medulla, respectively.

This means there are potentially four different items possible that could be asked.

Or the item could be reversed to couple the class of hormones with the appropriate region of the adrenal gland.

Item: Glucocorticoids (e.g. cortisol) are secreted by the __________________.

The answer in the marking key would be "zona fasciculata", and again this item could easily be repeated and test more knowledge by substituting another class of hormone for "Glucocorticoids (e.g. cortisol)" and changing the correct answer in the automated marking key to include the respective region whose cells secrete them.

Instead of "fill in blank" items the items could be presented as a repeated series of MCQs e.g.

Item: The general class of secretory products synthesised by the zona fasciculata are _______________.

A. Mineralocorticoids
B. Sex hormones
C. **Glucocorticoids**
D. Catecholamines
E. Peptides

Item: Glucocorticoids (e.g. cortisol) are secreted by the __________________.

A. Adrenal cortex
B. Zona glomerulosa
C. **Zona fasciculata**
D. Zona reticularis
E. Adrenal medulla

The correct answer is "C". Again, items could easily be repeated and test more knowledge by substituting "zona fasciculata" and "glucocorticoids" for the other correct matchings and changing the correct answer in the automated marking key. This means there are potentially four different items possible that you could pose.

This item could be substituted for an image-based item.

6.16.3 Level 3 Item (Testing Adrenal Gland Learning Outcome 7)

More "in depth" knowledge could be further tested, but the extent of the choices would depend on the specific curriculum offered by the histology teacher and whether that curriculum included a substantial content on "physiology".

Item: Functions of secretory products from the zona fasciculata include ____________.

A. Increasing heart rate and blood pressure.
B. Development of axillary and pubic hair in women.
C. **Promoting carbohydrate metabolism.**
D. Controlling electrolyte homeostasis.

A more difficult challenge is to label the zone on an image as shown for other items:

Item: Functions of secretory products from the region labelled **A** include __________.

A. Increasing heart rate and blood pressure.
B. Development of axillary and pubic hair in women.
C. **Promoting carbohydrate metabolism.**
D. Controlling electrolyte homeostasis.or

Item: Name the hormones secreted by the region of the adrenal gland labelled **A** in this medium magnification image.

Answer: Glucocorticoids, cortisol, corticosterone, androgens/gonadocorticoids, dehydroepiandrosterone/DHEA, DHEAS, androstenedione.

or

Item: Name the hormones secreted by the region of the organ labelled A in this medium magnification image.

Answer: Glucocorticoids, cortisol, corticosterone, androgens/gonadocorticoids, dehydroepiandrosterone/DHEA, DHEAS, androstenedione.

6.17 The "Assessments as a Learning Practice" Concept

Using the "Adrenal gland" as the example, there are a possibility of *at least 50 items* (some examples provided above) that can be devised for assessing the learning outcomes listed for the adrenal gland. As stated earlier, I clearly explain to the students how I design items for both their review quizzes and their final examination. I almost repeat to them what is described above. I tell them it is not possible to pose every type of item on the adrenal gland presented above–perhaps only 1 or 2 such questions and so I advise them to design the other possible items themselves. In fact, for 6 successive years of teaching the histology of human organs in a small class of 30 UCLA Extension students I included this as an online activity for students to share questions they designed. At the end of each "Weekly Graded Quiz", students were required to post alternative questions for each of the 20 questions in the "Weekly Graded Quiz" following the suggestions above for each of the 10 topics in the histology course.

Students regularly commented that **"assessments as a learning practice"** was a useful exercise for them to complete (Table 6.16: Item 11) and helped them in their learning and preparation for examinations (Table 6.16: Item 12).

In summary:

- Define the learning outcomes very carefully and list them clearly on the student LMS or at the commencement of each topic.
- Set assessments based on these expected learning outcomes.
- Create alternative versions of items (i.e. MCQ and "fill in the blank" formats).
- Encourage students to construct alternative items as part of their learning strategy.

Using the example given above for the "Adrenal gland" extensive histology quizzes (about 2500 questions!) are now available at https://www.meyershistology.com.

For images, the same image could be used for testing the other aspects of a concept assessed in a different item. It was easy to copy the questions and change the answer key and the image label. My policy was to take several images of various parts of the organ showing the key features that were being tested and use each one for the item options, and in a final examination, I would even use an image of a different histological section of the organ—again to reinforce the advice to correctly identify the organ.

Histology crosswords was also an assignment I often would ask students to complete. Crosswords are useful for students' problem solving and cognitive abilities. However, there are so many now on the internet, it is difficult to assume students did not consult these resources!

6.18 An Enhanced MCQ Format

The traditional MCQ format has been further enhanced on the "Thinkific" platform (https://www.thinkific.com) which enables MCQs to include any number of CORRECT answers rather than the single best option. This solves the problem of introducing confusion and ambiguity if there is a selection of choices as the alternative option (e.g.

(a) A, B, C and D are correct
(b) A, B and C only are correct
(c) A and B only are correct
(d) A only is correct
(e) D only in correct)

Table 6.14 An example of the unique components for quizzes in Meyer's Histology (https://www.meyershistology.com)

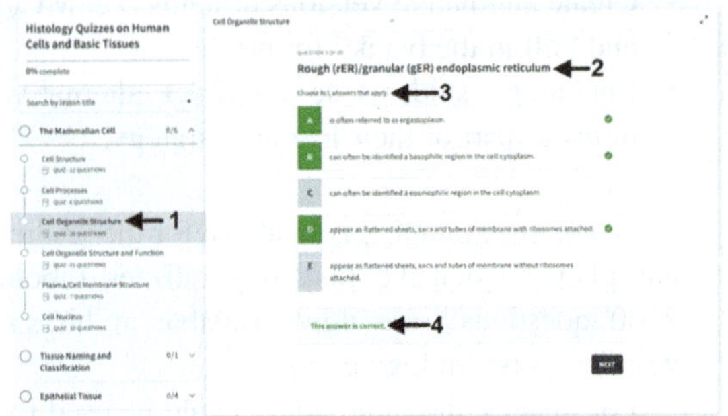

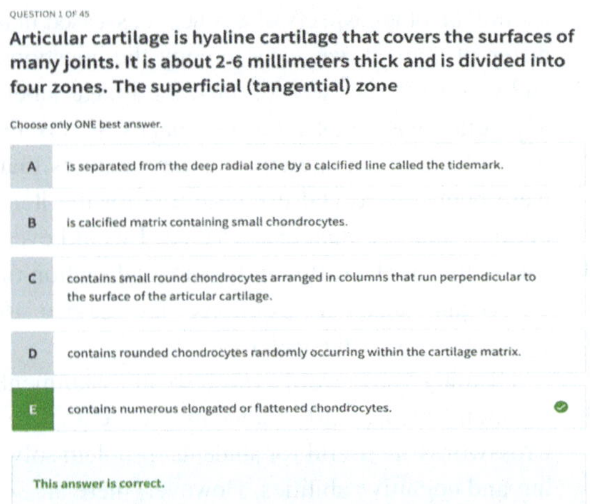

Unique components of quizzes

1. The database of items on "Cell Organelle Structure" contains 20 items. There is also the option of having items randomly selected from any of the total of 106 items from the 6 subheadings.
2. There are 2 formats used to load the item and answers.
3. After the item stem, there is the option to have more than one answer correct (Upper image). After the item stem there is also the option to have only one answer correct (Lower image).
4. Immediate feedback is provided which can also be accompanied by an explanation of the correct answer/s. Sometimes the item stem can be very short and simple (upper image) but often the stem contains reinforcing learning content before the item is posed (lower image).

Table 6.14 repeats an example of a quiz question from extensive quizzes now provided by (https://www.meyershistology.com) as described (Meyer (2023). There are over 2500 quiz questions. Again, quizzes are contained within the typical topics presented in a histology curriculum. In the example of a series of quiz items on "The Mammalian Cell" (Table 6.14), items are chosen from a database contained within appropriate subheadings.

Unique components of these quizzes are best summarised in point form with each point 1—4 labelled on the upper image in Table 6.14.

Apart from using annotated "still" images of components of tissue and organs for assessments, it was deemed important to test whether a student was competent in viewing a histological section and locating various histological characteristics of any tissue or organ (i.e. the practical examination of a section of a tissue or organ, like an assessment of correct use of the microscope done years ago) (Sect. 6.2.1). This would be an important skill to attain before studying pathology and being able to compare normal tissue or organ histology with a region of tissue indicating a pathological process.

6.19 Assessing Students' Ability to View Histological Sections and Identify/Locate Histological Features

6.19.1 Annotations on Digitised Slides

Digitised histology sections were once annotated (about 5 years ago), but we withdrew that facility because it was not seen (by academic staff—and students also, surprisingly) as simulating viewing a histological section in a traditional histology laboratory. Feedback from students both here at UWA as well as at the University of California Los Angeles (UCLA) where I continue to teach histology courses (completely online) preferred to view the histological structures on the interactive atlas then try and find them when viewing the slides without any annotations to prompt them (Table 6.16: Item 13). They found this to be a more positive learning experience. They found the annotations distracting and unnecessary (Table 6.16: Item 14). Also, the extensive videos (https://www.meyershistology.com), which are also being updated to provide the virtual microscopy experience, enabled students to very easily locate the required histological features when viewing the virtual microscopy database. But annotating digital slides was a desirable functionality to use in assessments.

6.19.2 Histology Practical Assignment

During each of the histology courses both at UWA and UCLA Extension, students were required to complete two assignments on "Practical histology". These assignments were introduced in 2008 when "screenshot" capture was available on mobile phones, computers, laptops etc. It was considered to be important for students to engage with the histological sections of human tissues and organs using virtual microscopy and view relevant histological characteristics of these organs. This facilitated students' achieving the skills required to recognise normal and abnormal structures within tissues and organs. It was also simulated using a light microscope.

The first assignment (Assignment 1) tasks were presented after completing 50% of the course topics:

1. The mammalian cell
2. Epithelium
3. Exocrine glands
4. Connective tissues
5. Cartilage

Students were presented a list of about 40 histological features to locate, capture and label and then students were prompted to submit the labelled images for assessment (Table 6.15). In addition, students were required to indicate on the histological section (either by submitting more labelled images or a X:Y coordinate of the histological feature) where they located the feature in the histological section. Students were *not permitted* to simply include any image that was in the resource https://www.histology-online.com or presented in any lecture or quiz. To locate the histological feature, they had to select the most appropriate histological section from the database that would clearly show the feature. Different histological features were included in the list each year for each course.

The second assignment (Assignment 2) was to be completed at the end of the course (before the final examination). The requirements for completing Assignment 2 were similar to Assignment 1 and covered the remaining topics of the histology course (e.g. Course 1—"Cells and Basic tissues", i.e.

1. Bone
2. Blood
3. Bone marrow
4. Peripheral nerve
5. Brain and spinal cord)

Evaluation reports from the students over five successive years (Table 6.16: Items 15 and 16) indicated they found it a very useful learning exercise, not too time consuming, and an enjoyable learning experience, which aided their ability to identify characteristic features of tissues and organs in histological sections.

Marking these assignments 1 and 2 was not a time-consuming task as I was very familiar with each histological section in the database and the most appropriate locations of all the histological entities listed in the assignments.

6.19.3 Histology Assignment: Use of Rubrics

A good rubric must have a list of specific criteria to be rated, and a scoring scale should include 3–5 levels of performance (e.g. Excellent/Good/Fair/

Table 6.15 Example of Assignment 1—list of histological features to be identified, imaged and labelled

The mammalian cell	Epithelium	Exocrine glands
1. Nissl substance 2. Nucleus 3. Nucleolus 4. Cytoplasm 5. Cell boundary **Connective tissues** 1. Elastic fibres 2. Collagen fibres 3. Reticular fibres 4. Macrophage 5. Fibroblast 6. Mast cell 7. Adipocyte 8. Dense regular connective tissue (Tendon) 9. Dense irregular connective tissue (Dermis) 10. Lamina propria	1. Simple squamous epithelium 2. Stratified squamous epithelium 3. Pseudostratified columnar epithelium 4. Neuroepithelium (olfactory epithelium) 5. Basal (stem cell) 6. Basement membrane 7. Cilia 8. Microvilli **Cartilage** 1. Hyaline cartilage 2. Chondrocyte 3. Lacuna 4. Perichondrium (chondrogenic layer) 5. "Territorial" matrix 6. Fibrocartilage 7. Isogenous group (of cartilage cells)	1. Mucous acinus 2. Serous acinus 3. Unicellular gland (goblet cell) 4. Centroacinar cell 5. Intercalated duct 6. Striated duct 7. Straight tubular gland 8. Myoepithelial cell 9. Coiled tubular gland 10. Secretory lobule

Poor) to indicate what level of expertise has been achieved for each skill, but in assignments 1 and 2 that I set for my students, they are merely a series of allocations of marks for tasks correctly achieved in identifying and labelling a histological structure.

1. Adequate magnification (1 mark)
2. Correct structure (1 mark)
3. Best example (i.e. showing most characteristic features) (1 mark)
4. Correct use of labels (1 mark)

Histology Assignments had only three levels of achievement (with an appropriate scale of marks) cited:

1. Very well done you know your stuff! (4 marks).
2. Okay but are you sure you have included everything? (2 marks).
3. More convincing details please (1 mark).

I prefer to use "numeric grades" to provide feedback to students although the University then condensed these into a grade as well when publishing final results.

6.20 There Is No Doubt Such Assessment Practices Are an Enormous Academic Workload: But It Paid off

In a research-intensive university such as UWA, allocating sufficient time away from my particular research interest to create the assessment packages as well as all the online learning resources initially was a challenge. It was a lot easier for me as time passed because I shifted from my research interest in reproductive biology to creating teaching and learning tools for eLearning and developing pedagogies for eLearning. It turned out to be a successful shift in interest as I was then able to receive teaching and learning grants and an Australian Teaching Fellowship. Development of the original "Histology Practical Assistant" was recognised by the Australian Government when in 1999 I was awarded an "Australian Award for university teaching". Since then, this resource has been extensively redeveloped (Meyer 2023) to contain all the necessary learning resources to complete the teaching of histology completely online. The

Table 6.16 Quantitative evaluation of assessment practices described in this chapter

Item	Question	Strongly agree (%)	Somewhat agree (%)	Neutral (%)	Somewhat disagree (%)	Strongly disagree (%)
1	You enjoyed collaborating with fellow students to complete learning exercises in the histology class	57	21	11	7	4
2	You enjoyed being "informally" assessed together as a student group and the informal competition	48	17	6	13	16
3	You enjoyed being "formally" assessed together as a student group for 10% of your final mark	11	9	7	28	45
4	You used the learning objectives as a guide in you preparation for formal examinations	44	31	10	9	6
5	Time required to complete learning and formative assessments for each topic was not overburdening?	56	29	7	5	3
6	The formative assessment each week motivated you to learn	55	20	15	6	4
7	I preferred to have "Pop-up" questions on the "Lecturio.com" platform appear during the lecture at the appropriate time	51	27	9	6	7
8	I preferred to defer completing the "Pop-up" questions on the "Lecturio.com" until the end of the lecture	41	36	11	6	6
9	"Pop-up" questions on the "Lecturio.com" platform made me concentrate and engaged me in the content being presented	46	24	10	9	11
10	The extended review content in the stem of questions in the formative assessments reinforced your learning	52	26	8	7	7
11	Using "assessments as a learning practice" was a useful exercise	54	14	11	10	11
12	Using "assessments as a learning practice" helped my learning and preparations for examinations	62	18	15	3	2
13	I preferred to view digital slides without annotations on the digital slides	62	21	10	4	3
14	Annotations on the digital slides were unnecessary for my learning and understanding of histological structures	57	22	14	3	4
15	Completing the assignments 1 and 2 helped my proficiency at viewing the histological sections	46	21	17	9	7
16	Completing assignments 1 and 2 helped my ability to locate characteristic features of tissues and organs when viewing digital slides	42	26	14	13	5

resource (https://www.histology-online.com) now includes an interactive histology atlas, virtual microscopy database, descriptive videos of all human cells, tissue and organs and an extensive collection (over 3500) of histology assessment items.

6.21 Assessment as Part of Learning Practices Improved Examination Performances: But that Comes with Issues

Since the implementation of all the learning resources described in Meyer's Histology (Meyer 2023) and particularly the use of assessments as part of the learning adventure, many more students have attained a "distinction" or "A" grade (<80%) or a "High Distinction" or A+ grade (>90%). The majority (almost 90%) of the 30 students completing histology courses at UCLA Extension (https://www.uclaextension.edu) each term gained either an "A" or "A+" grade. This was not unexpected because many of the students were of "mature age" and completing these courses to apply for entry into medical or dental school so they were highly motivated to gain good grades. However, marks/grades were more dispersed in the undergraduate histology courses I taught at UWA. But there was still a high proportion (greater than 65%) who attained a "A" or "A+" grade. Because of the students' success, one of the issues I faced was pressure (or more of a request—even command) to scale my marks to approximate a desired distribution spread of marks set by the university—but not a "Bell Curve".

There are many articles discussing the controversy of the use of a "Bell Curve" (Murad et al. 2021) and other forms of scaling marks, but my argument to my School and the Faculty was purely based on the hard work done by both myself (to create the assessment packages) and my students, most of whom engaged the "assessment for learning" practices I had encouraged. It was a common comment at student staff meetings held each semester that student representatives for the histology classes would ask other academic staff teachers (all most excellent teachers in their own unique ways) why they did not introduce similar assessments for learning practices in the other study units they were completing as part of their degree programme. Now because of the COVID-19 pandemic and the University's shift to an online learning format, staff are embracing the need for this paradigm shift to motivate and engage students more in their online learning platforms, and not without acknowledging the enormous workload this involves.

Over the past 5 years though the assignments have been omitted from the assessments for the medical and dental students as there is less emphasis on requiring them to identify each specific tissue and organ. However, in histology courses in Allied Health/Biomedical Sciences, they were retained because a desirable and important learning outcome was proficiency in identifying cells, tissues and organs. The newest innovation now available (released in December 2022) by SLICE makes completing assignments such as the one's described above much easier for the students—and histology instructor.

6.22 What Is the Latest Innovative Technology Enhancing Assessments in Histology?

6.22.1 SLICE

An extensive list of virtual microscopy sites available for use in histology education, and with free access, was provided by Chapman et al. (2020) and details of the use of virtual microscopy on an online platform for teaching and learning histology was described by Meyer (2023). One very popular online, virtual microscopy facility is SLICE, which promotes extensive visualisations for learning histology (and other image-based disciplines). Seven functionalities of SLICES's image-based online learning and teaching platform were described by Meyer (2023), but a brief review of SLICE is repeated here.

SLICE (https://www.best.edu.au/slice) is an Online Image Based Learning and Teaching Platform for viewing, annotating and interacting with images, as well as sharing knowledge with colleagues and students. SLICE hosts over 22,000 images, including virtual slides, from institutional collections which allows academics worldwide to choose any image that meets their needs for use in classroom activities in visually oriented biomedical and life science disciplines

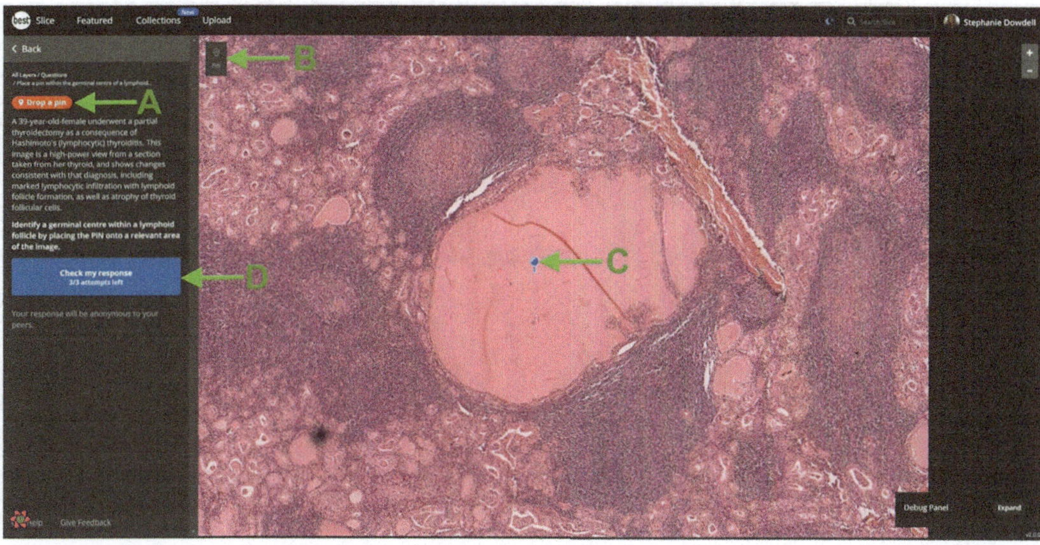

Fig. 6.6 The image-based question tool. Drop the pin

such as histology, anatomy, pathology and histopathology, haematology and botany. The Slice tools are designed specifically for education and include LMS integration, image annotation and image organisation. Cloud infrastructure provided by Amazon AWS has enabled SLICE's image-based learning and teaching platform to be shared cost-effectively by an unlimited number of students and their teachers.

Slice was awarded a Platinum medal in the 1EdTech 2022 Learning Impact Awards. 1EdTech is a non-profit collaboration of the world's leading universities, school districts, government organisations, content providers and technology suppliers, cooperating to accelerate learning technology interoperability, adoption and impact.

6.22.2 Image-Based Question Tool in SLICE

One essential learning outcome when studying histology is for the student to be able to identify cell and tissue components that characterise a particular human tissue or organ. In December 2022, a most useful innovation was added to the SLICE platform. The innovation was an image-based question tool that assesses students' ability to identify a histological feature on an image and then provides immediate feedback. It was the technological solution to my earlier years of setting student assignments whereby they searched histological sections using virtual microscopy to locate various histological structures—photographed them, labelled the histological features I had asked them to locate and submit to me for assessment. This question tool was something instructors over the past two decades could only imagine when using virtual microscopy and was dismissed as technologically impossible—until now.

6.22.3 Using the Question Tool

Students are provided a normal histology (or pathology) slide (Fig. 6.6). Descriptive text about each specific slide (e.g. the pathology) is provided in the left-hand column of the platform. In the example shown in Fig. 6.6, there is marked lymphoid follicle formation in this case of Hashimoto's (lymphocytic)thyroiditis. The item posed by the teacher/instructor appears immediately below this descriptive text and for this example, the student is required to identify a

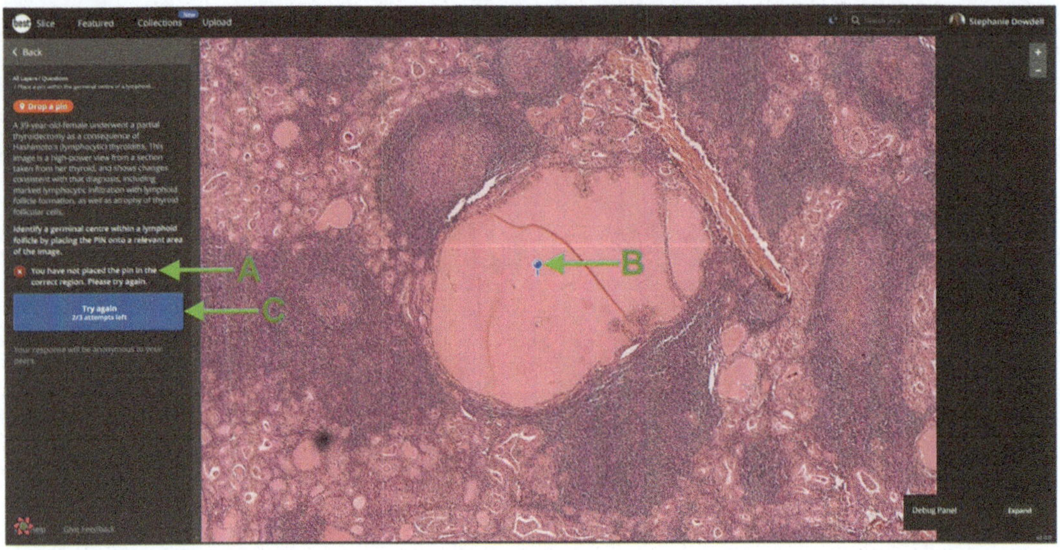

Fig. 6.7 The image-based question tool. Receive instant feedback

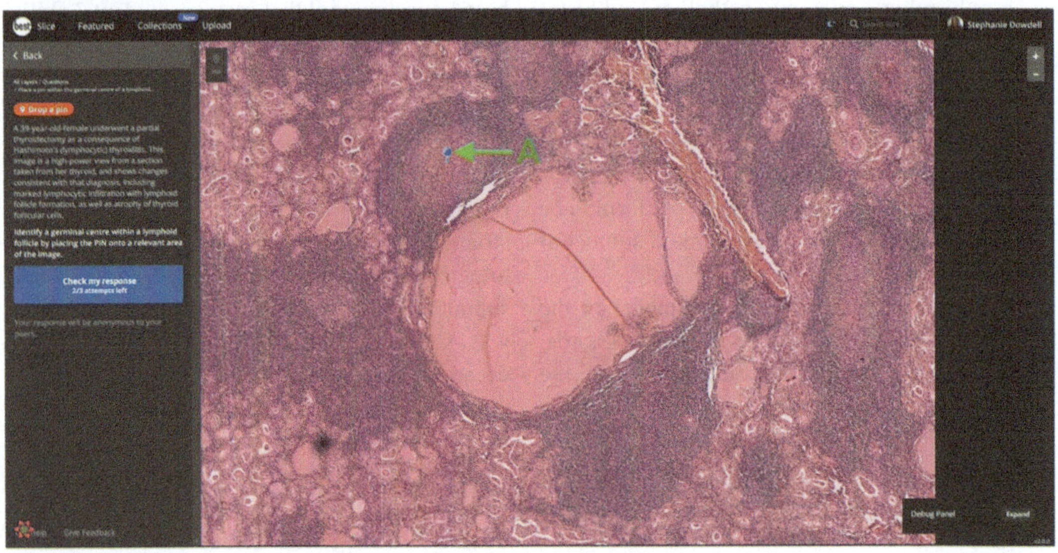

Fig. 6.8 The image-based question tool. Second attempt at dropping the pin

germinal centre in a lymphoid follicle by dropping a pin (Fig. 6.6a) in the correct area. After selecting the pin tool (Fig. 6.6b) the student places a pin into his/her chosen area of the image (Fig. 6.6c). Clicking on the "Check my response" button (Fig. 6.6d) the student receives immediate feedback as to whether he/she was correct or incorrect (Fig. 6.7a). This helps to alleviate the stress of marking and giving individualised feedback when the cohort size is large.

In the attempt shown the feedback was that the pin (Fig. 6.7a) was not placed in the correct area, and so the student can click the "try again" button (Fig. 6.7c) to have another attempt (Fig. 6.7b), knowing they only have three attempts at the exercise. In the second attempt, the student

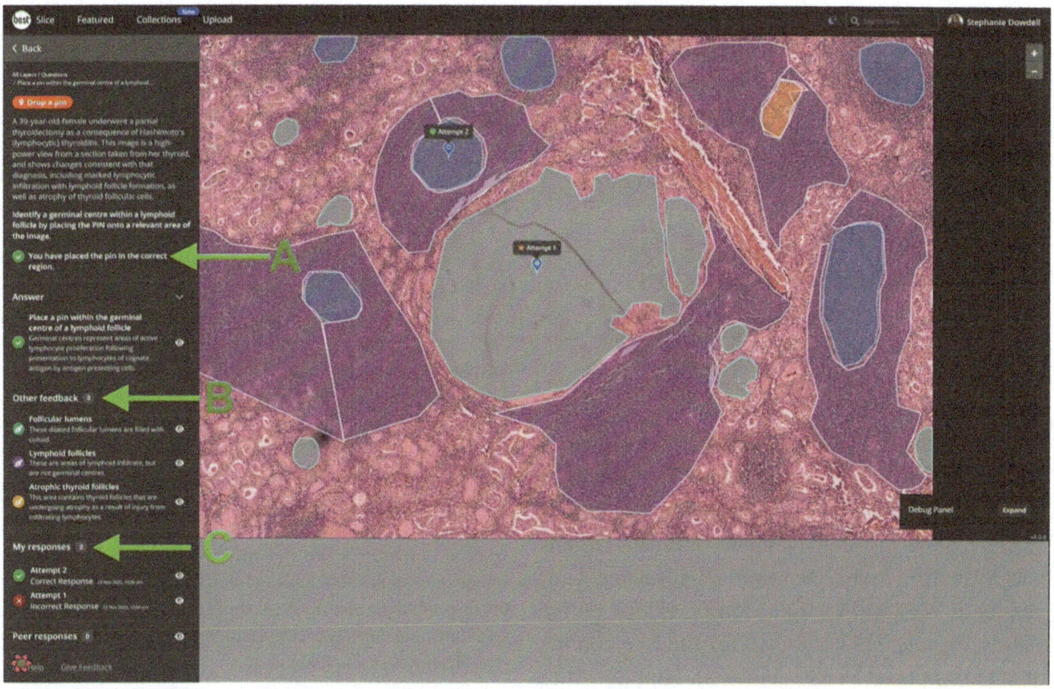

Fig. 6.9 The image-based question tool. Correct attempt at dropping the pin and additional feedback

drops the pin on a different region on the slide (Fig. 6.8a).

After correctly identifying the feature (Fig. 6.9), or using up the maximum number of attempts allowed, students are shown the correct region (Fig. 6.9a), along with any feedback annotations (Fig. 6.9b) to allow students to immediately remediate misconceptions when submitting their responses (Fig. 6.9c).

Some key features and benefits of this question tool are:

- Whilst using the question tool, students can interact with the digitised slide, and zoom in and pan across the histological slide.
- After placing a pin into their chosen area of the image, students receive immediate feedback as to whether they were correct or incorrect. This helps to alleviate the stress of marking and giving individualised feedback when the cohort size is large.
- When setting the question, the instructor has the option to either use the entire image or, if the feature is repeated multiple times across the image, a smaller field can be chosen and the question can be limited to only that smaller area of the image. Students are still able to zoom in and out and pan across the selected field of view.
- The instructor can review heat maps of student pins and filter for correct and incorrect responses to facilitate in-class discussion and address misconceptions in real time.
- The question tool has been designed to suit a variety of classroom uses, including both multiple- or unlimited-attempt formative classroom activities, and single-attempt, exam-style questions with no feedback provided to learners.

SLICE has reusable elements—so you save time whilst creating a question

One of SLICE's goals has been to save instructors time whilst creating questions. Over the last 10 years, registered users have collectively made almost two million annotations on images. The new question tool lets instructors use any of these previously made annotations or public layers when providing detailed feedback for a question.

As mentioned above, SLICE is not just a facility for storing anatomical, histological, or pathological images but all other image-based disciplines such as haematology and botany.

6.23 Quantitative Evaluation of Assessment Practices Described in this Chapter

At the end of each histology course from 2012 until 2016, an evaluation of assessment practices described in this chapter was provided for students to complete. Undergraduate students enrolled in a histology course "Human Organs and Systems" from 2012 until 2016 at the University of Western Australia completed the online survey. Results are combined for the five cohorts of students (Total = 680 students). Human ethics approval was obtained from the University of Western Australia (Reference number: RA/4/1/7150). The survey contained 33 questions, but only a summary of questions relating to assessment practices is provided in Table 6.16.

6.24 Conclusion

Over the past two decades, technology-enhanced and interactive learning strategies have revolutionised histology teaching to the extent that now histology teaching and learning can be delivered completely online. Technologies have enabled assessments to become an important tool for students' learning of histology—in particular, assessment packages that engage students to learn and those that return immediate feedback. These assessment strategies make histology learning efficient and significantly improve students' performance on examinations.

Acknowledgements The author would like to thank the UWA students who participated in the 2014 survey. The author is also grateful for the advice and assistance of Professor John Campbell, PhD, (deceased) and Professor Michael Hall, PhD, both from UCLA, in creating the unique learning resources. A University of Western Australia Teaching Fellowship, Australian Award for University Teaching, Australian Learning and Teaching Council Grant 00910-63001003 and Australian Learning and Teaching Council Fellowship ID11-1990 to the author provided financial support for the creation of all the resources used.

Conflicts of Interest *Geoffrey T. Meyer is the creator of all resources delivered by* https://www.histology-online.com *and* https://www.meyershistology.com *and a Co-Director of Histology-Online Pty Ltd that distributes these resources.* Students can subscribe individually for a very small fee (i.e. only about USD$14). Many universities subscribe (for a small fee) via their university library acquisition funds to enable their students to access the resources for 4 years. In some instances, *no* fee is charged, e.g. for developing nations characterised by a population with low and middle incomes, and other socio-economic indicators. This marketing of the resources is "non-profit" and essentially to provide an income to maintain the website, storage of large, scanned images and a reliable (but economical) Web-based membership management tool. Also, funds finance updating of software and any website changes etc., which require a web designer or IT specialist.

References

Chapman J, Lee L, Swailes N (2020) From scope to screen: The evolution of histology education. Biomed Vis Adv Exp Med Biol 1320

Gartner LP, Lee MJ (2022) Gartner & Hiatt's atlas and text of histology, 8th edn. Wolters Kluwer

Lowe JS, Anderson PG, Anderson SI (2018) Stevens & Lowe's human histology, 5th edn. Elsevier Health Sciences

Mescher A (2013) Junqueira's basic histology: text and atlas, 13th edn. McGraw-Hill Education

Meyer GT (2023) Biomedical visualisation – digital visualisation approaches in biomedical education. In: Border S, Rea PM, Keenan I (eds) Published in the book series Advances in experimental medicine and biology

Murad A, Jefri H, Le Ha P (2021) Why the bell curve system for giving grades needs reform. University World News

Nagandla K, Sulaiha S, Nalliah S (2018) Online formative assessments; exploring their value. J Adv Med Educ Prof 6(2):51–57

Ovalle WK, Nahirney PC, Netter FH (2013) Netter's essential histology. ClinicalKey 2012. Elsevier Saunders

Ross MH, Pawlina W (2016) Histology: a text and atlas. Wolters Kluwer Health

Rueshle S, Dorman M, Evans P, Kirkwood J, McDonald J, Worden J (1999) Critical elements: designing for online teaching. ASCILITE 99

Using Stereoscopic Virtual Presentation for Clinical Anatomy Instruction and Procedural Training in Medical Education

7

Edgar R. Meyer and Dongmei Cui

Abstract

This chapter begins by exploring the current landscape of virtual and augmented reality technologies in a post-pandemic world and asserting the importance of virtual technologies that improve students' learning outcomes while also reducing costs. Next, the chapter describes clinical anatomy instruction concepts in medical education, including applied anatomy content knowledge, pedagogical anatomy content knowledge, and virtual stereoscopic visualization studies that exemplify these concept areas, respectively. The chapter then explores the concept of procedural training with a specific emphasis on virtual stereoscopic anatomy visualization studies that exemplify or have implications for procedural training in medical education. Subsequently, the chapter discusses the benefits and challenges as well as the potential future positive and negative implications of virtual stereoscopic visualizations in medical education before finally concluding with some pensive considerations for the present and future of anatomy education and training using virtual technologies.

E. R. Meyer (✉) · D. Cui
Department of Advanced Biomedical Education, University of Mississippi Medical Center, Jackson, MS, USA
e-mail: emeyer@umc.edu; dcui@umc.edu

Keywords

Applied anatomy · Clinical anatomy · Medical education · Pedagogical anatomy · Procedural training · Virtual stereoscopic anatomy visualization

7.1 Introduction

American writer and philosopher Elbert Hubbard has been accredited with saying, "The world is moving so fast these days that the man [or woman] who says it can't be done is generally interrupted by someone doing it" (Esar 1951). Arguably, truer words have hardly ever been spoken or written given the pace of human technology and innovation advancements. As Hubbard's quote suggests, those who deny the possibility of achieving unfathomable feats are inevitably proven wrong, outcompeted, and outperformed. In capitalistic economies like the United States, the mere potential of acquiring wealth quickly can be a powerful motivating factor for entrepreneurs and investors to take risks on starting businesses and engaging in global market trades, thus surpassing others' triumphs and successes. Evidence of this reality manifested with the proliferation of new online businesses and digital services, especially during the global COVID-19 pandemic. Unfortunately, this seemingly inequitable phenomenon became catastrophic for many small business owners during the pandemic, but it also

D. Cui et al. (eds.), *Biomedical Visualisation*, Advances in Experimental Medicine and Biology 1431,
https://doi.org/10.1007/978-3-031-36727-4_7

provided a setting for exponential growth in digital enterprises.

One such digital enterprise that ultimately benefited from the pandemic was the virtual reality (VR) market whose global revenue reached 9.3 billion United States dollars (USD) by 2021 and achieved a 2.67-billion USD increase between 2021 and 2022 with a projected total of 28.84 billion USD by 2026 (Alsop 2022). Despite initial pandemic-related supply-and-demand setbacks in 2020, another report projected even higher revenue values reaching as high as 227.34 billion USD by the end of the current decade (FBI 2022a). Unsurprisingly, the growth of the VR market in healthcare specifically is expected to reach 6.2 billion USD by the end of the same time period due to its high demand (FBI 2022a, 2022b), and when considering both VR and augmented reality (AR) revenues in healthcare, the combined total is projected to amount to 9.79 billion USD over the next 5 years (BW 2022). While these fiscal values sound promising for companies that manufacture and distribute virtual hardware and software to health science instructors and students, they raise the question of which entities are funneling this level of revenue into the market. In other words, who is mainly footing the bill?

According to Perkins Coie, VR and AR technologies pose the greatest economic disruption worldwide to the gaming and entertainment industries, followed by healthcare and medical device (38%) and education (28%) sectors (2020). Just as individual entrepreneurs and innovators must adapt to the changing landscape of technology, so too must firms and institutions within these sectors respond to the changing consumer demands. Disruption can be positive or negative for individuals or organizations, depending on the circumstances. This level of volatility does warrant time for consideration of cost-benefit analyses regarding various factors pertaining to virtual technologies, such as their purpose, affordability, utility, validity, reliability, and effectiveness at improving outcomes measures. For instance, when considering the use of virtual technologies in the teaching of anatomy to health professional students, their usage might be more sustainable and affordable than human donors, but if the technological devices do not enable students to achieve desired learning outcomes, then they essentially just become cool toys. Therefore, when health science educators and trainers are contemplating purchasing and implementing these devices to improve their students' learning or their trainees' procedural skills, respectively, then they must offer their due diligence in reviewing within the literature experimental findings that corroborate these products' effectiveness, if any. Moreover, these educators and trainers should conduct beta or pilot tests on the products to ascertain the level of their potential, especially if no formal studies on the products exist. After all, health science professionals have a logical obligation to their students and trainees, their patients, their colleagues, and/or their own faith in the scientific method to follow evidence-based practices in any scholarly or diagnostic endeavor in pursuit of learning outcomes that are most efficient and cost effective for all stakeholders involved.

The previously mentioned fiscal values include the sales of VR and AR products not only used in the care, treatment, and diagnosis of patients but also in the education and training of future and current healthcare professionals (e.g., medical students, residents, and physicians). This chapter will focus on an overview of the use of virtual stereopsis in medical education for clinically relevant anatomy and procedures both in the classroom and laboratory settings for pre-clinical students and in the clinical or clinically simulated settings for clinical students and trainees. This overview will entail a brief description of several examples of virtual stereoscopic visualization techniques and implications in clinical anatomy instruction and procedural training in medical education and a discussion of the benefits and challenges of using virtual stereoscopic visualizations in medical education.

7.2 Clinical Anatomy Instruction and Procedural Training in Medical Education

7.2.1 Clinical Anatomy Instruction

The discipline called clinical anatomy is relatively new among the biomedical sciences. Given the fact that novel discoveries in the gross, or macroscopic, human body have been relatively minimal and controversial in their importance over the course of the current decade (Kumar et al. 2018; Kwon 2020), novel research has gravitated toward clinical applications of gross anatomy to surgical or other medical procedures as well as to innovations in anatomical sciences education. Clinical anatomy captures both the applied content knowledge of how human anatomy can inform and improve the clinical practice of health care professionals and the pedagogical content knowledge of how clinically relevant human anatomy can be taught more effectively and efficiently to health science professional students (e.g., medical, dental, physical therapy, occupational therapy, physician associate, optometry, nurse anesthesia, chiropractic, podiatric, mortuary science, etc., students).

This burgeoning shift in the advancement of the anatomical sciences beyond just the bench to also the clinic and the classroom has warranted the development and growth of master and doctoral programs designed specifically for training the next generation of anatomy educators. Most of these new programs include Master of Science and/or Doctor of Philosophy in Clinical Anatomy programs. Due to factors like higher rates of faculty retirement versus lower rates of students entering the profession, there is an evident shortage of anatomy educators in places such as the United States, Canada, and the European Union (EU), including the United Kingdom (UK) given the fact that this first study was published before the UK's exit from the EU (Wilson et al. 2020, 2021; Edwards et al. 2022). As a result, these new programs have the potential to restore a healthy supply of anatomy educators to institutions in need of anatomy faculty members. Many of these programs also offer different research tracks that allow students to explore clinically relevant research, educational research, or a combination of both in their master theses or dissertations. A few examples of these different tracks in the United States and Canada are discussed in the following sections.

7.2.1.1 Applied Anatomy Content Knowledge

Ever since the sixteenth century, anatomy has been a foundational discipline for medical education in the Western world, but it has been a foundational discipline arguably even longer in the Middle and Far East, especially in Islamic (Alghamdi et al. 2017) and ancient Chinese (Shaw et al. 2022) cultures, respectively. In fact, if one were to review most medical education curricula, one would easily find, regardless of whether curricular disciplines were integrated or not, that normal and/or abnormal anatomy comprise a majority of medical students' education. Unsurprisingly then, 80% of the content on Step 1, the first part of the United States Medical Licensing Examination (USMLE), at most, can pertain to normal or abnormal anatomy (i.e., gross anatomy, embryology, histology and cell biology, and pathology) (FSMB and NBME 2023). As a result, one cannot disregard the importance of the anatomical sciences in medical education. However, anatomical science knowledge on the USMLE Step 1 is not usually assessed in isolation, but often in the context of clinical vignettes, or items with a stem that includes a patient case or scenario and a question followed by a series of answer choices (FSMB and NBME 2022). Therefore, future health science professionals, including future health care practitioners and future anatomy educators, should be educated in clinical contexts. Future health science professionals should be educated in clinical contexts not only because their examinations contain items in a similar contextual format but also because requiring students to synthesize multi-disciplinary content in the contexts in which or for which they will be working or teaching in the future is more pragmatic and realistic.

Many medical schools throughout the United States and Canada and throughout the world are located at academic health science centers where biomedical and clinical research and patient care are occurring simultaneously. This proximity allows medical students and future medical educators to learn more conveniently in the contexts in which they will be living out their careers. Even at medical schools that are not necessarily associated with an academic health science center, there are still clinicians present who are often involved in medical students' instruction, even in the pre-clinical stages of their education. Many of the courses or modules in medical curricula are often taught by a team of faculty and led by at least one basic science educator with a PhD or equivalent degree and one clinical educator with a medical doctorate (e.g., MD, Dr. MuD, Dr. Med, and DO) or equivalent degree in other countries (e.g., MBBS, BMBS, MBChB, MBBCh, B.Med, MB, BM, BS, and B. Surg). This organization in the anatomical sciences, for instance, allows medical practitioners such as radiologists and surgeons to deliver clinical correlation lectures and perhaps administer skills, such as basic suturing, during their anatomy courses or integrated, systems-based modules, thus allowing medical students to experience the relevance of the foundational anatomy knowledge they are learning.

Examples of Virtual Stereoscopic Visualizations for Applied Anatomy This section will continue to showcase this contextualized applied anatomy instruction through virtual stereoscopic anatomical visualizations, using several examples in the United States and Canada. The first example features the Corps for Research of Instructional and Perceptual Technologies (CRIPT) lab (Fig.7.1), the first laboratory in the world to study and explore human body three-dimensional (3D) modeling with the mission to emerge as a global leader in digital anatomy education and evaluation research (2016a, b). Since the lab's inception at the University of Western Ontario in 2006, its founder, Dr. Timothy Wilson, and his research team have generated a considerable number of publications, many of which feature stereoscopic visualization studies on applied anatomy concepts, such as surgery (de Ribaupierre and Wilson 2012; Roach et al. 2012), laparoscopy (Mistry et al. 2013; Roach et al. 2014; Dawidek et al. 2016), and dental implant preparation (Massey et al. 2013). Many of the co-authors on these studies were master or PhD students at the University of Western Ontario who are now currently serving as faculty members at medical centers or practicing as healthcare professionals in various fields. Therefore, virtual stereoscopic visualization studies generated from the CRIPT lab have influenced how anatomical knowledge is applied not only to the instruction of future healthcare professionals but also to the active research conducted by students in the health sciences who will inevitably become the health science educators and practitioners of the future.

Secondly, collaborations between researchers involved in projects executed in the CRIPT lab inspired the establishment of a similar lab governed by Dr. Dongmei Cui at the University of Mississippi Medical Center (UMMC) (2023). The ultimate goal of this lab called the Clinical Anatomy Research and Scholarship Laboratory (CARSL), or commonly known at the institution as the 3D Virtual Anatomy Lab, is to create novel, innovative instructional tools that augment clinical anatomical knowledge acquisition and application in the health sciences (Fig. 7.2). These tools incorporate virtual stereoscopic visualization techniques, and they have been featured in several publications focused on applied anatomy concepts. These concepts include the implications of virtual stereoscopic anatomical models for procedural training, such as in the interventional delivery of anesthetics (Chen et al. 2020), development of valid and reliable virtual stereoscopic anatomical models for medical education and training (Meyer et al. 2018; Meyer 2019), and the implementation of those models in educational sessions demonstrating their surgical applications (Meyer 2019). The stereoscopic projection system used in the CARSL includes 3D software (Amira® software), dual projectors, a silver screen, a well-equipped computer

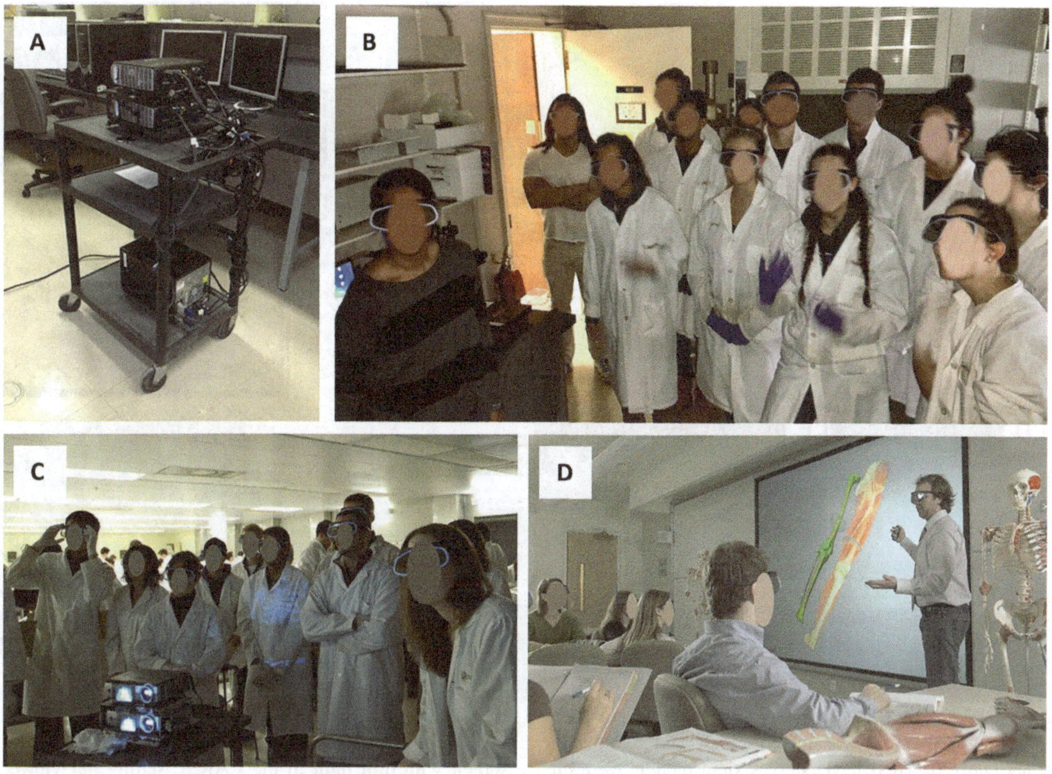

Fig. 7.1 CRIPT Laboratory and Anatatorium. (**a**) depicts a mobile cart that allows the portability of a typical dual projection system necessary for presenting stereoscopic visualizations. (**b**) shows students observing these visualizations in the CRIPT laboratory while using the glasses with polarized lenses to appreciate the added depth of stereopsis; such learning sessions are ideal for teaching anatomical content and regions that are too difficult and/or time consuming for dissection in compact curricula with reduced time. (**c**) provides an example of how portable hardware can also be used in the cadaveric laboratory to supplement medical students' learning of more complex regions of anatomy. (**d**) captures Dr. Tim Wilson's demonstrating bone and musculature models of the lower limb in a stereoscopic display using a commercial projection system in a virtual learning lab space called the Anatatorium that initiated his extensive line of virtual anatomy research and education. This space was utilized in courses in which students, such as those in nursing, kinesiology, and other health sciences, did not have access to cadavers for learning anatomy. The photos in this figure have been included with the permission of Dr. Tim Wilson

workstation, and a high-speed Dell Precision T7600 computer (Cui et al. 2017). Stereoscopic visualization is achieved via two eyes seeing slightly different images so that when the brain puts these two images together, a profound impression of three-dimensionality results (Poggio and Poggio 1984). Individuals wear 3D glasses that have polarized lenses that match the polarization axes of the projector filters of the projector (Cui et al. 2017). Similar to the Canadian lab example previously described, the CARSL has impacted clinical anatomy education not only for medical students in their first year of education but also for the advancement of the primary authors of these studies into their current professional practitioner and educator roles, respectively.

Thirdly, a considerable amount of novel research concerning applied anatomical visualizations has been contributed by investigators in the lab of Dr. Anne Agur at the University of Toronto. Virtual stereoscopic visualization studies have offered implications in such applied anatomy concept areas as muscle loading

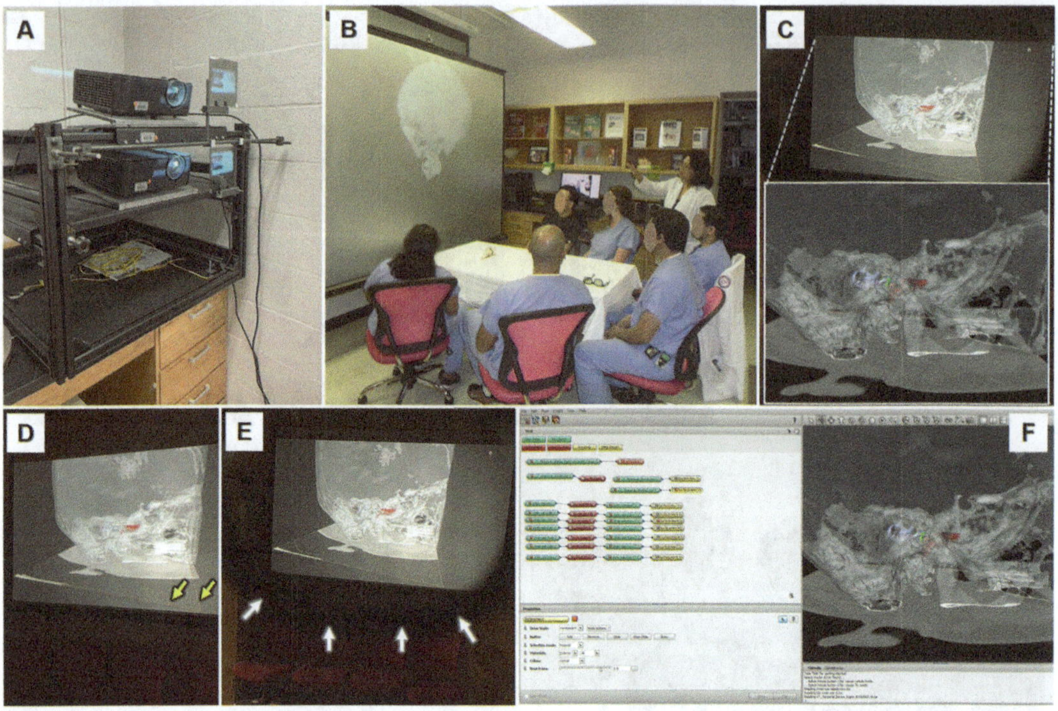

Fig. 7.2 Clinical Anatomy Research and Scholarship Laboratory (CARSL). 3D stereoscopic models were created in the Clinical Anatomy Research and Scholarship Laboratory (CARSL), and these models can be used in lectures or stereoscopic presentations within the laboratory. (**a**) shows the projection system, including the high-definition LCD (Liquid Crystal Display) dual projectors which are controlled by the Dell T7600 workstation (not pictured). Two linear polarizing filters were placed in front of the projectors to produce the stereoscopic displays of the models. (**b**) depicts Dr. Dongmei Cui's (standing) demonstrating a virtual model of the skull displayed on a silver screen as collaborator Dr. Jian Chen (seated, black shirt) and other medical residents and students view the model in a stereoscopic format while wearing the polarized 3D glasses. (**c**) shows a stereoscopic presentation of the model skull (temporal bone) with the transparent bone view feature applied to reveal deeper middle and inner ear structures and their 3D relationships to the overlying bone. The model is displayed on a large silver screen with dim light in the CARSL setting but viewed through a lens of the polarized 3D glasses. A more magnified view of the transparent skull (temporal bone) model is also included as an inset outlined in white with white segmented leader lines connecting it to the more distant screen view. (**d** and **e**) compare images of the transparent skull (temporal bone) model viewed without 3D glasses and viewed with 3D glasses, respectively, so that the reader can appreciate the dual images (yellow arrows) superimposed on one another in (**d**), causing the display to appear out of focus when in reality, the dual images create the added depth of stereopsis as viewed in (**e**) when the viewer wears 3D glasses with polarized lenses. The rim (white arrows) of the lens frame for the right eye is visible in this image. (**f**) provides an example of the Amira® software platform used to generate the virtual anatomical model of the skull (temporal bone)

diagnostics (Falcinelli et al. 2018) and denervation in the treatment of pain (Tran et al. 2022). Although some other published research from the Agur lab does not explicitly state the use of virtual stereopsis, it does involve sophisticated digitization and computational modeling techniques ultimately used in the construction of 3D virtual models—processes first described in the article by Ravichandiran et al. (2009). This research entails applied anatomy concepts, such as treatment of dysphagia (Shaw et al. 2017), peripheral nerve blocks and denervation in the treatment of pain (Davies et al. 2012; Tran et al. 2019a, b), neurosurgery (Doglietto et al. 2017), and reconstructive surgery (Fung et al. 2009; Chang et al. 2013; Walton et al. 2015; Peer et al. 2022). The visualization of these virtual models using stereopsis has the potential to confer additional benefits to medical students' and trainees' learning.

Although this section only showcases three examples of labs that have engaged in applied anatomy research involving virtual stereoscopic visualizations, there are additional examples throughout the literature. A number of these studies, in addition to emphasizing clinical applications, also explore pedagogical concepts, especially since the massive increase in the development and implementation of virtual anatomical models has compelled educator scholars to determine whether these models actually improve students' learning.

7.2.1.2 Pedagogical Anatomy Content Knowledge

Many basic and clinical scientists alike would assert the importance of understanding the content knowledge specific to their respective disciplines. After all, having a solid understanding of foundational concepts in a particular field allows one to navigate successfully and confidently the routine demands of that field's research and/or healthcare delivery. This firm grasp of content knowledge, reinforced further by its habitual use, also affords researchers and clinicians mental freedom for creativity and innovation through the implementation of novel research that generates new knowledge and/or the development of novel devices and procedures that improve standards of patient care.

Nevertheless, many basic and clinical scientists, especially at academic health science centers, are also educators of health science students in various disciplines. Thus, scientists and clinicians who teach students cannot rely on content knowledge alone, but also on the pedagogical content knowledge of their disciplines. In other words, pedagogical content knowledge includes the concepts and skills that educators need to know in order to instruct students in their disciplines efficiently and effectively so that their students will also learn to master discipline-specific content knowledge in the same way as the educators have mastered their discipline-specific content knowledge. Pedagogical anatomy content knowledge is no different as anatomists or anatomy educators must apply to their teaching the basic learning theories, teaching methods, and instructional strategies that ideally foster the learning of the anatomical sciences among their respective student cohorts. Appropriate application of pedagogical anatomy content knowledge often entails an iterative, self-reflective process in which the educator is engaging in practices that ultimately facilitate and improve students' learning of anatomy. These practices mainly include developing learning objectives, designing and implementing discrete learning sessions and assessments aligned with those objectives, and adjusting instruction as necessary based on observations, formative assessments, students' performance, and students' course and faculty evaluations. But how do anatomy educators decide which instructional strategies or teaching methods to incorporate into their lessons in order to achieve improved student learning outcomes? And better yet, how do anatomy educators learn how to implement these instructional strategies or teaching methods correctly and effectively? The answer to both of these questions lies at the heart of educational or pedagogical research and scholarship in the anatomical sciences.

Scholarly Obligations of Anatomy Educators Educators in the anatomical sciences then must commit to three basic scholarly obligations to both their students and their profession. *First*, anatomy educators must be willing to read the literature pertaining to their respective fields and specialty interests on a regular basis to remain up to date on evidence-based practices so that they can be sure to implement the same or similar practices within their classrooms and labs. In addition, anatomy educators should read the literature regularly, paying attention to any novel, innovative teaching techniques that hold instructional potential and that might be worthy of implementing with their students for one, two, or even several learning sessions.

Secondly, anatomy educators must be willing to conduct novel educational research of their own. That research could involve educators testing a published, evidence-based strategy on their own students to determine if the learning gains exhibited in the student cohorts of the published study are generalizable to their own students. That research could involve educators' developing and implementing with their students a completely new instructional strategy never attempted or simply implementing with their students a strategy never used in anatomy instruction. Moreover, that research could involve educators' exploring one or more manipulatable variables in the educational environment and their effects on one or more responding variables in the students. Nonetheless, that research could take on a variety of purposes, though the ultimate goals should be the increase of positive outcomes and the decrease of negative outcomes. Regardless of the type of research that is conducted, educator researchers should foremost analyze, interpret, and draw conclusions from data for the quality improvement of their courses, if nothing else. However, educator researchers only become educator scholars when they commit to sharing their findings with the rest of their own scientific community and the general public through publications and/or scientific meetings, thus allowing other educators to learn insights that might be beneficial for their own learners. In this way, anatomy educators add to the existing literature by offering new discoveries and new knowledge that further advance the scholarly enterprise of their profession. Current anatomy educators have asserted that established independent educational research is recommended for current trends in the training of future anatomy educators (Schaefer et al. 2018).

Thirdly, anatomy educators must be willing to don the mantle of mentor for one or more students in their careers. Teaching students how to learn in the anatomical sciences using various instructional strategies is a commendable feat. Teaching students why they are learning specific content using contexts applicable to their future professions—contexts such as the applied anatomy examples described in the previous section—is also a noteworthy endeavor. However, teaching students how to teach the anatomical sciences they have learned to others is arguably an even more laudable accomplishment. This practice of teaching others to confirm a deeper understanding of content is certainly a routine that anatomy educators can model and encourage among their students in and outside the classroom. But anatomy educators can extend this modeling role further perhaps by even training some or many of their students in the practices of conducting applied and/or educational anatomy research, thus producing anatomy educator scholars for the future. This privilege should also be an honor that basic science or clinical educators in any discipline relish as it offers them an opportunity to influence others not only just to carry on their work but also to make it even better than before. There is considerable evidence that reveals that mentorship programs in the health sciences can positively impact scholarly productivity and career satisfaction among junior faculty (Woods et al. 1997; Jackson et al. 2003; Menon et al. 2016). For example, at UMMC, the development of a Clinical Anatomy division/program proposed a structured career development path for anatomy educator scholars in effort to train a new generation of clinical anatomists (Cui and Lehman 2014). Today, this program is successful in preparing this new generation of anatomy educator scholars, now with eight graduates since 2018 and more to come.

Examples of Virtual Stereoscopic Visualizations for Pedagogical Anatomy This section will continue to showcase these pedagogical anatomy concepts specifically through virtual stereoscopic anatomical visualizations, using several examples in the United States and Canada. Studies from the CRIPT lab at the University of Western Ontario, for example, explored the relationship between learners' visual spatial ability (their ability to mentally rotate 2D and 3D objects) and their performance of surgical-related tasks (Roach et al. 2012, 2014; Mistry et al. 2013). Another study explored a range of factors that influence spatial anatomy comprehension before and after instruction with different computer visualizations (Nguyen et al. 2012). A study also explored the relationship between spatial ability and various modalities of 3D digital anatomy models (Labranche et al. 2022). In fact, earlier work from the CRIPT lab led to the development of a validated Spatial Anatomy Test (SAT), versions of which have been used in explorations of cerebral blood flow (Loftus et al. 2017, 2018). Moreover, the researchers in the lab have explored the importance of interactivity in learning with virtual stereoscopic models such as in the development of an interactive 3D eye model (Allen et al. 2015) as well as the importance of haptics and the development of the Haptics Abilities Test (HAT) (Sveistrup et al. 2023). Furthermore, a considerable amount of research regarding eye tracking, which was also used with the HAT, has provided insights into relationships between eye movements and spatial learning (Roach et al. 2016, 2017a, b, 2019).

Research from the CARSL at UMMC has also explored comparisons of monoscopic and stereoscopic visualization formats as well as comparisons of virtual stereoscopic lab and cadaveric dissection lab learning environments. From this lab, a study evaluated the effectiveness of 3D vascular stereoscopic models versus 2D flat-screen (monoscopic) models on first-year medical students' learning, given their differing spatial visualization abilities (Cui et al. 2017). In a study comparing the short- and long-term retention of middle and inner ear anatomy among medical students' viewing either monoscopic or stereoscopic virtual models, there was no significant difference in students' performance on short- and long-term post-3D tests between learning groups (Meyer 2019). Nevertheless, despite the fact that there was a significantly lower difference between the short- and long-term post-3D test scores, the post-3D test scores were still significantly higher than the pre-3D test scores (Meyer 2019). Moreover, the percentage drop in retention between immediate (short-term) post-3D and one-month (long-term) post-3D tests was considerably lower (Meyer 2019) than the average percentage drop in retention of learned meaningful information reported in the literature (Ebbinghaus 1966; Custers 2010). These results were more positively different from an earlier 2016 (published in 2020) pilot study of medical students' short- and long-term retention of virtual stereoscopic pelvic anatomy during and after a cadaveric gross anatomy course in which pelvic dissections and prosections were implemented (Meyer et al. 2020). Results in this pilot study revealed a significantly lower difference between students' pre-3D test and one-month long-term post-test scores and no significant difference between students' pre-3D test and six-month long-term test scores (Meyer et al. 2020). Since the publication of these studies, more research has been conducted on retention of anatomical information from virtual learning experiences.

Both of these CARSL studies also utilized the incorporation of radiographic features with the middle and inner ear and pelvic models. This feature in Amira® allows radiographic slices to interact with a 3D model generated from the same radiographic data. Each radiographic slice can be superimposed on the 3D model, accurately indicating the location of the structures on the radiographic images and their 3D relationships with the model (Cui et al. 2016). These interactions have the potential to improve students' and trainees' knowledge of anatomy interpreted from radiographic imaging modalities.

In some of these example studies, researchers were able to explore both applied anatomy concepts as well as pedagogical anatomy concepts, especially since a regular pedagogical practice in the anatomical sciences is to educate students using clinical applications. Similarly, when learners are medical residents or fellows gaining additional training beyond their years of undergraduate or graduate medical education, an expected pedagogical practice is to train medical residents and fellows to perform clinical procedures.

7.2.2 Procedural Training

Through procedural training techniques, medical residents and fellows are expected to master the performance of procedures in a gradual process. The final goal of achieving procedural performance mastery for medical trainees is initiated through habitual practice using virtual and/or physical models that simulate the procedural tasks and conditions. When trainees have gained mastery over performing the simulated procedure using virtual and/or physical models, they usually progress through additional simulations of increasing gradations of sophistication and realism. For instance, many medical schools incorporate into resident training the use of mannequins and high-fidelity mannequins which can move and respond to trainees with the intent of mimicking patients. In addition, many medical schools also recruit the use of standardized patients who offer an even higher level of realism in simulating patients when medical residents are practicing performing standard patient-interaction procedures, such as taking a medical history and performing a physical exam.

Examples of Virtual Stereoscopic Visualizations for Procedural Training This section will focus solely on examples in which virtual stereoscopic visualizations have been used in the procedural training of medical students, medical residents, or other health professional students. Many of the applied clinical anatomy examples discussed in an earlier section also incorporate virtual stereoscopic visualizations for procedural training purposes. This fact is not surprising since many medical procedures involve the application of clinically relevant anatomy knowledge.

From the CRIPT, several studies explored virtual simulations of surgical procedures. One study piloted a virtual stereoscopic model of the skull and cerebrum in which structures, such as the temporal lobe and amygdala–hippocampus complex were removed in a stepwise fashion to simulate a neurosurgical procedure for medical student and resident viewers (de Ribaupierre and Wilson 2012). Another study from the same lab examined four counterbalanced groups of health professional students' performance of one of two different surgical flap procedures after observing videos of the respective procedure in monoscopic (2D) and stereoscopic (true 3D) visualization formats (Roach et al. 2012). Yet two other studies explored laparoscopic procedural skills in first- and second-year medical students across monoscopic and stereoscopic visualization modalities (Mistry et al. 2013; Roach et al. 2014). The final study mentioned from the CRIPT lab confirmed the validity of measurements from a virtual micro-CT scan reconstruction image of the mandible for planning dental implant preparation (Massey et al. 2013).

The CARSL at UMMC featured one study that described the safest needle trajectories for performing trigeminal ganglion block injections using virtual stereoscopic images of the ganglion and surrounding structures, such as muscles of mastication, the parotid gland, and nearby arteries and veins (Chen et al. 2020). Medical residents viewing such a model repeatedly before actually performing a trigeminal nerve block procedure could potentially have a more reduced risk of damaging essential structures during the actual procedure than medical residents attempting the procedure without any prior familiarity with the safest injection pathways.

Similarly, from Dr. Anne Agur's lab at the University of Toronto, the studies modeling muscle and nerve fiber patterns and pathways using 3D visualization with stereoscopic capabilities can also be viewed multiple times by medical residents prior to their performance of the actual procedures involving the relevant anatomical structures depicted. Such practice could potentially reduce medical residents' errors in the diagnosis of abnormal muscle loading in craniomaxillofacial structures (Falcinelli et al. 2018) and in the performance of wrist nerve denervation procedures to treat chronic wrist pain (Tran et al. 2022). Reduction of error among residents and physicians in conducting clinical diagnoses and procedures is one of the many benefits of virtual stereoscopic learning.

7.3 The Benefits and Challenges of Virtual Stereopsis in Medical Education

When educators are considering incorporating virtual stereoscopic visualizations into medical education and training, there are a variety of benefits and challenges to ponder (Table 7.1). Conducting an analysis of the pros and cons of utilizing such virtual technologies in the specific contexts of each different educational learning environment or format both within and across institutions is highly recommended. Such an approach allows educators to make decisions that are aimed at improving students' and trainees' learning outcomes, the ultimate goal of any educational intervention or innovation.

7.4 Potential Positive and Negative Implications of Virtual Stereopsis in Medical Education

The current benefits and challenges of virtual stereopsis in medical education have the potential to ripple into positive and negative implications for medical education and training in the future (Table 7.2). Many of these implications are dependent on the potential intersection and overlap of other innovations, hot topics, or new and revised standards for medical education. Reading and reflecting over these implications might foster discussions among educators regarding the explorations of similar concepts at their own institutions. In addition, such reflection might even encourage educators to engage in the development and implementation of novel ideas or projects that benefit their own students and trainees as well as educators, students, and trainees at other institutions if they decide to publish and share the fruits of their labors.

7.5 Conclusions

In the current age of the ever-growing and ever-expanding development and usage of virtual and digital technologies in health sciences education, educators are challenged to consider its pros and cons for their students. Virtual stereoscopic presentation in medical education is discussed in this chapter as one example of a visualization format that has been used in the instruction and training of medical students and residents, among other health science students and trainees. Virtual stereoscopic presentation is different from the 2D flat screen, or monoscopic, presentation of lectures in the regular classroom, given the depth cues it affords viewers. In this regard, it has the potential to offer students additional learning benefits. Namely, stereoscopic anatomical models can be used to teach anatomy and evaluate students' retention (Meyer 2019; Meyer et al. 2020) and spatial abilities (Luursema et al. 2006, 2008; Hilberlink 2009; Nguyen and Wilson 2009; Aziz et al. 2002; Brewer et al. 2012; Anderson et al. 2013; Roach et al. 2014; Cui et al. 2017). This chapter has described clinical anatomy instruction and procedural training in medical education and compared the benefits and challenges as well as the potential positive and negative implications of virtual stereopsis in medical education. In addition, several examples of virtual stereoscopic visualization techniques used in clinical anatomy instruction and procedural training and of different research and educational tracks in the United States and Canada were provided. The authors hope that this chapter will provide useful information that benefits educators in their teaching and career development, especially as they consider decisions regarding the incorporation of virtual stereoscopic anatomy visualizations in their learning environments.

Acknowledgments The authors extend much appreciation and gratitude to Dr. Tim Wilson who generously provided copious amounts of photos of his students actively engaged in learning and research and of the research equipment in his lab. A few of these photos were used with permission in Fig. 7.1 of this chapter.

Table 7.1 Benefits and challenges of virtual stereopsis in medical education

Benefits	Challenges
Many students and trainees find the learning format to be engaging and exciting.	Some students experience dizziness, nausea, headaches, or other negative side effects.
Some institutions find virtual technology learning environments more cost effective and sustainable than cadaveric laboratories.	Many institutions with established body donation programs find virtual technology to be an added cost needing further exploration.
If financial resources are available, institutions can create stationary and portable learning environments to benefit students.	Even if financial resources are available, students might have limitations on their personal devices that hinder virtual learning.
As virtual stereoscopic technologies have evolved, some applications enable viewing visualization on personal devices.	Many students and educators are unaware of which applications are the most accurate and/or are the easiest to use.
Many students experience learning gains when virtual stereopsis is incorporated into self-paced or guided instructional formats.	Some students and educators experience a steeper learning curve when trying to learn how to use new technologies.
The learning format has shown evidence of particular learning gains among many students and trainees with lower spatial abilities, given the added depth cues.	The learning curve, in general, and the specific virtual model content with the added depth cues can also increase some students' cognitive load, thus reducing learning gains.
When interactivity is applied to virtual stereoscopic visualizations, many students find these experiences more realistic and congruent with learning real human anatomy.	Some students find the combination of interactivity and virtual stereoscopic visualization to be overwhelming in their contribution to cognitive overload.
This learning format, when used on personal devices especially, can be used repetitively, as self-paced modules can be repeated as many times as desired by student and trainee users.	This learning format, when provided in a physical laboratory space without portability or continuous access, might not be as useful to students desiring rehearsal practice.
This learning format has evidence to show that it can help many students learn anatomy, especially when it is used as a supplement to traditional, cadaveric anatomy learning.	This learning format, especially in the absence of cadaveric donors at institutions, cannot mimic or replicate the experience of learning with real human tissues and organs.
This learning format, when repeated exposure and use are available for students and trainees, can improve students' and trainees' mastery of procedural skills in simulated formats before they are practiced on real patients, thus aiding in the reduction of medical errors and of harm to patients.	Despite the repeatability of this learning format and the mastery of procedural skills it can afford in simulated conditions, students and trainees are still tasked with the transference of this knowledge to scenarios involving real people which virtual simulations cannot ideally duplicate.

Note: This table does not present an exhaustive list of the benefits and challenges of using virtual stereopsis in medical education and training. Readers may think of additional benefits and challenges from their own past, current, and future research and existing prior knowledge

Table 7.2 Potential positive and negative implications of virtual stereopsis in medical education

Positive implications	Negative implications
There is substantial potential for many advances in anatomical sciences and medical education, in general, with the increased implementation of this learning modality.	Many virtual technology companies could seek to make exorbitant profits off of students, educators, and institutions who are new to the virtual technology market.
There is great potential for institutions using this modality, especially those with the financial resources, to enable positive learning outcomes among students.	There will inevitably be an inequity in the paucity of learning gains among students at institutions that have limited financial resources around the globe.
With insights from combined eye tracking and spatial learning research, there is potential for providing visual cues to learners with lower spatial abilities to help improve their learning of anatomy (Roach et al. 2019).	Further research that corroborates the benefits of visual cues for low-spatial ability learners in other cohorts with other anatomy content knowledge will compel anatomy educators to change paradigms of how they teach content.
With applications of interactivity and haptic feedback to this learning modality, there is potential for the modification of the Multimedia Learning Theory (MMLT) to assert the need for touch, even in virtual learning, for the benefit of maximizing student learning (Wilson 2021; Sveistrup et al. 2023).	There is potential for future research to show that virtual learning alone, without the addition of haptic feedback to visual and auditory input, is not fully adequate in fulfilling students' learning needs in the anatomical sciences in which touch is an equally important sense to utilize in addition to sight and sound.
The quickly evolving nature of virtual technology can ensure that this particular modality only becomes ever-increasingly realistic in its application to learning anatomy.	As more and more technological updates are applied, there will be a seemingly endless need for institutions to pursue the next-best virtual tool, perhaps at even greater expenses.
Increasing realism in virtual technology can enhance the students' and trainees' learning experiences, especially when simulations are involved, thus augmenting simulation learning beyond current standards.	Despite the number of advancements in the realism of simulations incorporating virtual stereoscopic visualizations, students and trainees will still be tasked with transferring skills to interactions with real people.
Interprofessional education is becoming an increasingly hot topic in medical education within the United States and Canada, especially since interprofessional education is included in the Liaison Committee on Medical Education (LCME) standards (AAMC 2023, p. 14), so there is unique potential for the incorporation of this learning modality into interprofessional interactions involving learners from multiple health science professions.	The potential for increased complexity with interprofessional learning experiences incorporating virtual stereoscopic learning modalities poses unique barriers for institutions where coordinating and implementing interprofessional education across various different schools and programs is already cumbersome and challenging enough without the added considerations of virtual technology and the learning gains it can provide.
Justice, equity, diversity, and inclusion (JEDI) are additional hot topics in medical education within the United States and Canada, especially since cultural competence and health inequities are included in the LCME standards (AAMC 2023, p. 16), so there is unique potential for the incorporation of this learning modality into student interactions that promote justice, equity, diversity, and/or inclusion (JEDI) among learners and the patients with whom they interact.	The concepts of JEDI pose a problematic paradox when virtual technologies, such as virtual stereoscopy, are considered, namely in regard to equity because of financial inequalities that might exist between institutions and individual learners. In addition, learning differences, such as spatial abilities, will make the incorporation of this virtual modality with equitable considerations for students and the patients with whom they interact more challenging.

Note: This table does not present an exhaustive list of the potential positive and negative implications of virtual stereopsis in medical education and training. Readers may think of additional implications from their own past, current, and future research and existing prior knowledge

References

Alghamdi MA, Ziermann JM, Diogo R (2017) An untold story: The important contributions of Muslim scholars for the understanding of human anatomy. Anat Rec 300:986–1008. https://doi.org/10.1002/ar.23523

Allen LK, Bhattacharyya A, Wilson TD (2015) Development of an interactive anatomical three-dimensional eye model. Anat Sci Educ 8:275–282. https://doi.org/10.1002/ase.1487

Alsop T (2022) Consumer and enterprise VR market size worldwide 2021–2026. Statista, New York. https://www.statista.com/statistics/1221522/virtual-reality-market-size-worldwide/#statisticContainer. Accessed 4 Jan 2023

Anderson P, Chapman P, Ma M et al (2013) Real-time medical visualization of human head and neck anatomy and its applications for dental training and simulation. Curr Med Imaging Rev 9:298–308. https://doi.org/10.2174/15734056113096660004

Association of American Medical Colleges (AAMC) (2023) LCME functions and structure of a medical school. Liaison Committee on Medical Education (LCME), Association of American Medical Colleges (AAMC). Available at https://lcme.org/publications/

Aziz MA, McKenzie JC, Wilson JS et al (2002) The human cadaver in the age of biomedical informatics. Anat Rec 269:20–32. https://doi.org/10.1002/ar.10046

Brewer DN, Wilson TD, Eagleson R et al (2012) Evaluation of neuroanatomical training using a 3D visual reality model. Stud Health Technol Inform 173:85–91. https://doi.org/10.3233/978-1-61499-022-2-85

Business Wire (BW) (2022) Global AR-VR in healthcare market, by component (hardware, software), by device type, by application, by end use, by region, competition, forecast & opportunities, 2017-2027F. Global AR-VR in healthcare market. Business Wire, Dublin. https://www.businesswire.com/news/home/20220719005693/en/Global-9.79-Bn-AR-VR-in-Healthcare-Market-to-2027-by-Component-Device-Type-Application-End-Use%2D%2D-ResearchAndMarkets.com. Accessed 4 Jan 2023

Chang Y, Cantelmi D, Wisco et al (2013) Evidence for the functional compartmentalization of the temporalis muscle: a 3-dimensional study of innervation. J Oral Maxillofac Surg 71:1170–1177. https://doi.org/10.1016/j.joms.2013.04.005

Chen J, Prem A, Crim R et al (2020) Identification of the preferred interventional approach to trigeminal ganglion using three-dimensional reconstruction stereoscopic model. Radiol Med Diagnost Imaging 3:6 pages. https://doi.org/10.31487/j.RDI.202.02.01

Corps for Research of Instructional and Perceptual Technologies (CRIPT) (2016a) History. OceanWP Theme by Nick. https://anatatorium.com/history/. Accessed 4 Jan 2023

Corps for Research of Instructional and Perceptual Technologies (CRIPT) (2016b) The C.R.I.P.T. Anatatorium. University of Western Ontario, London. http://www.anatatorium.com/CRIPT/Home.html. Accessed 4 Jan 2023

Cui D, Lehman (2014) Development of a clinical anatomy division for anatomical education. Annals of anatomy Volume 196:214. Poster presented at the 25th International Symposia on Morphological Sciences (ISMS) meeting, Xian, China. July 26 to July 30

Cui D, Lynch JC, Smith AD et al (2016) Stereoscopic vascular models of the head and neck: a computed tomography angiography visualization. Anat Sci Educ 9:179–185. https://doi.org/10.1002/ase.1537

Cui D, Wilson TD, Rockhold RW et al (2017) Evaluation of the effectiveness of 3D vascular stereoscopic models in anatomy instruction for first year medical students. Anat Sci Educ 10:34–45. https://doi.org/10.1002/ase.1626

Custers EJFM (2010) Long-term retention of basic science knowledge: a review study. Adv in Health Sci Educ 15: 109–128. https://doi.org/10.1007/s10459-008-9101-y

Davies JC, Fattah A, Ravichandiran et al (2012) Clinically relevant landmarks of the frontotemporal branch of the facial nerve: a three-dimensional study. Clin Anat 25: 858–865. https://doi.org/10.1002/ca.22162

Dawidek MT, Roach VA, Ott MC et al (2016) Changing the learning curve in novice laparoscopists: incorporating direct visualization into the simulation training program. J Surg Educ 74:30–36. https://doi.org/10.1016/j.jsurg.2016.07.012

de Ribaupierre S, Wilson TD (2012) Construction of a 3-D anatomical model for teaching temporal lobectomy. Comput Biol Med 42:692–696. https://doi.org/10.1016/j.compbiomed.2012.03.005

Doglietto F, Qiu J, Ravichandiran et al (2017) Quantitative comparison of cranial approaches in the anatomy laboratory: a neuronavigation based research method. World J Methodol 7:139–147. https://doi.org/10.5662/wjm.v7.i4.139

Ebbinghaus H (1966) Über das gedächnis. Untersuchungen zur experimentellen psychologie. Nachdruk der Ausgabe Leipzig 1885. Amsterdam, Netherlands. English edition: Ebbinghaus H (1966) Memory: a contribution to experimental psychology (trans: Ruger HA, Bussenius CE). Columbia University, New York. http://psychclassics.yorku.ca/Ebbinghaus/index.htm

Edwards DN, Meyer ER, Brooks WS et al (2022) Faculty retirement will likely exacerbate the anatomy educator shortage. Anat Sci Educ. Advance online publication. https://doi.org/10.1002/ase.2217

Esar E (1951) The Treasury of humorous quotations. Phoenix House Publishing, LLC, Chicago, p 103

Falcinelli C, Li Z, Lam et al (2018) Diffusion-tensor imaging versus digitization in reconstructing the masseter architecture. J Biomech Eng 140:111010 (6 pages). https://doi.org/10.1115/1.4041541

Federation of State Medical Boards (FSMB), National Board of Medical Examiners (NBME) (2022) Sample test questions: Step 1. United States Medical Licensing Examination (USMLE). https://www.usmle.org/sites/

default/files/2021-10/Step_1_Sample_Items.pdf. Accessed 4 Jan 2023

Federation of State Medical Boards (FSMB), National Board of Medical Examiners (NBME) (2023) Step 1 content outline and specifications. United States Medical Licensing Examination (USMLE). https://www.usmle.org/prepare-your-exam/step-1-materials/step-1-content-outline-and-specifications. Accessed 4 Jan 2023

Fortune Business Insights (FBI) (2022a) Virtual reality market size, share & COVID-19 impact analysis, by component (hardware, software, and content), by device type (head mounted display, VR simulator, VR glasses, treadmills & haptic gloves, others), by industry (gaming, entertainment, automotive, retail, healthcare, education, aerospace & defense, manufacturing, others), and regional forecast, 2022–2029. Market Research Report. Fortune Business Insights, Maharashtra. https://www.fortunebusinessinsights.com/industry-reports/virtual-reality-market-101378. Accessed 4 Jan 2023

Fortune Business Insights (FBI) (2022b) Virtual reality (VR) in healthcare market size, share & COVID-19 impact analysis, by component (hardware, software, and content), by application (pain management, education & training, surgery, patient care management, rehabilitation & therapy procedures, and others), and regional forecast, 2022–2029. Market Research Report. Fortune Business Insights, Maharashtra. https://www.fortunebusinessinsights.com/industry-reports/virtual-reality-vr-in-healthcare-market-101679. Accessed 4 Jan 2023

Fung L, Wong B, Ravichandiran et al (2009) Three-dimensional study of pectoralis major muscle and tendon architecture. Clin Anat 22:500–508. https://doi.org/10.1002/ca.20784

Hilberlink AJ (2009) A measure of the effectiveness of incorporating 3D human anatomy into an online undergraduate laboratory. Br J Educ Technol 40:664–672. https://doi.org/10.1111/j.1467-8535.2008.00886.x

Jackson VA, Palepu A, Szalacha L (2003) "Having the right chemistry": a qualitative study of mentoring in academic medicine. Acad Med 78:328–334. https://doi.org/10.1097/00001888-200303000-00020

Kumar A, Ghosh SK, Faiq MA et al (2018) A brief review of recent discoveries in human anatomy. QJM 112: 567–573. https://doi.org/10.1093/qjmed/hcy241

Kwon D (2020) New discoveries in human anatomy. The Scientist. https://mail.google.com/mail/u/0/?zx=62xdpmn49ig4#inbox/FMfcgzGsmDmpTXBFmQbXgmbMBbvGSPTg?projector=1&messagePartId=0.1. Accessed 15 Apr 2023

Labranche L, Wilson TD, Terrell M et al (2022) Learning in stereo: the relationship between spatial ability and 3D digital anatomy models. Anat Sci Educ 15:291–303. https://doi.org/10.1002/ase.2057

Loftus JJ, Jacobsen M, Wilson TD (2017) Learning and assessment with images: a view of cognitive load through the lens of cerebral blood flow. Br J Educ Technol 48:1030–1046. https://doi.org/10.1111/bjet.12474

Loftus JJ, Jacobsen M, Wilson TD (2018) The relationship between spatial ability, cerebral blood flow and learning with dynamic images: a transcranial doppler ultrasonography study. Med Teach 40:174–180. https://doi.org/10.1080/0142159x.2017.1395401

Luursema JM, Verwey WB, Kommers PAM et al (2006) Optimizing conditions for computer-assisted anatomical learning. Interact Comput 18:1123–1138. https://doi.org/10.1016/j.intcom.2006.01.005

Luursema LM, Verwey WB, Kommer PAM et al (2008) The role of stereopsis in virtual anatomical learning. Interact Comput 20:455–460. https://doi.org/10.1016/j.intcom.2008.04.003

Massey ND, Galil KA, Wilson TD (2013) Determining position of the inferior alveolar nerve via anatomical dissection and micro-computed tomography in preparation for dental implants. J Can Dent Assoc 79:d39. https://jcda.ca/article/d39. Accessed 4 Jan 2023

Menon V, Muraleedharan A, Bhat BV (2016) Mentoring for junior medical faculty: existing models and suggestions for low-resource settings. Asian J Psychiatr 19:87–88. https://doi.org/10.1016/j.ajp.2015.12.013

Meyer ER (2019) Validated virtual three-dimensional stereoscopic middle and inner models for examining first-year medical students' short- and long-term retention. Dissertation, University of Mississippi Medical Center

Meyer ER, James AM, Cui D (2018) Hips don't lie: expert opinions guide the validation of a virtual 3D pelvis model for us in anatomy education and medical training. HAPS Educator 22:105–118. https://doi.org/10.21692/haps.2018.023

Meyer ER, James AM, Sinning et al (2020) A pilot study of the impact of three-dimensional stereoscopic models of pelvic anatomy on short- and long-term retention in first-year medical students. HAPS Educator 24:7–22. https://doi.org/10.21692/haps.202.021

Mistry M, Roach VA, Wilson TD (2013) Application of stereoscopic visualization on surgical skill acquisition in novices. J Surg Educ 70:563–570. https://doi.org/10.1016/j.jsurg.2013.04.006

Nguyen N, Wilson TD (2009) A head in virtual anatomy: development of a dynamic head and neck model. Anat Sci Educ 2:294–301. https://doi.org/10.1002/ase.115

Nguyen N, Nelson AJ, Wilson TD (2012) Computer visualizations: factors that influence spatial anatomy comprehension. Anat Sci Educ 5(2):98–108. https://doi.org/10.1002/ase.1258

Peer M, Tran J, Li et al (2022) Parametric multi-scale modeling of the zygomaticus major and minor: implications for facial reanimation. J Craniofac Surg 33:701–706. https://doi.org/10.1097/SCS.0000000000008128

Perkins Coie (2020) 2020 augmented and virtual reality survey report. Perkins Coie, LLP, Washington. https://www.perkinscoie.com/images/content/2/3/231654/2020-AR-VR-Survey-v3.pdf. Accessed 4 Jan 2023

Poggio GF, Poggio T (1984) The analysis of stereopsis. Ann Rev Neurosci 7:379–412. https://doi.org/10.1146/annurev.ne.07.030184.002115

Ravichandiran K, Ravichandiran M, Oliver ML et al (2009) Determining physiological cross-sectional area of extensor carpi radialis longus and brevis as a whole and by regions using 3D computer muscle models created from digitized fiber bundle data. Comput Methods Prog Biomed 95:203–212. https://doi.org/10.1016/j.cmpb.2009.03.002

Roach VA, Brandt MG, Moore CC et al (2012) Is three-dimensional videography the cutting edge of surgical skill acquisition? Anat Sci Educ 5:138–145. https://doi.org/10.1002/ase.1262

Roach VA, Mistry MR, Wilson TD (2014) Spatial visualization ability and laparoscopic skills in novice learners: evaluating stereoscopic versus monoscopic visualizations. Anat Sci Educ 7:295–301. https://doi.org/10.1002/ase.1412

Roach VA, Fraser GM, Kryklywy JH et al (2016) The eye of the beholder: can patterns in eye movement reveal aptitudes for spatial reasoning? Anat Sci Educ 9:357–366. https://doi.org/10.1002/ase.1583

Roach VA, Fraser GM, Kryklywy JH et al (2017a) Different perspectives: spatial ability influences where individuals look on a timed spatial test. Anat Sci Educ 10:224–234. https://doi.org/10.1002/ase.1654

Roach VA, Fraser GM, Kryklywy JH et al (2017b) Time limits in testing: an analysis of eye movements and visual attention in spatial problem solving. Anat Sci Educ 10:528–537. https://doi.org/10.1002/ase.1695

Roach VA, Fraser GM, Kryklywy JH et al (2019) Guiding low spatial ability individuals through visual cueing: the dual importance of where and when to look. Anat Sci Educ 12:32–42. https://doi.org/10.1002/ase.1783

Schaefer AF, Wilson AB, Barger BB et al (2018) What does a modern anatomist look like? Current trends in the training of anatomy educators. Anat Sci Educ 12(3):225–235. https://doi.org/10.1002/ase.1806

Shaw SM, Martino R, Mahdi et al (2017) Architecture of the suprahyoid muscles: a volumetric musculoaponeurotic analysis. J Speech Lang Hear Res 60:2808–2818. https://doi.org/10.1044/2017_JSLHR-S-16-0277

Shaw V, Diogo R, Winder IC (2022) Hiding in plain sight-Ancient Chinese anatomy. Anat Rec 305:1201–1214. https://doi.org/10.1002/ar.24503

Sveistrup MA, Langlois J, Wilson TD (2023) Do our hands see what our eyes see? Investigating spatial and haptic abilities Anat Sci Educ (published before print). https://doi.org/10.1002/ase.2247

Tran J, Peng P, Agur A (2019a) Evaluation of suprascapular nerve radiofrequency ablation protocols: 3D cadaveric needle placement study. Reg Anesth Pain Med 44:1021–1025. https://doi.org/10.1136/rapm-2019-100739

Tran J, Peng PWH, Gofeld et al (2019b) Anatomical study of the innervation of posterior knee joint capsule: implication for image-guided intervention. Reg Anesth Pain Med 44:234–238. https://doi.org/10.1136/rapm-2018-000015

Tran J, Ho L, von Schroeder et al (2022) Anatomical study of the innervation of triangular fibrocartilage complex and distal radioulnar and radiocarpal joints: implications for denervation. J Hand Surg Am 47:843–854. https://doi.org/10.1016/j.jhsa.2022.05.008

University of Mississippi Medical Center (UMMC) (2023) Faculty profiles: Dongmei Cui, MD (hon), PhD. Advanced Biomedical Education. University of Mississippi Medical Center, Jackson. https://www.umc.edu/som/Departments%20and%20Offices/SOM%20Departments/Advanced-Biomedical-Education/About-Us/Faculty1/Dongmei-Cui-MD-PhD.html. Accessed 4 Jan 2023

Walton C, Li Z, Pennings et al (2015) A 3-dimensional anatomic study of the distal biceps tendon: implications for surgical repair and reconstruction Orthop J Sports Med 3:2325967115585113 (6 pages). https://doi.org/10.1177/2325967115585113

Wilson TD (2021) Visualization technologies—I can see clearly now but the feel is gone. Invited commentary on: Bogomolova et al (2021) Stereoscopic three-dimensional visualization technology in anatomy learning: a meta-analysis. Med Educ 55:285–288. https://doi.org/10.1111/medu.14448

Wilson AB, Notebaert AJ, Schaefer AF et al (2020) A look at the anatomy educator job market: anatomists remain in short supply. Anat Sci Educ 13:91–101. https://doi.org/10.1002/ase.1895

Wilson AB, Kaza N, Singpurwalla DJ et al (2021) Are anatomy PhDs nearing extinction or adapting to change? United States graduate education trends in the anatomical sciences. Anat Sci Educ 14:432–439. https://doi.org/10.1002/ase.2013

Woods ES, Reid A, Arndt A et al (1997) Collegial networking and faculty vitality. Fam Med 29:45–49

Creating Virtual Models and 3D Movies Using DemoMaker for Anatomical Education

8

David L. Miles and Dongmei Cui

Abstract

Three-dimensional (3D) anatomy models have been used for education in health professional schools globally. Virtual technology has become more popular for online teaching since the COVID-19 pandemic. This chapter will describe a project in which a series of virtual anatomical models of organs and structures were developed for educational purposes, and it will describe in detail how to build three-dimensional (3D) movies using DemoMaker. Although setting up the 3D system was complicated and challenging, the process of reconstructing 3D models from radiographic images and the steps of creating animations and 3D movies are exponentially simpler. These efforts require minimal training, thus allowing most people to be able to engage in this modeling process and utilize the moviemaking steps. Amira® software and computed tomographic angiography (CTA) data were used to create 3D models of the lungs, heart, liver, stomach, kidney, etc. The anatomical locations of these structures within the body can be identified and visualized by recording information from multiple CTA slices using volume and surface segmentation. Ultimately, these virtual 3D models can be displayed via dual projectors onto a specialized silver screen and visualized stereoscopically by viewers as long as they wear 3D polarized glasses. Once these 3D movies are created, they can be played automatically on a computer screen, silver screen, 3D system playback screen, and video player, and they can be embedded into PowerPoint lecture slides. Both virtual models and movies are suitable for self-directed learning, face-to-face class teaching, and virtual anatomy education. Model animations and 3D movie displays offer students the opportunities to learn about anatomy and the anatomical positions of organs in the body and their 3D relationships to one another. By observing and studying these 3D models, students have the potential to be able to compartmentalize the anatomical information and retain it at a higher level than students learning corresponding anatomy without similar resources.

Keywords

Stereoscopic models · Three-dimensional (3D) models · Three-dimensional (3D) movies · Anatomy · Education · Surface

D. L. Miles
School of Graduate Studies in the Health Sciences, University of Mississippi Medical Center, Jackson, MS, USA

D. Cui (✉)
Department of Advanced Biomedical Education, University of Mississippi Medical Center, Jackson, MS, USA
e-mail: dcui@umc.edu

D. Cui et al. (eds.), *Biomedical Visualisation*, Advances in Experimental Medicine and Biology 1431,
https://doi.org/10.1007/978-3-031-36727-4_8

rendering · Computed tomographic angiography (CTA)

8.1 Introduction

The use of stereoscopic virtual models has the possibility to benefit individuals who utilize them by improving their anatomical education, enhancing their knowledge of three-dimensional (3D) anatomical relationships, and expanding their basic anatomical knowledge to clinically relevant contexts. The creation of 3D models allows users to view the desired organ(s) at a higher quality and level of realism than that conveyed by two-dimensional (2D) images in textbooks and online resources. While textbooks and online resources are helpful, there is potential that the use of Amira® software (Thermo Fischer Scientific, Waltham, MA) which allows the creation of stereoscopic virtual models (Stalling et al. 2005) will have the effect of increasing content learning, comprehension, and retention. The potential significance of learning gains from stereoscopic model visualizations is based on the difference between learning via 2D images or 3D models viewed and/or rotated in a 2D plane and learning via 3D models viewed in a stereoscopic format. With 2D images and 3D models viewed on 2D computer screens, the viewer is typically "detached" from the information. There is almost an impersonal barrier between the viewer and the data as if the viewer cannot rationalize or comprehend the images or models as being authentic, accurate representations of genuine anatomical structures. With 3D models, realism is increased because human beings with two eyes naturally visualize their surroundings using stereopsis. The enhanced realism offered by virtual stereoscopic 3D models can help reduce the barrier between viewers and potentially help the viewer to take the next step in comprehending the information. Not only are the virtual stereoscopic 3D models more visually stimulating, but they can also be physically manipulated. This manipulation allows the model to be rotated 360 degrees in any direction to allow a full view of all sides of the model. The ability to engage interactively in the dynamic control and movement of the models allows viewers to see desired organs and structures from any perspective that they wish, instead of the immovable 2D perspectives used in textbooks and static online images.

8.1.1 Literature Review

The use of 3D modelling in anatomical education and research has been studied in multiple papers (Ackerman 1998; Adams and Wilson 2011; Anderson et al. 2013; Brewer et al. 2012; Brown et al. 2012; Foo et al. 2013; Gary et al. 1999; Luursema et al. 2006, 2008; Nguyen and Wilson 2009; Nguyen et al. 2012; Sergovich et al. 2010; Cui 2015; Cui et al. 2016, 2017). There are many areas in which digital rendering can be used, ranging from the entertainment industry with video games to the healthcare industry with medical imaging and anatomical modeling. Recent studies and other scholarly work from the Cui lab found that 3D modelling and rendering has the possibility to play a positive role in anatomy education and enhance the quality of clinical training (Cui et al. 2017, 2019; Yang et al. 2019; Chen et al. 2017; Meyer and Cui 2020).

Regarding the forms of digital display, this study utilized monoscopic and stereoscopic imaging. Monoscopic displays allow the display and rotation of 3D images, but on a 2D screen. This modality is utilized widely with virtual products such as the Anatomage Table (Anatomage, Santa Clara, CA), Essential Anatomy 5 (3D4 Medical from Elsevier, RELX Group, London, UK), and 3D Anatomy Software (BodyViz, Clive, IA). Stereoscopic displays allow the projection of an image that will appear in a realistic 3D format as though they are "popping out" of the screen when the user views them while wearing specialized glasses. Stereoscopic views are common in civilian use, such as in watching 3D movies on a silver screen at the theater, where the viewer must wear specialized glasses to view the movie in 3D.

Amira® is a program that allows the creation of 3D models and their stereoscopic display.

Amira® software has been used in multiple studies regarding anatomical model creation and the human body. Examples of such studies include those pertaining to the paranasal sinuses and cervical vertebrae (Chen et al. 2017), the female pelvis (Sergovich et al. 2010), the male pelvis (Meyer et al. 2018), the head and neck (Nguyen and Wilson 2009; Brewer et al. 2012; Cui et al. 2016), and the larynx (Hu et al. 2009). A study conducted on creating 3D models of the cerebral ventricular system (Adams and Wilson 2011) found success in the use of Amira®. Amira® also has the capability to produce animations and movies through the use of the DemoMaker utility. This tool allows the creation of animations detailing the creation of the data in their 3D format. The focus of this paper is to explore the generation of 3D models of selected thoracic and abdominal organs and structures, generate movies regarding the creation of these models, and propose the future use of 3D anatomical models in the field of education.

8.1.2 Stereopsis and 3D Models

One of the forms of 3D viewing is known as stereopsis. Stereopsis is a term that translates into "three-dimensional sight." In humans, stereopsis is caused by two slightly different images being projected on the retinas so that a 3D image is produced. The 3D models were created by stacking slices of data corresponding to the desired organ. In this project, a 3D projection system was used to project the stereoscopic models on a silver screen. In order to do this, two specialized projectors attached to two linearly polarized filters were used to project two different images of the models from a high-speed computer with an NVIDIA Quadro K6000 video card (NVIDIA Corporation, Santa Clara, CA) (Cui 2015). Viewers wore specialized 3D glasses with two different lenses—each one filtering out a different projected image and essentially ensuring that each of the viewers' retinas is only receiving one of the two projected images. Thus, the viewers visualized and perceived the combined dual virtual images as a single 3D model hovering and rotating in 3D space. While the 3D model was projected, viewers could use the 3D mouse to control these 360-degree rotational movements of the model in all axes, move the model closer to or further away from view, and move the model higher or lower in elevation, allowing the Amira® program user more dynamic flexibility in interacting with and viewing any models reconstructed by the program and presented by the dual projection system.

8.2 Data Used for the Creation of Stereoscopic Models

The majority of 3D models in anatomy education are created by using reconstructions of slices from computed tomography (CT) and magnetic resonance imaging (MRI) scans. The multiple individual slices from these scans are the sources from where the data is collected using a software program. CT scans depend heavily on the different densities of structures in the body. Due to this density dependence, the volume rendering of bones and other dense structures is easier with CT scans. In CT scans, denser objects will appear whiter or brighter than less dense objects. For example, bones appear more strikingly white than other organs. However, organs that have similar densities are very difficult to differentiate from one another. This lack of differentiation makes volume rendering much more inaccurate when reconstructing organs with similar densities. As a result, the surface rendering technique is a viable and popular choice for reconstructing organs and other soft tissue structures from CT scans although the surface rendering process is time consuming and reliant on the anatomical knowledge of the individual or group conducting the process (Cui et al. 2019). Nevertheless, one of the disadvantages of CT scans is their poor capture of less dense soft tissue, making MRI scans more suitable for visualizing organs and other soft tissue structures.

MRI scans are another source of data for the reconstruction of 3D anatomical models. MRI scans offer an advantage over CT scans in having a greater capability of imaging soft tissues. On

MRI, organs with different densities are easily distinguishable, as well as structures with different intensities. However, MRI is inferior to CT scans in imaging hyperdense structures such as bones or structures filled with air or with very low densities such as the lungs. Therefore, MRI is mainly used for imaging structures with similar densities but slightly varying soft tissue details, such as those within the brain, while CT scans are used for imaging strongly contrasting organs, such as the lungs, liver, and bones.

Computed tomography angiography (CTA) imaging is commonly used for mapping the anatomy of the vascular tree network of different blood vessels in the body. This imaging technique can be used in clinical settings to visualize vascular networks in the surgical planning process prior to endovascular procedures or high-risk procedures, such as coronary artery bypass grafting (CABG); to diagnose vascular diseases and conditions, such as atherosclerosis or other vessel occlusions, vessel enlargements like aneurysms, and congenital heart disease; and to manage vascular emergencies, such as clotting (thrombosis) and internal bleeding (hemorrhaging) (Baliyan et al. 2019). This type of imaging was used in the project described in this chapter (see Fig. 8.1). The use of CTA imaging allowed for the complete mapping and visualization of the aorta and other structures. CTA scanning is similar to CT scans, but with the aid of iodinated contrast material, it has the increased capability for mapping and viewing blood vessels and blood-rich organs, such as bones, the heart, the liver, and the spleen.

Magnetic resonance angiography (MRA), like CTA, employs the aid of a contrast agent, such as gadolinium, to map and image vascular tree networks of blood vessels; however, unlike CTA, MRA is typically noninvasive, and it does not require the use of radioactive contrast agents. Moreover, some MRA techniques do not require a contrast medium at all. MRA is also very similar to MRI as they both rely on the magnetic properties of protons in the body to image the body's structures. There are many useful aspects of MRA ranging from detecting aneurysms to mapping. On the other hand, MRA is more specialized toward imaging different types of blood vessels, and it is especially useful for detecting and diagnosing vascular abnormalities, such as aneurysms, arteriovenous malformations, aortic coarctation or dissection, cerebral stroke, coronary artery disease, peripheral atherosclerosis, and pulmonary embolism as well as for screening and monitoring purposes (de Leucio and de Jesus 2021). Overall, MRA/MRI and CTA/CT scans each have their strengths and weaknesses that can determine the most effective imaging processes to use for certain structures or regions in the body (Yang et al. 2019). Therefore, using combined data from all of these imaging techniques would have greatly enhanced this project.

8.3 Available Techniques for Constructing Stereoscopic Models

8.3.1 Volume Rendering

A common way to construct stereoscopic models is through the process of gathering data via segmentation. In segmentation, the data is gathered slice-by-slice from the CT/CTA or MRI/MRA scan. In order to do this, the desired organ must be identified, and the data on its shape and location are slowly gathered from every slice in which it is present. One of the most common ways to collect segmented data is by a process known as volume rendering. Volume rendering is a form of assisted segmentation, a process where the desired data does not have to be manually traced, but instead, tools used in Amira® (e.g., Brush, Magic Wand, and Threshold) will automatically acquire large sets of data based on certain criteria in the data. Through this process, data is gathered based on the relatively similar densities of structures. A specified intensity threshold is set for the desired data and all structures that have that similar density will then be selected by volume rendering techniques. Many organs, tissues, blood vessels, etc., have different densities, thus allowing a variety of structures to be mapped with volume rendering techniques (see Fig. 8.2).

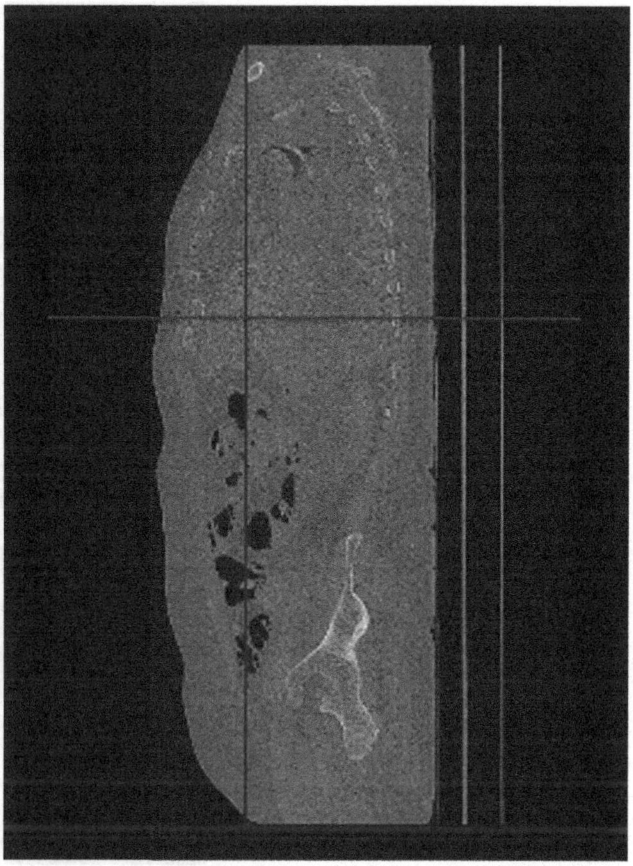

Fig. 8.1 An example of a CTA slice (Sagittal Scan of CTA Imaging) was used to create a 3D stereoscopic model, which focusing on a Portion of the Liver

8.3.2 Surface Rendering

The surface rendering method is a much more manual route in the process of creating stereoscopic models. In this process, the data must be gathered manually from every slide where it is present. This data is collected either by one of the lassos, brush, or blow tools. After the structure's data have been fully collected, the software generates it into a 3D model. Through the use of different surface rendering modules, each structure can be individually added and/or separated from the rest of the data (see Fig. 8.3). The disadvantage of surface rendering is the sheer amount of time required to successfully acquire the data. As a result, the volume rendering method is typically the preferred method. However, surface rendering can produce much more accurate reconstructed products than volume rendering. Surface rendering also allows individual structures to be created and to be included in the overall model in separate layers so that structures can be individually added or removed.

8.3.3 Combined Rendering

Both volume rendering and surface rendering techniques have their distinct advantages and disadvantages. The semi-auto combined rendering process utilizes both of these techniques in the rendering of 3D models (Chen et al. 2017). Since volume rendering techniques work best on structures that have higher densities than the rest of the structures in digitized scans, but not so well when the structures in the scans become harder to discern from each other, volume rendering techniques can be used to quickly make a model. After the initial model is made, surface rendering techniques can be used to make any

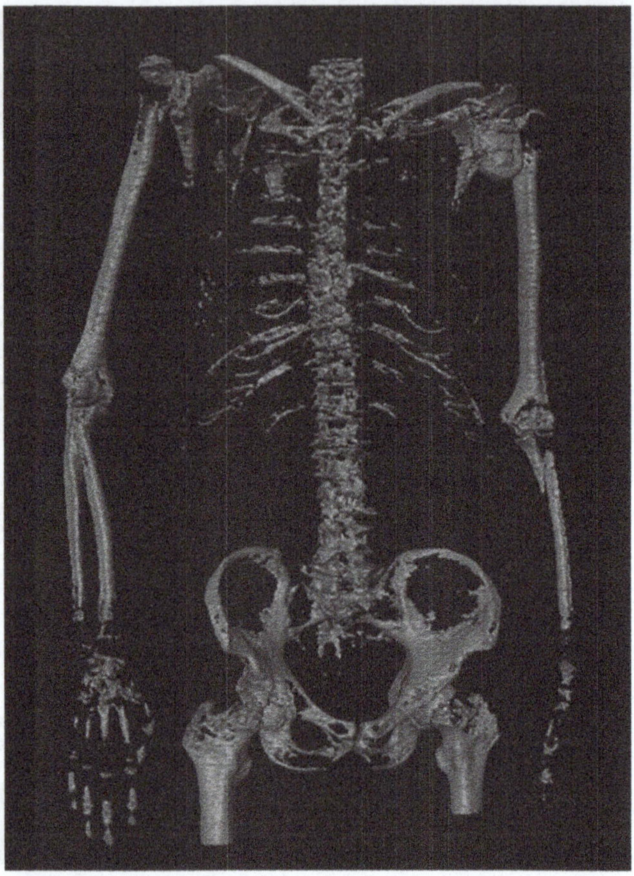

Fig. 8.2 Model of a portion of the skeleton created by volume rendering

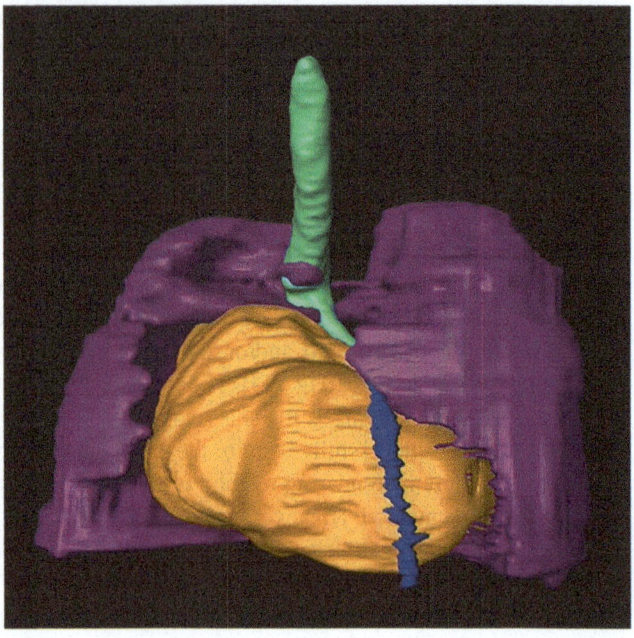

Fig. 8.3 Surface rendered models of thoracic organs

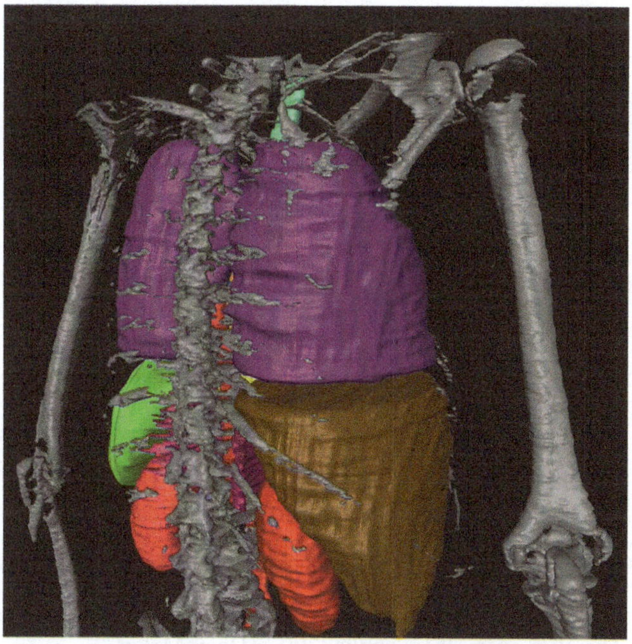

Fig. 8.4 Combination of surface-rendered and volume-rendered models

necessary adjustments, such as additions or cancellations to the model. These techniques can increase the speed of making models. The speed of volume rendering techniques mixed with the specificity of surface rendering techniques allows for more accurate models to be made than with a single approach alone (see Fig. 8.4). Although this combined process takes longer than usual because it involves two processes instead of one, it still has the potential to be a very useful resource in the future. For clarity, it must be noted that Fig. 8.4 is not an example of semi-auto combined rendering; it is simply the combination of products from surface rendering and volume rendering. Semi-auto combined rendering was not explicitly used during this study, but its use is still worth mentioning in the list of available methods for the construction of stereoscopic models.

8.4 Procedural Steps of Segmentation and Display

The process of creating accurate 3D models began with the download of the CTA imaging data into the desired Amira® network. This network will act as a host for all future relevant data in the study. Once the CTA scan is acquired, its data can be viewed under the Segmentation Editor tab. This tab allows users to scroll through the CTA scan and find the desired organ of their choosing. Once the desired organ is identified, its data can be acquired. As previously mentioned, there are multiple ways to acquire data (surface rendering, volume rendering, and combined rendering). All "softer," less dense organs, were reconstructed using surface rendering techniques, but the bones and other denser organs were reconstructed using volume rendering techniques. The bones were displayed through the Volren module. Regarding surface rendering, the lasso tool in Amira® was used to trace the softer, less dense organs in every slice in which they were present. These data were accumulated into a rough 3D model of the organs of interest. This model was created by applying the associated data to the Surface Generator module connected to each of the desired organs within the module. Once the data was applied, a Surface module would appear, and a Surface View module could be created. The created Surface View module would finally display the data. After the creation of the first 3D model, more data could be collected from each organ again, but from a different plane of view (e.g., coronal instead of frontal). These iterative processes created a more accurate model. After the model was made sufficiently more accurate through these processes, a smoothing feature in Amira® was used to level a majority of the inaccuracies and rough edges of a model of an organ within the overall model. These new data were again displayed via the Surface View module to produce the finalized organ. This process was repeated for every desired organ.

The creation of animations/movies was utilized to showcase the insertion of the models into their respective locations of the body. This was completed by transferring the gathered data into a new network. A Bounding Box module was then attached to the data, allowing the body data to be placed into a 3D box to show its boundaries. The data were then applied to a DemoMaker module which then created a MovieMaker module. This module allowed the creation of an animation of the 3D models in the body. The animation could be edited and adjusted in MovieMaker to create the animation to the user's liking (see Fig. 8.5).

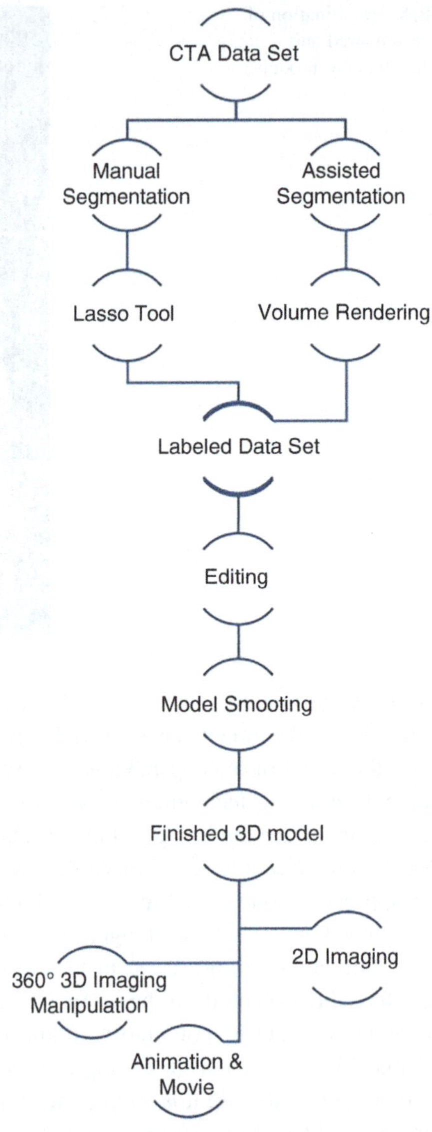

Fig. 8.5 Overview of procedural steps for 3D model's segmentation and display

8.5 Creation of 3D Movies

8.5.1 DemoMaker

The DemoMaker of Amira® allowed the gathered data associated with 3D models to be conveyed in an animated movie. It is a useful tool that has the potential to advance education in the anatomical sciences beyond the scope of this study. The movie created from the DemoMaker has a self-moving OrthoSlice plane that will move through the entire body represented in the radiographic scans. As it moves through the body, it will insert the 3D models into their respective locations. After the models are inserted, the skeleton appears in its location around the organs. Finally, we designed the movie so that after all data is inserted, the camera viewing the models will rotate around all organs in the model to give a 360-degree view.

The process by which the movies were created began with the data module for each desired 3D model being placed into a collective network. The surface modules of the data were then extracted and connected to surface view modules. A Bounding Box was then placed around the 3D models. At this point, multiple aesthetic modules could be added: OrthoSlice, Volren, Bounding Box, Camera Rotate, etc. These all added their own unique additions to the future movies. The

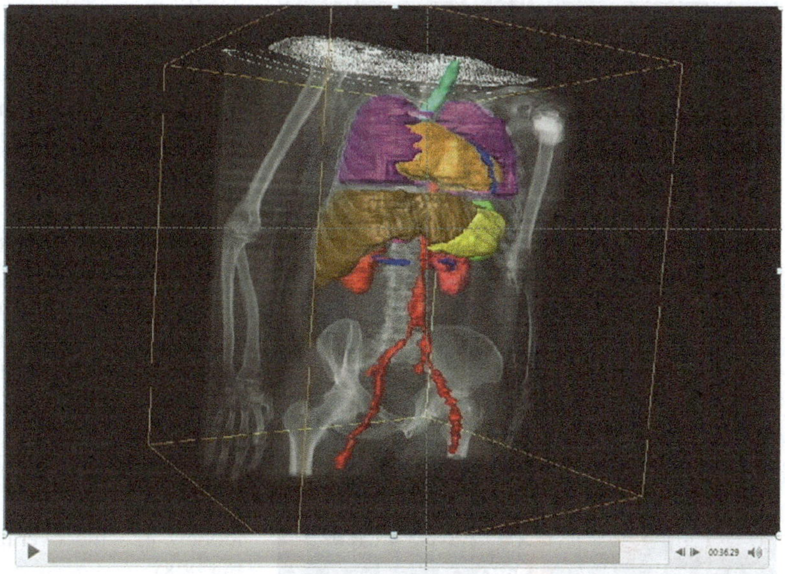

Fig. 8.6 Screenshot from 3D movie display

DemoMaker module was what made the movies, and through DemoMaker the user can access DemoDirector which allows the editing of the movies. Through DemoDirector, the length of the movie, the timed insertion of the 3D models, the insertion of other modules, and more were accessed and controlled. Thus, this series of controls through DemoMaker were used to make the movies in this project (see Fig. 8.6).

In order to successfully implant the data for the 3D models, a few steps are required. First, the surface module of the desired data must be acquired, along with its surface view. These data are then transferred to the DemoDirector portal. Once the data are in the portal, they can be modified so that the corresponding models appear at certain times and are displayed in ways according to the user's choosing. The Volren module can be utilized in the same fashion. To display the skeleton, the Volren module is attached to the main database. Afterward, the module is applied to the DemoDirector through which it is adjusted to the user's discretion. The Camera Rotate feature allows a 360-degree pan of the 3D model. Once this feature is applied to the DemoDirector, it can be edited to rotate the model at certain speeds and/or times.

8.5.2 DemoMaker Use

The use of DemoMaker expands the realism of created 3D models by allowing their creation to unfold before the viewers' very eyes. This capability could potentially help increase students' spatial awareness of organs represented in the models, increase their knowledge of those organs, help decrease likelihood of their losing interest in the anatomical information, and more. The demos and movies created can be edited in a variety of ways to the creators' liking. For example, annotations can be added to the models in the movies. The annotations can either be in a 2D or 3D stereoscopic format. The 2D annotations are only useful in a 2D plane in which they can be placed over the desired area requiring the annotative labels. However, when the models are rotated or manipulated, the 2D annotations' positions are no longer relevant to their original locations. This problem is corrected through the 3D annotations. 3D annotations have the ability to remain fixed to their relevant positions when the models are rotated and manipulated. This 3D annotation application, therefore, serves as a more accurate tool to use in the labeling of 3D models generated from 3D data (see Fig. 8.7).

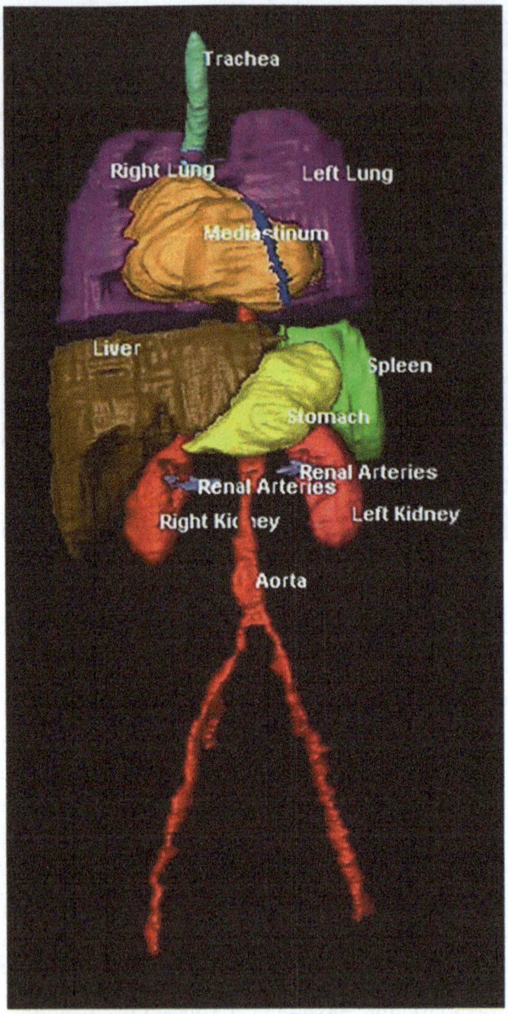

Fig. 8.7 Complete 3D model with annotations

8.6 Results

8.6.1 Completion of Models, Movie, and Annotation

Stereoscopic models were developed from multiple different organs imaged by the CTA scans, such as the associated skeletal structures, two lungs, one heart, one liver, one spleen, one stomach, one trachea, two kidneys, and selected blood vessels.

The study resulted in the successful completion of multiple 3D models of bodily organs and the completion of an animation/movie which details the creation of the models. The list of models created by surface rendering techniques includes the left lung, the right lung, the liver, the right kidney with its renal arteries, the left kidney with its renal arteries, the stomach, the middle mediastinum (and some of its component structures), some lower thoracic cage muscles, the spleen, the trachea, and the aorta. The skeletal system was also modelled, but by using the volume rendering technique instead of the surface rendering technique (see Table 8.1). These models are created based on real anatomy structures, and they have the potential to be useful in future educational use. The created movie was also a success. It detailed the accurate insertion of the skeletal structure models and other organ models into the overall constructed model (see Fig. 8.8).

8.6.2 Stereoscopic Visualization

The stereoscopic projection system requires certain features and materials to run properly and function at its full capacity. An example of the preferred computer software would be a Dell Precision T7600 computer workstation (Dell Inc., Round Rock, TX) with an NVIDIA Quadro K6000 video card (NVIDIA Corp., Santa Clara, CA). Dual projectors are the features required to display the data. These projectors work together to help display the virtual 3D models. Along with the dual projectors, a silver (Da-Lite Model C 10000 diagonal, silver matte finish; Da-Lite Corp., Warsaw, IN) projection screen on which the virtual 3D models are projected is needed. Finally, 3D glasses are required to view the finalized data in its desired virtual 3D stereoscopic format. The explanation of how this visualization format is achieved was described earlier in Sect. 8.1.2.

There are also several additional features that can be incorporated into the presentation of a virtual 3D model. The use of OrthoSlice allows a radiographic slice to be inserted into the overall model. This slice accurately conveys the modeled structures' relevance in position and size to the actual scanned image source data from which the

Table 8.1 List of organ models created and their respective characteristics

Organ models	Methods of segmentation	Time required for model completion	Number of CTA scan slices comprising the models (frontal plane)	Number of CTA scan slices comprising the models (transverse plane)
Lungs (2)	Surface rendering	3 h each	166	177
Heart (1)	Surface rendering	3 h	93	116
Liver (1)	Surface rendering	4 h	201	176
Spleen (1)	Surface rendering	1 h	88	113
Stomach (1)	Surface rendering	30 min	52	86
Kidneys with renal arteries (2)	Surface rendering	Kidneys (2 h each) Renal arteries (1 h each)	55 32	120 27
Middle Mediastinum (1)	Surface rendering	3 h	93	116
Trachea (1)	Surface rendering	45 min	65	135
Aorta (1)	Surface rendering	1 h	26	286
Common iliac arteries (2)	Surface rendering	45 min each	56	239
Skeletal structures	Volume rendering	5 min	N/A	N/A

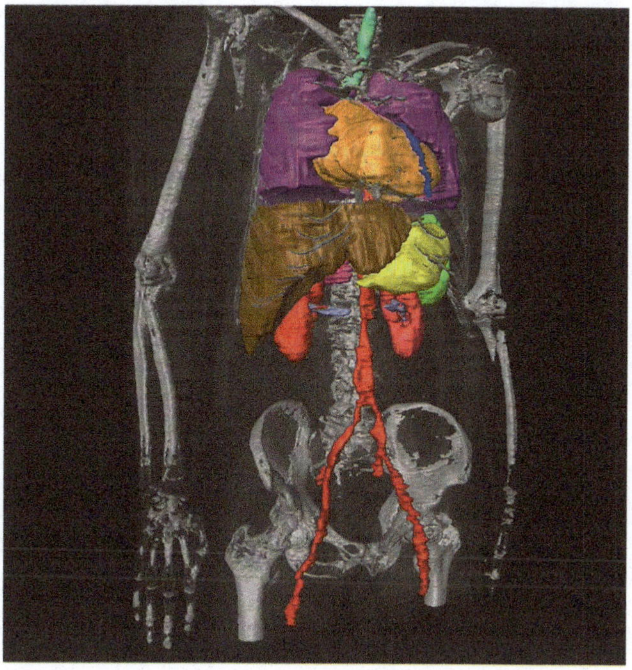

Fig. 8.8 Model of organs acquired by volume rendering (skeleton) and surface rendering (remaining organs)

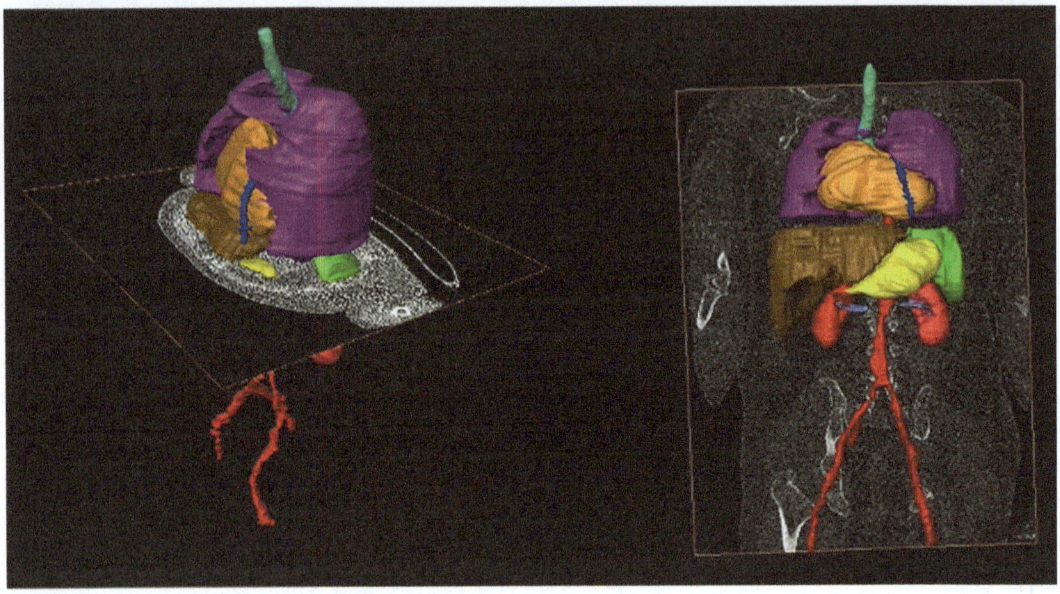

Fig. 8.9 Transverse and coronal CTA slices inserted into the 3D model

structures were generated. The OrthoSlice radiographic insertion has the capability to be freely moved across the overall model through a sagittal, coronal, or transverse plane. Other features contributing to the presentation of the virtual 3D model are the ability to take a snapshot image of the desired virtual 3D model for incorporation into other educational products, to create movie clips that detail the creation of the virtual 3D model in an animated sequence as previously described, and to toggle the display format of the virtual 3D model between monoscopic and stereoscopic formats, allowing viewers to visualize the virtual 3D model in the absence of additional equipment necessary for achieving stereopsis or with the aid of such equipment, respectively. As models with an inserted CTA images in transverse and coronal planes are displayed monoscopically or stereoscopically, they can be rotated in 360-degree angles to provide the viewer a full understanding of the 3D orientations and relationships of the anatomical structures depicted in the model (see Fig. 8.9). The stereoscopic format though provides viewers with a more realistic 3D display of the models as if they were actually physically present floating in mid-air, though they are virtually present in real space.

8.6.3 Implementation

There are many opportunities for advancing education through the use of virtual 3D anatomical models. Examples of areas where these models can be implemented are training sessions for medical residents, physicians, and other health science professionals as well as learning sessions for health professional students (Rozen et al. 2008, 2009; Bloch et al. 2015; Schoenthaler et al. 2016; Unger et al. 2016; Chen et al. 2017). Depending on their desired purpose, virtual 3D stereoscopic models of anatomical structures can have a wide range of uses. For students, these 3D models offer a realistic representation of the organ or structure which they are studying. More lifelike models have the potential to enhance students' ability to retain the associated anatomical information. The zoom, rotational, and other interactive features programs, such as Amira®, allow users to engage dynamically with virtual 3D models. This level of interactivity with virtual 3D anatomical models can also potentially enhance students' retention of anatomical information. In addition, there is potential that students who study with virtual 3D anatomical models compared to students who do not will achieve higher performance on examinations. However, this claim extends past the bounds of this project.

8.7 Discussion

8.7.1 Benefits

There are many benefits that accompany the use of virtual 3D stereoscopic anatomical models. These models can be used by a diverse group of learners, such as anatomy educators, research scientists, healthcare professionals (e.g., physicians, dentists, occupational therapists, and physical therapists), medical residents, college students, graduate students, health professional students (e.g., medical students and dental students), and the general public. These models can be used in a variety of settings such as classrooms, laboratories, or home environments for a variety of purposes, such as education, research, and quality control. The virtual 3D models described in this project can be displayed in one-on-one, small-group, or large-group learning sessions in locations ranging from small lab spaces to auditorium-style lecture halls, either under guided facilitation by a skilled instructor or during self-study opportunities. This type of model can be used in anatomy education, and use un exams or pre-test and post-tests of educational research studies (Cui et al. 2017; Meyer et al. 2018; Meyer and Cui 2020); the 3D stereoscopic models can also be used in surgical simulation and clinical training in regional anatomy (Chen et al. 2017). There are many different features that accompany the 3D models, such as snapshot features, a 2D and 3D measurement capability, and the stereoscopic visualization feature which displays the data in 3D. Studies have shown that the use of 3D models in education can help students increase their performance in their studies of complex anatomical structures of human anatomy (Luursema et al. 2006, 2008; Meyer 2019; Nicholson et al. 2006; Nguyen et al. 2014). Furthermore, the fast-paced advancement of virtual anatomical visualization technologies has the potential to compel researchers toward the next steps in updating their capability to produce even more accurate, realistic, and effective virtual 3D anatomical models.

8.7.2 Limitations

Regarding limitations, the data that was utilized in creating 3D stereoscopic models had a few problems that caused some inaccuracies in the project. There was a pacemaker implanted into the heart of the individual, thus making obtained CTA images difficult to distinguish the structures in the thoracic cavity region, which vary from light to extremely strong for these slice images. This interference looked more like white noise static present throughout the scans, resulting in the data taken from the lungs and heart having some inaccuracies. To help counter these inaccuracies, extra care was taken during the modeling of the lungs and heart to make them as accurate as possible. Nevertheless, some inaccuracies may have still persisted in the models. Another limitation is the reduced capability of CTA in imaging soft tissue structure and the ensuing difficulty in differentiating different soft tissue structures from one another during the model creation.

8.7.3 Future Directions

The use of the virtual 3D anatomical models generated in this particular project can play an important role in aiding students in their understanding of human anatomy. Future steps that can be taken could be such actions as improving the ability to create more accurate models, discovering ways to further increase the usefulness of these virtual 3D anatomical models in an educational environment, and determining how effective the use of these virtual 3D anatomical models in an educational environment really are.

References

Ackerman MJ (1998) The visible human project: a resource for anatomical visualization. Stud Health Technol Inform 52:1030–1032. https://doi.org/10.1109/ITAB.1997.649393

Adams CM, Wilson TD (2011) Virtual cerebral ventricular system: an MR-based three-dimensional computer model. Anat Sci Educ 4:340–347. https://doi.org/10.1002/ase.256

Anderson P, Chapman P, Ma M et al (2013) Real-time medical visualization of human head and neck anatomy and its applications for dental training and simulation. Curr Med Imaging Rev 9:298–308

Baliyan V, Shaqdan K, Hedgire S et al (2019) Vascular computed tomography angiography technique and indications. Cardiovasc Diagn Ther 9:S14–S27. https://doi.org/10.21037/cdt.2019.07.04

Bloch E, Uddin N, Gannon L et al (2015) The effects of absence of stereopsis on performance of a simulated surgical task in two-dimensional and three-dimensional viewing conditions. Br J Ophthalmol 99: 240–245. https://doi.org/10.1136/bjophthalmol-2013-304517

Brewer DN, Wilson TD, Eagleson R et al (2012) Evaluation of neuroanatomical training using a 3D visual reality model. Stud Health Technol Inform 173:85–91

Brown PM, Hamiliton NM, Denison AR (2012) A novel 3D stereoscopic anatomy tutorial. Clin Teach 9:50–53

Chen J, Smith AD, Khan MA et al (2017) Visualization of stereoscopic anatomic models of the paranasal sinuses and cervical vertebrae from the surgical and procedural perspective. Anat Sci Educ 10:598–606. https://doi.org/10.1002/ase.1702

Cui D (2015) Development and evaluation of 3D stereoscopic models of vasculature of head and neck for anatomy education. UMI Dissertation Publisher, Ann Arbor, MI

Cui D, Lynch JC, Smith AD et al (2016) Stereoscopic vascular models of the head and neck: a computed tomography angiography visualization. Anat Sci Educ 9:179–185. https://doi.org/10.1002/ase.1537

Cui D, Wilson TD, Rockhold RW et al (2017) Evaluation of the effectiveness of 3D vascular stereoscopic models in anatomy instruction for first year medical students. Anat Sci Educ 10:34–45

Cui D, Chen J, Meyer E et al (2019) Anatomy visualizations using stereopsis: current methodologies in developing stereoscopic virtual models in anatomical education. Adv Exp Med Biol 1156:49–65. https://doi.org/10.1007/978-3-030-19385-0_4

de Leucio A, de Jesus O (2021) MR angiogram. [Updated 2022 Jul 25]. In: StatPearls [Internet]. StatPearls Publishing, Treasure Island

Foo JL, Martinez-Escobar M, Juhnke B et al (2013) Evaluating mental workload of two-dimensional and three-dimensional visualization for anatomical structure location. J Laparoendosc Adv Surg Tech A 23: 65–70

Gary A, Norman GE, Spero L et al (1999). Do virtual computer models hinder anatomy learning? Acad Med 74:S87–589

Hu A, Wilson T, Ladak H et al (2009) Three-dimensional educational computer model of the larynx: voicing a new direction. Arch Otolaryngol Head Neck Surg 135(7):677–681. https://doi.org/10.1001/archoto.2009.68

Luursema JM, Verwey WB, Kommers PAM et al (2006) Optimizing conditions for computer-assisted anatomical learning. Interact Comput 18:1123–1138

Luursema LM, Verwey WB, Kommer PAM et al (2008) The role of stereopsis in virtual anatomical learning. Interact Comput 20:455–460

Meyer ER (2019) Validated virtual three-dimensional stereoscopic middle and inner models for examining first-year medical students' short- and long-term retention (Doctoral dissertation). Retrieved from the University of Mississippi Medical Center

Meyer ER, Cui D (2020) Anatomy visualization using stereopsis: assessment and implication of stereoscopic virtual models in anatomical education. Adv Exp Med Biol 1235:117–130. https://doi.org/10.1007/978-3-030-37639-0_7

Meyer ER, James AM, Sinning A et al (2018) A pilot study of the impact of three-dimensional stereoscopic models of pelvic anatomy on short- and long-term retention in first-year medical students. HAPS Educator 24:7–22. https://doi.org/10.21692/haps.2020.021

Nguyen N, Wilson TD (2009) A head in virtual anatomy: development of a dynamic head and neck model. Anat Sci Educ 2:294–301. https://doi.org/10.1002/ase.115

Nguyen N, Nelson AJ, Wilson TD (2012) Computer visualizations: factors that influence spatial anatomy comprehension. Anat Sci Educ 5(2):98–108. https://doi.org/10.1002/ase.1258

Nguyen N, Mulla A, Nelson AJ et al (2014) Visuospatial anatomy comprehension: the role of spatial visualization ability and problem-solving strategies. Anat Sci Educ 7:280–288

Nicholson DT, Chalk C, Funnell WR et al (2006) Can virtual reality improve anatomy education? A randomized controlled study of a computer-generated three-dimensional anatomical ear model. Med Educ 40:1081–1087

Rozen WM, Ashton MW, Grinsell D et al (2008) Establishing the case for CT angiography in the preoperative imaging of abdominal wall perforators. Microsurgery 28:306–313. https://doi.org/10.1002/micr.20496

Rozen WM, Ashton MW, Whitaker IS et al (2009) The financial implications of computed tomographic angiography in DIEP flap surgery: a cost analysis. Microsurgery 29:168–169. https://doi.org/10.1002/micr.20594

Schoenthaler M, Schnell D, Wilhelm K et al (2016) Stereoscopic (3D) versus monoscopic (2D) laparoscopy: comparative study of performance using advanced HD optical systems in a surgical simulator model. World J

Urol 34:471–477. https://doi. org/https://doi.org/10.1007/s00345-015-1660-y

Sergovich A, Johnson M, Wilson TD (2010) Explorable three-dimensional digital model of the female pelvis, pelvic contents, and perineum for anatomical education. Anat Sci Educ 3(3):127–133. https://doi.org/10.1002/ase.135

Stalling D, Westerhoff M, Hege H (2005) Amira: a highly interactive system for visual data analysis. The visualization handbook. Elsevier, pp 749–767

Unger B, Tordon B, Pisa J et al (2016) Importance of stereoscopy in haptic training of novice temporal bone surgery. Stud Health Technol Inform 220:439–445. https://doi.org/10.3233/978-1-61499-625-5-439

Yang GC, Wilson TD, Lehman MN et al (2019) Comparison of magnetic resonance angiography and computed tomography angiography stereoscopic cerebral vascular models. Adv Exp Med Biol 1205:1–9. https://doi.org/10.1007/978-3-030-31904-5_1

Teaching Cellular Architecture: The Global Status of Histology Education

9

Michael Hortsch, Virgínia Cláudia Carneiro Girão-Carmona, Ana Caroline Rocha de Melo Leite, Ilias P. Nikas, Nii Koney-Kwaku Koney, Doris George Yohannan, Aswathy Maria Oommen, Yan Li, Amanda J. Meyer, and Jamie Chapman

Abstract

Histology or microanatomy is the science of the structure and function of tissues and organs in metazoic organisms at the cellular level. By definition, histology is dependent on a variety of microscope techniques, usually light or more recently virtual, as well as electron microscopy. Since its inception more than two centuries ago, histology has been an integral component of biomedical education, specifically for medical, dental, and veterinary students. Traditionally, histology has been taught in two sequential phases, first a didactic transfer of information to learners and secondly a laboratory segment in which students develop the skill of analyzing micrographic images. In this chapter, the authors provide an overview of how histology is currently taught in different global regions. This overview also outlines which educational strategies and technologies are used, and how the local and cultural environment influences the histology education of medical and other students in different countries and continents. Also discussed are current trends that change the teaching of this basic science subject.

M. Hortsch (✉)
Departments of Cell and Developmental Biology and of Learning Health Sciences, University of Michigan Medical School, Ann Arbor, MI, USA
e-mail: hortsch@umich.edu

V. C. C. Girão-Carmona
Department of Morphology, Federal University of Ceará, Fortaleza, Brazil
e-mail: virginia.girao@ufc.br

A. C. R. de Melo Leite
University of International Integration of the Afro-Brazilian Lusophony, Redenção, Brazil
e-mail: acarolmelo@unilab.edu.br

I. P. Nikas
School of Medicine, European University Cyprus, Nicosia, Cyprus
e-mail: i.nikas@euc.ac.cy

N. K.-K. Koney
Department of Anatomy, University of Ghana Medical School, University of Ghana, Korle Bu, Ghana
e-mail: nkkoney@ug.edu.gh

D. G. Yohannan · A. M. Oommen
Government Medical College, Thiruvananthapuram, Kerala, India

Kerala University of Health Sciences, Thrissur, Kerala, India

Y. Li
Department of Human Anatomy, Histology and Embryology, Fudan University, Shanghai, China
e-mail: liyan@shmu.edu.cn

A. J. Meyer
Department of Anatomy, Physiology, and Human Biology, School of Human Sciences, The University of Western Australia, Perth, WA, Australia
e-mail: amanda.meyer@uwa.edu.au

J. Chapman
Tasmanian School of Medicine, University of Tasmania, Hobart, TAS, Australia
e-mail: Jamie.Chapman@utas.edu.au

D. Cui et al. (eds.), *Biomedical Visualisation*, Advances in Experimental Medicine and Biology 1431,
https://doi.org/10.1007/978-3-031-36727-4_9

Keywords

Histology · Microanatomy · Medical education · Technology-enhanced instruction · Microscopy · Virtual microscopy

9.1 A Short History of Histology

Histology or microanatomy has been a core biomedical science teaching topic for approximately two centuries. It addresses the structural description of metazoic organisms at the cellular level and how different types of cells are arranged in tissues and organs. Most importantly, these structural arrangements and cellular specializations are closely linked to the physiological and biochemical functions (physiology and biochemistry) that are carried out in various organs and organ systems. Histology also serves as the foundational science for the microscopic analysis of diseased tissues and organs, known as pathology. Therefore, medical, dental, veterinary, and other biomedical students usually participate in a histology course or component during their preclinical education.

As the cellular dimension is beyond the resolution of the human eye, the histological analysis relies on optical instruments, specifically different types of microscopes for magnifying these structures and making them visible. It was in the seventeenth century when craftsmen in Europe were able to produce quality glass lenses that enabled the construct of single lens and compound microscopes. Among others, several Dutchmen, including Zacharias Janssen (1585–bef 1632), Cornelius Drebbel (1572–1633), and Antoni van Leeuwenhoek (1632–1723), and two Italian scientists, Francisco Fontana (1580/90–1656) and Galileo Galilei (1564–1642), were among the first investigators to create and use microscopes for scientific studies (Chapman et al. 2020). This enabled the next generations of scientists to make foundational observations and define central concepts in the field of microscopic biology, later called histology. The Englishman Robert Hooke (1635–1703) introduced the term "cell" from the Latin *cella* for small room or chamber when he observed cork bark under a primitive compound microscope (Hooke 1665). Based on Hooke's observation, it would take another one and a half centuries for the concept of the biological cell as the fundamental biological unit to develop. Theodor Schwann (1810–1882), Matthias Jakob Schleiden (1804–1881), and contemporaries formulated the cell theory of life in the first half of the nineteenth century (Schwann 1838a, b, c; Schleiden 1838). Around 1800, the Frenchman Xavier Bichat (1771–1802) coined the term "tissue" from the old French word *tissu*, which itself is derived from the Latin word *texere* meaning to weave. He defined 21 different tissues in the human body and also developed the foundational concept of pathology as the science of diseased tissues (Bichat 1800; Shoja et al. 2008). The term histology or *Histologie* in German, as a contraction of the Greek words *στός* or *histos* for "web/tissue" and *λογία* or *logia* for "science/knowledge," was introduced in 1819 by the German Carl Mayer (1787–1865) (Mayer 1819).

The nineteenth century saw several technological advancements that enabled additional histological discoveries. First was a continued improvement of light microscopes, which included the invention of achromatic lenses by Joseph Jackson Lister (1786–1869) (Lister 1830) and the condenser and apochromatic lenses by Ernst Abbe (1840–1905) (Volkmann 1966). As a consequence, a large number of high-quality light microscopes were manufactured and became available to researchers and educators. In addition, supplementary instruments, like the microtome, were invented and used for cutting thin sections of embedded tissues for viewing under a microscope (Purkyně 1834; His 1870). In parallel, a variety of fixation and staining methods for biological specimens were developed, which further improved the quality of the structural analysis by light microscopes (Griffith 1864; Titford 2005, 2006).

As the theoretical resolution of light microscopes is limited by the wavelength of the radiation used and the numerical aperture of the optical instrument, most subcellular structures cannot be imaged by traditional optical light microscopy (TM). This limitation was overcome by the invention of the electron microscope (EM). Working at the Technical University Berlin, in 1933 Ernst Ruska (1906–1988) and Max Knoll (1897–1969) built the first functional electron microscope opening the subcellular dimension for analysis by cell biologists. That development resulted in the discovery and photographic documentation of new, so far unidentified subcellular organelles and compartments, viruses, and large biomolecules (Haguenau et al. 2003).

Thus, from its beginning the scientific field of histology depended on different microscopy technologies. More recently, additional optical and non-optical technologies and experimental approaches have complemented traditional microscopy for the scientific analysis of cell structure and function, morphing histology into the new field of cell biology (Bechtel 2006; Scott and Logan 2004).

9.2 A Short History of Histology Education

The first documented histology/microscopy courses were offered around 1830 to medical students at German universities (Tuchman 1993). In the second half of the nineteenth century, many Central European medical schools offered basic histology and pathology courses to their students that were often supplemented with microscope laboratory exercises (Anonymus 1875; Bohm et al. 1904; Cotter 2001; Kölliker 1867; Orth 1878). In the following decades, histology became widely accepted as a component of preclinical medical education and incorporated into the biomedical curricula at most European and North American universities (Hortsch 2023a, Fig. 9.1). Developing from these beginnings, histology education maintained a two-part structure, consisting of a didactic and a laboratory segment. The didactic part, traditionally in the form of a lecture or more recently a flipped classroom modality, conveyed the histological terminology and basic fact knowledge about cells, tissues, and organs to learners. The second part consisting of laboratory instruction originally involved light microscopes and dye-stained tissue on glass slides. Students were required to find and recognize cells and cellular structures in histological preparations that had been previously introduced in a didactic segment. Early histology laboratory manuals often included detailed staining instructions (Huber 1892, 1900) and technical details and the physics of microscopy (Schaffer 1920), suggesting that the preparation and staining of sectioned tissues by learners were often required before they could be investigated with a microscope (Fig. 9.2). Histology laboratory manuals from the 19th and early twentieth century also contain empty pages for students to

Fig. 9.1 Laboratory classwork in mammalian anatomy and histology at the University of Michigan (appr. 1893). Photograph by J. Jefferson Gibson; Source: University of Michigan Bentley Historical Library

Fig. 9.2 Detail from a photograph depicting histology laboratory instruction at the University of Michigan Medical School (appr. 1893). Photograph by J. Jefferson Gibson; Source: University of Michigan Bentley Historical Library

document their observations in the form of drawings (Huber 1892, 1900). This artistic approach is still used for histology education and assessment in some countries (Yohannan et al. 2019; Cracolici et al. 2019) and is correlated with a better understanding of tissue organization and examination performance (Balemans et al. 2016; Cogdell et al. 2012). However, with decreasing time dedicated to histology instruction, drawing exercises play a diminishing part in today's histology education and at many universities are no longer used.

These two main components of histology education have different learning goals that can be described by the dual-processing theory of learning (De Neys 2006; Evans 2003; Evans and Stanovich 2013). The first didactic segment of histology teaching provides nomenclature and fact knowledge that a learner can memorize, sometimes without a deeper understanding of the material. In contrast, histology laboratory sessions require students to identify microscopic structures by using an analytical thought process. However, this second step involving higher reasoning is based on and requires the theoretical knowledge that was acquired in the first step. Therefore, the two main components of histology education are sequential and interdependent, and both are required for teaching learners an advanced mastery of the field (Hortsch 2023a). If one step, like laboratory instruction, is removed from the learning process, many students will achieve a lower level of histology competency (Gribbin et al. 2022).

With few exceptions, this two-phase structure has been the central tenet of histology education from its beginning up to the present day. However, the educational approaches and technologies used for most of the last two centuries have changed, especially over the last

30 years. Up to the turn of the millennium, live synchronous lectures were the main didactic format for delivering informational content to students and light microscopes and stained tissues on glass slides preparations were used for histology laboratory sessions. More recently, alternative forms of information delivery have been tested for histology instruction and often found equivalent or better in comparison to the traditional lecture modality. These teaching strategies include the flipped classroom approach (Li and Guo 2014; Gilliland 2017; McLean 2018; MacDonald et al. 2021; Zhong et al. 2022), educational gaming approaches (Felszeghy et al. 2019), as well as team- and problem-based teaching (Dickerson and Kubasko 2007; Goldberg and Dintzis 2007; Triola and Holloway 2011; McLean 2018). Although these educational modalities are currently not widely used, they might gain more popularity with the general change from teacher- to learner-centered education (Hannafin and Land 2000; Bloodgood 2012; Khalil et al. 2013; Jurjus et al. 2018). Another recent change is the use of technology-based learning approaches for histology. Reports have been published about the use of e-books (Young et al. 2013; Pawlina and Ross 2018; Lowrie 2020; Gartner and Lee 2022), websites (Sander and Golas 2013; Sorensen and Brelje 2023; Hortsch 2023b), recorded lecture videos (Cardall et al. 2008; Holaday et al. 2013; Selvig et al. 2015), Massive Open Online Courses (MOOC) (Multon et al. 2018; Zhang et al. 2018), podcasts (Beylefeld et al. 2008; Sakthi-Velavan and Zahl 2023), Internet-accessible video demonstrations (Yohannan et al. 2022; Chapman 2023), e-learning platforms (Sander and Golas 2013; Drees et al. 2020), online tutorials (Rosenberg et al. 2006), mobile applications (Hortsch 2016; Ostrin and Duschenkov 2016), self-directed learning modules (Chimmalgi and Hortsch 2022; Schoenherr et al. 2022), and social media (Maske et al. 2018; Essig et al. 2020) for histology education. Again, these new technologies are often only used locally or by individual learners.

Starting shortly after the year 2000, the laboratory teaching of histology changed considerably with the introduction of virtual microscopy (VM), replacing the use of traditional light microscopes and glass slides for histology instruction in many countries (Hortsch et al. 2023) (Fig. 9.3). During the last decade of the twentieth century, the use of personal computers for medical education, the connectivity of the Internet, and the ability to store large image files on computer servers laid the foundation for this development (Gu and Ogilvie 2005). VM requires a special microscope setup with an automated movable stage and a digital camera that is able to scan regular histological glass slides for the creation of high-resolution digital image files that can be stored on a central computer server. Similar to the strategy used for Google Earth (Gorelick et al. 2017), users only download small segments of the large image files via the Internet or from a local computer server to their personal computer station that allows them to view selected areas of the histological preparation at any magnification of their choice (limited only by the original resolution of the image file) (Fig. 9.3). VM users can pan over any scanned region of the digitized slide and zoom in or out for high or low magnification views of areas of interest. Usually, only high-quality glass slides are digitized providing all learners with the same selected image material. Virtual slide files can easily be shared via electronic databases like the Virtual Microscopy Database (Lee et al. 2018) or viewed on histology websites (Hortsch 2023b; Sorensen and Brelje 2023; Takizawa 2022). Another advantage of VM is that rare specimens are available to all students in a histology class. The increasing use of VM for histology teaching goes hand in hand with its use for telemedicine and the pathological analysis of patient-derived material (Hortsch et al. 2023). VM enjoys wide popularity for histology education in the most affluent global regions. But its use in developing countries is quickly growing, a process that was further accelerated by the recent COVID-19 pandemic (Hortsch et al. 2023) and the Virtual Microscopy Database (Lee et al. 2018).

The other major change in how histology is taught worldwide is curricular in nature. Until 20–30 years ago, histology education was usually organized in stand-alone courses, independent

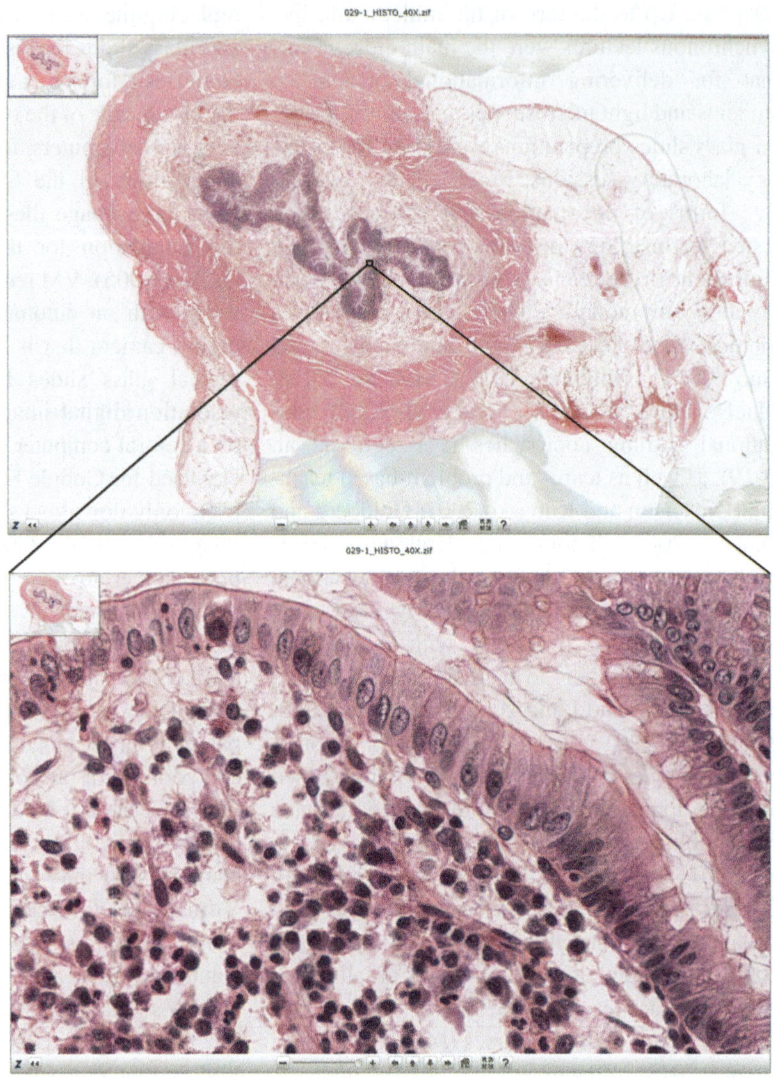

Fig. 9.3 A virtual microscopy slide from the University of Michigan collection (Hortsch 2023b) depicting a Hematoxylin and Eosin-stained cross-section of the small intestine. The top panel shows the slide at the lowest magnification, whereas the lower panel depicts a section of the slide at the highest magnification

and not coordinated with other topics and courses in biomedical curricula. In most developed countries, this has changed with the introduction of integrated curriculum structures (Abali et al. 2014; Bolender et al. 2013; Brauer and Ferguson 2015; Daniel et al. 2021; Eisenstein et al. 2014; Kozu 2006; Shankar et al. 2014; Sherer et al. 2014). After the initial introduction of basic tissues and histological concepts, organ system histology is taught at the same time as the corresponding organ gross anatomy, physiology, biochemistry, pharmacology, and pathology. Thus the structure–function relationship at the cellular level and the importance of organ organization become more apparent to biomedical learners. Again, the incorporation of histology into integrated curricular structures has not progressed evenly and in some local educational systems histology is still organized in a separate course (Table 9.1).

In summary, the way histology is taught to biomedical students is constantly evolving and different global regions are at different stages in this process. The following segments will summarize how histology is currently taught in different countries and on different continents and will

Table 9.1 Predominant histology teaching modalities on different continents

	North America	South America	Europe	Africa
Curricular structure of histology education	Mostly integrated	Mostly integrated	Integrated, as well as stand-alone courses	Mostly integrated and some stand-alone courses
Didactic instruction	Live and recorded lectures	Live and recorded lectures, expositive classes, and seminars	Live and recorded lectures	Live lectures and some online courses
Laboratory instruction	Online or in-person	Mostly in-person, some online	Mostly in-person, some online	*In-person*
Microscopy format	Mostly VM (often including EM micrographs)	TM or TM in combination *with VM*	VM or TM exclusively, but also VM combined with TM	TM and some VM
Other common learning resources/ modalities	LMS, websites, textbooks, e-books, mobile apps, podcasts	Inverted class model, clinical case studies, digital platforms, e-books	Websites, textbooks, journals, mobile apps, videos, podcasts, games	Websites, textbooks, e-books, YouTube
Primary assessment modalities	Predominantly MCQs, when possible case-based	Mixed formats including MCQs, semi-structured and open questions	MCQs, essays, spot tests, and clinical cases	MCQs, steeple chase, short answers questions

	South Asia	East Asia	Australia
Curricular structure of histology education	Mostly stand-alone courses	Integrated, as well as stand-alone	Mostly integrated
Didactic instruction	Live and recorded lectures	Live and recorded lectures	Live and recorded lectures
Laboratory instruction	In-person	Mostly in-person, some online	Mostly in-person
Microscopy format	Mostly TM	VM or TM exclusively, some VM combined with TM	Mostly VM
Other common learning resources/modalities	Textbooks, mobile apps	Websites, textbooks, mobile apps	LMS, websites, textbooks, YouTube
Primary assessment modalities	Short notes, identification of tissue glass slides, drawing diagrams	Various types of question formats	Various types of question formats

LMS, learning management system; MCQs, multiple-choice questions; TM, traditional microscopy; VM, virtual microscopy

address general and local factors that impact its teaching to biomedical students and sometimes impede the introduction of novel technologies and teaching approaches.

9.3 Histology Education in Different Global Regions

Histology is important for understanding the cellular basis of many human diseases and serves as the foundation of pathology. Consequently, it is an established component of medical, dental, and veterinary education at most universities worldwide. The following chapter segments provide overviews of how histology/microanatomy is currently being taught in different countries and global regions. The information that is discussed includes the curricular format of histology instruction, either stand-alone courses or curricular integration and whether histology teaching follows the traditional pattern of two modalities, a didact information transfer (lecture, flipped classroom, or other) and laboratory learning (either in-person or online) (Table 9.1). The predominant type of microscope technology used for histology laboratory instruction, either TM using light microscopes or VM, is also addressed, as well as the use of other e-learning resources (websites, video recording, podcasts, social media, mobile apps, and more) and novel educational approaches (flipped classroom, case- and team-based learning, etc.) (Table 9.1). Although the recent COVID-19 pandemic often accelerated changes in histology teaching that were already in progress, the status of histology education still diverges greatly in different countries and geographic areas.

9.3.1 The State of Histology Education in North America

Basic science education is similar in US and Canadian medical, dental, and veterinarian schools to that in other highly industrialized global regions, such as Europe and Australia. This also applies to many Caribbean medical schools that act as feeder schools for the postgraduate clinical system on the North American continent (Norcini et al. 2006). Professional schools in North America usually offer a 4-year educational plan awarding an MD (Doctor of Medicine), DO (Doctor of Osteopathic Medicine), DVM (Doctor of Veterinary Medicine), DDS (Doctor of Dental Surgery), DMD (Doctor of Dental Medicine), or other doctoral degrees. Relatively few students in biomedical PhD (Doctor of Philosophy) programs will take a histology class and if they do, often as an elective course or to support an educational career choice. Most students enter a professional biomedical program with a bachelor's or master's degree after completing a 4-year college education, sometimes with additional graduate school experience. A minority of these students have taken a proper histology course in college (Johnson et al. 2015; Selvig et al. 2015) and thus the subject of histology and its challenges are new to these students when they start their basic biomedical training.

For more than a century, histology instruction has been part of the preclinical phase of medical/dental/veterinary school education in North America, usually in the first and/or second year (Burk et al. 2013; Hortsch 2023a). Until the turn of the millennium, histology was usually taught as a stand-alone independent course (Painter 1994; Hightower et al. 1999). However, with the introduction of an integrated biomedical curriculum structure, histology lessons are now often part of organ-based courses and coordinated with the other basic sciences and clinical specialties like pathology (Jonassen et al. 2016). In 2017, 98% of US allopathic medical schools reported that histology instruction was part of an integrated curriculum (McBride and Drake 2018). At some, but not all North American schools, histology and its clinical counterpart pathology are now taught together or at least in temporal proximity (Shaw and Friedman 2012).

The biphasic structure of histology education is still the predominant form of microanatomy instruction in most schools on the North American continent (Hortsch 2023a). Lectures are often recorded, and lecture attendance is not mandatory at many schools, allowing students to view lecture videos asynchronously (Selvig et al. 2015; Zureick et al. 2018), an option that turned into a positive during the COVID-19 pandemic when all educational instructions were converted to an online mode. A successful implementation of the flipped classroom strategy for histology didactic instruction has been reported in several North American programs (Gilliland 2017; MacDonald et al. 2021). However, this option has not been adopted by most other histology teachers on the continent. Several other new teaching strategies and e-learning technologies for the teaching of histology have been tested and are sporadically used in the USA and Canada. These include e-textbooks (Young et al. 2013; Pawlina and Ross 2018; Lowrie 2020; Gartner and Lee 2022), websites (Hortsch 2023b; Sorensen and Brelje 2023; Takizawa 2022), social media (Essig et al. 2020; Rosenberg et al. 2006), and mobile applications (Hortsch 2016; Ostrin and Duschenkov 2016).

Over the past decades, the laboratory segment of histology teaching has undergone two significant changes at North American schools, the first being a transition from the use of TM with histological glass slides to VM. As of 2017, 67% of

US allopathic medical schools reported the exclusive use of VM for histology laboratory teaching, 10% employed a combination of both TM and VM, 10% used only TM, and 13% only static images (McBride and Drake 2018). Normally, institutional virtual slide collections are accessible to students via the Internet, allowing them to perform the laboratory exercises at a time and location of their own choosing. This appeals to today's generation of learners, who want to have control over where and when they are studying. Several articles from US schools reported about the replacement of in-person teaching events for histology with online instruction, either replacing the didactic or laboratory sessions or the entire histology course (Thompson and Lowrie 2017; Yen et al. 2014; Barbeau et al. 2013; Gadbury-Amyot et al. 2013). As a drawback, this approach to histology learning places a lot of responsibility back onto learners who may not be ready to handle the self-discipline and time-management skills required for this approach. The lack of personal teacher-to-learner interactions not only correlated with significantly lower learning outcomes (Selvig et al. 2015), but may also make it more difficult to identify students who struggle with the subject and need timely academic support (Hortsch and Mangrulkar 2015).

The second recent major change in North American histology education is a precipitous reduction of curricular time dedicated to histology laboratory teaching (Hightower et al. 1999; Gartner 2003; Drake et al. 2009, 2014; McBride and Drake 2018; Hortsch 2023a). At a few medical schools, this has resulted in the complete removal of scheduled time for histology laboratory instruction (Daniel et al. 2020). A recent report indicates that the cancelation of official histology laboratory time is correlated with a statistically highly significant drop in students' histology learning outcomes (Gribbin et al. 2022).

Assessment of histology and of most other core science topics at North American schools is usually based on multiple-choice questions (MCQs). In contrast, many gross anatomy courses/components still have a prosection/dissection-based laboratory assessment component. The majority of multiple-choice histology questions are image based (Zaidi et al. 2017). As histology addresses the structure–function relationship of health or normal tissues and organs, it is challenging to create good histology MCQs that are clinical in nature and address higher levels in Bloom's taxonomy pyramid (Zaidi et al. 2017; Fales-Williams 2016; King et al. 2019). With the integration of more clinical relevance into preclinical curricula, there are calls for making histology assessment questions more medically relevant, rather than solely serving a basic science role (King et al. 2019; Mantovani et al. 2019; Sherer et al. 2014). In addition, following the curricular integration of basic and clinical subjects, histology assessment questions now have to be more multi-dimensional by incorporating the other basic sciences and clinical aspects (Thompson et al. 2023). For the first US professional medical examination, the USMLE Step 1 examination, histology is already tested together with cell biology, its modern counterpart.

In the spring of 2020, due to the emerging COVID-19 pandemic, most educational events, including histology instruction, had to shift to an online-only modus. The pre-COVID-19 pandemic state of histology instruction at North American schools already incorporated many online components, and these resources often allowed for a rapid response to the arising emergency situation. Nevertheless, many instructors had to learn the efficient use of lecture recording technologies and had to modify their teaching approach in order to capture students' interest and keep learners engaged (Sweetman 2021). It is still too early to comment on whether North American universities will retain all or some of the COVID-19-mandated changes to histology instruction or whether schools will return to the pre-pandemic teaching modus.

9.3.2 The State of Histology Education in South America

In South America, basic science instruction at the university level is an important part of the education of students in different health science areas such as veterinary medicine (Jiménez et al. 2020), nursing (Muñoz 2018), dentistry (Fernández et al. 2022), and medicine (Mantovani et al. 2019). Basic science subjects, like histology, usually complement other preclinical and clinical curricular components (Araya et al. 2021). In a majority of South American countries most health science degrees require 5–6 years of study (Biondi and Cortese 2020; Camacho et al. 2022).

Histology is seen as one of the basic science topics that are fundamental for the education of health science professionals (Garrido et al. 2021). In South America, histology teaching has been included in higher biomedical education since the second half of the nineteenth century (Torres et al. 2010). However, only during the twentieth century, a separate histology course was included in biomedical curricula together with other biological science subjects (Calado 2019). Presently, histology is offered in the first semesters of health science curricula at most South American universities (Mantovani et al. 2019; Fernández et al. 2022; Rojas et al. 1999). However, there have been experimental efforts to exclude histology from the curricular basic science framework and to teach it during the clinical phase (Parra 2022).

In most South American countries, histology classes are taught in Spanish or Portuguese, and the workload distribution varies from 64 to 128 h, generally scheduled in one or two terms. Usually, these classes include laboratory activities using TM with in-person instruction. Only a few universities offer VM for histology instruction.

In some South American countries, for example, in Ecuador, histology teaching is still based on a traditional approach. The teaching and learning process is often teacher-centered using an expository class strategy that minimizes the involvement of students (Chóez et al. 2018). Even with the introduction of student seminars, the passivity of students in the teaching process remains a problem that sometimes results in an unsatisfactory learning outcome (Chóez et al. 2018). Jiménez and coworker, surveyed the University of Veterinary Medicine and Zootechnique of the National University of Loja (Ecuador) students, who had traditional veterinary histology classes, about their interest in using a flipped classroom model (Jiménez et al. 2020). The results showed that students and professors knew about this type of teaching method and were interested in implementing it for histology teaching. The reported results encourage the use of the flipped classroom strategy for histology education at South American universities (Pavanelo and Lima 2017).

Traditional histology teaching in a course for nursing students at the University Las Americas in Chile was analyzed by Munõz (2018). At this institution, human histology and embryology were taught with slides and images in a didactic lecture style. Subsequently, practical laboratory instruction used TM with glass histology slides followed by a discussion of the micrographic images. However, the limited time for laboratory activities suggested the use of VM to improve students' academic performance. Comparing results from before and after the switch to VM showed higher student performance with VM, indicating that VM is an equivalent or even superior educational tool when compared to TM (Muñoz 2018).

A study of students at the Medical Faculty of the University of Chile, who used digitized images in their histology course, showed that this strategy also supports the learning process and helps them to identify histological structures when using TM (Rojas et al. 1999). An article from the National University of Córdoba in Argentina reports on the use of a free website called ODONTOWEB for histology and embryology teaching (Avila and Samar 2011). This website was specifically designed for dentistry students and contained digital still images from the digestive, respiratory, and urinary systems with a short explanatory text.

A survey about social networks at the Universidad Nacional de Colombia collected opinions about the use of traditional classes and activities in a microscopy laboratory (Parra 2022). This study included teachers who adopted technological tools, such as videos, interactive platforms, and virtual laboratories in their teaching methodology. Students reported that these didactic resources made it easier to remember the information and increased learner motivation. In general, students felt that histology has the potential to contribute to a better understanding of other disciplines and clinical concepts and can support the development of deductive thinking (Parra 2022).

In Brazil, a variety of new educational strategies are used for teaching histology, like delivering didactical material via social media (Vulcani et al. 2020), reviews by student course assistants (Silva and Belo 2012), websites (Nóbrega et al. 2014), and digital books (Silva et al. 2020). The use of digital technologies for histology education at Brazilian public universities is still at an early stage. In fact, most Brazilian universities do not have a dedicated infrastructure for teaching histology and/or histopathology, specifically for the use of VM. Some universities offer histology websites with theoretical concepts, clinical cases, and banks of static images. However, most of these resources only address the identification of important cellular structures without allowing a progressive enlargement and panning of the images that are central features of VM.

A study concerning the use of VM at a Brazilian dental school showed that this technology improved the interaction with students and that the high quality of the virtual slides allowed students to identify histological structures faster (Fonseca et al. 2015). Other authors demonstrated that students in Brazil who were exposed to VM during the COVID-19 pandemic reported subjective improvements in learning efficiency. However, no difference in academic performance was observed when students using VM were compared with students exposed to only TM (dos Santos et al. 2021).

In South America, the assessment of histology learning success is usually based on examinations with MCQs and a practical test with glass slides. Most questions test the recognition of structures and the classification of tissues and organs based on optical light microscope observations. Some South American universities use assessments with open questions and photomicrographs of cells, tissues, and organs.

A study from the National University of the Centre of the Province of Buenos Aires in Argentina reports on the assessment in a veterinary medicine histology course centering mainly on mistakes, however, without discussing them properly (Teruel et al. 2017). The teaching strategy of the course was changed to provide feedback for practical training sessions, and evaluations and a mixed assessment model was adopted that was composed of multiple-choice, semi-structured, and open questions. As a result, the authors noticed a significant increase in student approval and satisfaction with teachers' performances, the teaching material adopted, and the learning environment.

At the Regional Blumenau University in Brazil, the use of deliberate pedagogy, a strategy of repetition and refinement, was used to improve medical students' performance in a morphological microscopy course (Sant'Anna et al. 2022). The authors examined the influence of the microscope technology on students' identification of specific structures. Their data indicated that students taught by the deliberate practice strategy performed better in finding specific structures on histological slides when compared to students who were taught with educational games relating to the theoretical content. The authors concluded that deliberate practice is an effective teaching method for histology (Sant'Anna et al. 2022).

Histology education in South American countries is very variable and has evolved over the years, from the use of TM and students drawing in notebooks to the reading of textbooks and atlases and the incorporation of information and communication technologies, such as VM (Calado 2019). This evolution was accelerated during the global COVID-19 pandemic which interrupted face-to-face educational activities in most countries. COVID-19 forced the adoption of new teaching resources for the teaching of histology in South America, either in a synchronized or an asynchronized format, specifically increasing the use of flipped classroom approaches and VM (dos Santos et al. 2021).

Many of the microscopy platforms and histology resources are in languages other than Spanish or Portuguese with a majority only available in English (Joaquim et al. 2022). Nevertheless, these English language teaching and learning tools were used in many South American countries during the pandemic, and some are still being used today for histology instruction (Vijayan 2021). However, the lack of such resources in the local languages may be one reason why educators often prefer to hold on to TM and old-fashioned teaching strategies for histology and pathology education. After restrictions enforced by the pandemic ended and students returned to classrooms, the use of many teaching adaptations and e-learning tools was unfortunately discontinued, specifically for histology teaching in Brazil. For the Brazilian medical education system, the upcoming challenges of incorporating new technologies are daunting (Carvalho et al. 2020), and this applies specifically to histology and/or histopathology teaching. However, these changes are important for improving the quality of the learning process and education in general (Wanderer et al. 2020).

9.3.3 The State of Histology Education in Europe

Most Universities in Europe offer 5- or 6-year medical degrees to undergraduate students, who usually enter medical school aged 18–20 years. Only a few schools provide 4-year postgraduate degrees, which are similar to the North American medical education system (Campos-Sanchez et al. 2012; Gatumu et al. 2014; Zaletel et al. 2016; Smith and Pawlina 2021; Nikas et al. 2022). Various types of curricular structures are used for medical education, including traditional, systems-based, and problem-based (Quirk and Harden 2017; Smith et al. 2021). To harmonize and ensure the quality of higher education in Europe, the European Higher Education Area (EHEA) was established, including both European Union (EU) and non-EU member countries (Bergan and Matei 2020). However, program details and assessment formats are determined by national agencies that are responsible for quality assurance and accreditation. All courses within each curriculum are associated with a pre-defined number of European Credit Transfer System (ECTS) points, which are transferable between European Universities. To achieve accreditation as a medical provider, a specific number of ETCS points have to be collected by a candidate, for instance, 360 ETCS points for a 6-year medical degree (Patrício et al. 2008; Murtomaa 2009; Zaletel et al. 2016; Panteli and Maier 2021).

Histology is a traditional component of medical, dental, and veterinary education at European Universities, as well as other programs such as biology and sports science (Gatumu et al. 2014; Garcia et al. 2019; Saverino et al. 2022). It is usually taught in the first preclinical years of medical school, either as a stand-alone course or as part of an integrated curriculum (Gatumu et al. 2014; Mione et al. 2016; Moxham et al. 2017; Darici et al. 2021; Nikas et al. 2022). According to data from 13 European universities, histology is usually taught in the first 2 years of medical studies (Moxham et al. 2017). Only at the Sorbonne University in France, histology was taught during the first 3 years of the medical curriculum. A significant heterogeneity exists regarding the histology teaching hours at different universities (Moxham et al. 2017). The total histology instruction hours ranged from 30 (at the University of Padova, Italy) to 180 (at the University of Niš, Serbia), while 6 out of 13 universities schedule more than 100 h devoted to histology instruction in their curriculum. Furthermore, Sander and Golas reported (2013) 164 histology instructional hours at Aarhus University, Denmark, whereas at the Universities of Turku in Finland and of Ghent in Belgium 54 and 75 h, respectively, are dedicated to histology education (Mione et al. 2013; Nivala et al. 2013). Notably, after switching from TM to VM, Ghent University significantly reduced the time for histology teaching by 30% for histology lectures and 75% for laboratory sessions (Mione et al. 2013). In a recent study presenting data from the UK and Ireland, medical school histology teaching on average (with a range of 2–104 h) accounts for 23.6 h, which is significantly less than the 51 h reported for US medical schools (McBride and

Drake 2018; Smith et al. 2021). In summary, whereas many European universities still run stand-alone histology courses with a relatively high number of teaching hours, others have reduced the contact hours for histology instruction in their integrated curricula.

Similar to other parts of the world, histology instruction at European universities is based on a combination of lectures and laboratory sessions. Lectures are usually conducted before laboratory sessions, yet some have successfully applied a flipped classroom approach (Mione et al. 2011; Nikas et al. 2022). As students may not appreciate the clinical importance of histology, many educators integrate clinical examples for motivating learners and for demonstrating the significance of histology in everyday clinical practice (Moxham et al. 2017). Textbooks, journals, websites, applications, videos, podcasts, and gaming strategies such as Kahoot! (Kahoot! A.S.A., Oslo, Norway) are resources often used by European schools for histology instruction (Moxham et al. 2017; Schmidt 2013; Felszeghy et al. 2019; Tauber et al. 2019; Darici et al. 2021; Nikas et al. 2022). Notably, a few instructors still ask their students to draw their microscopic observations (Zaletel et al. 2016), while others include student-led presentations in their histology courses to facilitate active learning (Tauber et al. 2019).

In contrast to many schools in North America, most European schools still offer on-campus histology laboratory sessions, with many making these sessions compulsory for their students (Nikas et al. 2022; Campos-Sanchez et al. 2012; Gatumu et al. 2014; Darici et al. 2021). On-campus laboratory sessions often start with a short lecture-style introduction that is followed by the examination of glass and/or virtual histology slides (Tauber et al. 2019; Zaletel et al. 2016; Nikas et al. 2022). Whereas some programs use a more traditional approach, others apply an independent, student-centered methodology with instructors acting as facilitators (Felszeghy et al. 2019). Interestingly, Campos-Sánchez and coworkers (2012) reported that students at the University of Granada in Spain favored the traditional "reception learning" over an independent, "self-discovery learning" modality. They were more motivated to participate in traditional histology sessions for obtaining important diagnostic, therapeutic, and research skills. The flipped classroom model of instruction has also been successfully used in European histology teaching programs. In the report by Mione and coauthors, medical students independently and asynchronously studied a selected set of histological virtual images before attending a 75-min on-campus session that discussed the self-study material (Mione et al. 2016).

Among European medical schools, there is a significant heterogeneity regarding the use of VM or TM glass slides for histology laboratory instruction. Some educators from affluent countries such as Belgium, Denmark, and Finland reported on the exclusive use of VM in their courses (Mione et al. 2013; Vainer et al. 2017; Felszeghy et al. 2019). Others offered a combination of glass and virtual slides to their learners (Gatumu et al. 2014; Tauber et al. 2019; Nikas et al. 2022). European students often prefer VM over TM, especially for their own independent studies (Boutonnat et al. 2006; Schmidt 2013; Tauber et al. 2021b). A blended approach with both VM and TM is also popular with some educators and students (Tauber et al. 2021a; Nikas et al. 2022). Publications from universities in many European countries have reported on the successful implementation of VM for their histology courses, including Germany (Schmidt et al. 2011; Schnackenberg 2013; Darici et al. 2021), Switzerland (Glatz-Krieger et al. 2006), Denmark (Sander and Golas 2013; Vainer et al. 2017), Finland (Felszeghy et al. 2019), UK (Campbell et al. 2010), and Poland (Filipiak et al. 2011). Most schools use their own virtual slide collections and host them either on dedicated websites (Glatz-Krieger et al. 2006; Schmidt 2013; Darici et al. 2021) or commercial platforms such as the PathXL (Cirdan Imaging Ltd., Lisburn, UK) (Vainer et al. 2017) or Aiforia (Aiforia Technologies Oyj, Helsinki, Finland) (Felszeghy et al. 2019). However, according to Moxham et al. (2017), many European universities still rely mainly or exclusively on TM for histology instruction. In contrast to medical schools, histology laboratory instruction within biology programs often focuses more on hands-on histological techniques (e.g., fixation, staining, and embedding) and the proper use of laboratory equipment such as light microscopes (Garcia et al. 2019). In other fields of education, such as sports science, histology laboratory sessions are not offered at all (Saverino et al. 2022).

There are many differences in how histology knowledge is assessed at different European universities. Some European medical schools exclusively use an MCQ format that includes questions with and without images (Smith et al. 2021; Mione et al. 2013; Darici et al. 2021; Nikas et al. 2022). Other schools also use other assessment methods, such as essays, oral examinations, clinical cases, and traditional or digital spot tests (Moxham et al. 2017; Vainer et al. 2017; Smith et al. 2021). In integrated examinations, histology-related questions usually constitute a small fraction of the whole examination content. Educators have expressed concern that students might focus less on learning the histology content and more on other topics tested in the same examination and that students' academic performance in subsequent medical school classes, specifically pathology, might suffer as a consequence (Nivala et al. 2013; Moxham et al. 2017).

During the recent COVID-19 pandemic, schools worldwide had to shift to online instruction (Cuschieri and Calleja Agius 2020; Longhurst et al. 2020; Torda 2020). Notably, this transition was smoother for histology and pathology than for other subjects, mostly due to the previous adoption of VM which facilitated the immediate conversion to online instruction (Darici et al. 2021). Some European educators published their experiences of teaching histology or pathology during the COVID-19 pandemic, many relying primarily on synchronous sessions that were conducted online (Darici et al. 2021; Ishak et al. 2022; Nikas et al. 2022; Saverino and Zarcone 2022). Darici et al. (2021) at the Westfälische-Wilhelms-University in Münster, Germany utilized their "Virtuelle Mikroskopie" histology software for online histology teaching and reported that students' examination scores remained stable at pre-pandemic levels. Livestreamed microscopic sessions using a camera attached to the light microscope were offered to histology learners by Saverino and Zarcone (2022) at the University of Genoa in Italy. During their sessions, these authors examined histology glass slides at different magnifications and captured static images that were shared with all course participants using Microsoft Teams (Microsoft Corporation, Albuquerque, NM, USA). Their report concludes that online histology laboratory sessions were appreciated, and personal interactions with the teaching staff were missed by their students. Lastly, during the COVID-19 pandemic, Nikas et al. (2022) offered their histology and pathology sessions at the European University of Cyprus synchronously with concurrent recordings. They reported that most histology students had a positive virtual learning experience, although not as favorable as more senior students attending online pathology sessions. Furthermore, female students, advanced degree holders, students attending classes in their home country, learners receiving high examination scores, and students reporting lower self-assessed stress levels also expressed more favorable opinions about online histology instruction. Interestingly, students' Internet connection quality affected online examination scores (Nikas et al. 2022).

In conclusion, many differences exist regarding the curricular structure, instruction, use of TM versus VM, and assessment of histology at different universities within the EHEA. Despite implementing significant curricular changes and emerging technologies, many European universities still keep a traditional education approach and rely on on-site teaching, with a predominant or exclusive use of TM and relatively many hours of histology instruction. However, others have followed a trend similar to North American schools with less instruction hours dedicated to histology education while focusing largely on VM and independent learning.

9.3.4 The State of Histology Education in Africa

Medical education in Africa is diverse and varies significantly in different geographic regions. North African countries bordering the Mediterranean, as well as South Africa, have more advanced university systems and access to modern technologies than the rest of the African continent. Histology education at universities in Ghana is fairly representative of most sub-Saharan countries and will be used as an example in this segment. Medical education in Ghana currently runs both undergraduate and graduate school programs, and students graduate with either a medical (MBChB) or a dental degree (BDS). Based on its colonial past (Monekosso 2014), it used to follow the British educational system, with high school graduates entering an undergraduate medicine program. Recently, many schools have adopted the North American system of medical education, with students holding an undergraduate degree when applying to a graduate medical program (Marsden 2006). The adoption of a graduate-entry program led to a change in the way medical school courses are taught. A modular system of medical education was implemented together with a problem-based (Amoako-Sakyi and Amonoo-Kuofi 2015) and/or team-based learning approaches (Agyei-Nkansah et al. 2020). Over time, more of the undergraduate programs at public and newly established private medical schools are adopting this system of medical education. Medical schools in Ghana offer a 6-year program for undergraduates or a 4-year program for graduate students leading to a MBChB degree (Bachelor of Medicine and Bachelor of Surgery) or BDS (Bachelor of Dental Surgery).

Histology education is not limited to medical students in Ghana. In addition to medical school, histology is also taught to medical laboratory science undergraduate and master's students in anatomy programs. Histology is usually offered as a stand-alone course or as part of a modular course. In most curricula, students are introduced to the four basic tissues before the various organ systems are taught.

As described for histology education, in general, the teaching of histology in African countries involves two phases: didactic lectures and practical laboratory sessions (Fig. 9.4). During the laboratory sessions, students usually use light microscopes and printed micrographs to reinforce the topics learned in the preceding lectures. Four to five students usually share a single microscope and slide set. Many African schools often lack light microscopes and glass

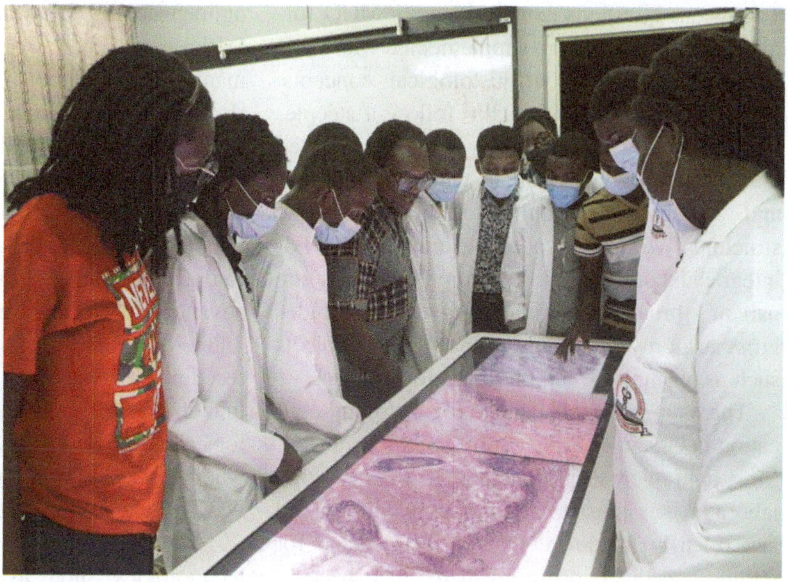

Fig. 9.4 Dr. N. Koney teaching histology to a group of medical students by using virtual histology slides displayed on an Anatomage table (Anatomage, Santa Clara, CA, USA) at the University of Ghana

slide sets and use VM websites and micrographs to supplement histology instruction.

The current generation of students in Africa is technologically inclined and prefers online resources to augment what they learn in class. For histology, students at the University of Ghana use online resources such as the Michigan Histology Website (Hortsch 2023b), the Histology Guide website (Sorensen and Brelje 2023), and the Histology@Yale website (Takizawa 2022). Student usage of these websites is dependent on their lecturer's or peers' recommendations or based on their own Internet searches. Histology learners in Ghana also like to use YouTube resources. During histology laboratory sessions, students often use mobile phone cameras to take pictures of the microscope slides and to label them later on their personal computers or tablets. Most African histology students prefer e-books for their personal studies or alternatively use histology textbooks with an atlas section (histological images) for their laboratory work. E-books are user friendly and have the advantage that students do not need to purchase an expensive paper textbook.

Depending on the school and the curriculum, different assessment formats are used in Ghana to judge students' learning success in histology. In stand-alone histology courses, students are assessed solely on theory and practical skills. The theory questions usually have an MCQ format and occasionally are supplemented with short answer questions about histological concepts. Practical examinations usually follow a steeplechase format where histology slides are placed under microscopes together with micrographs, and students are required to identify these structures within a specific time frame. For integrated courses, the histology questions are part of a broader examination that also includes gross anatomy, embryology, and other subjects such as physiology and pharmacology.

There are many challenges for histology education in Ghana and other sub-Saharan countries. Most light microscopes that are used in histology laboratory sessions at African universities are old and often break down. Due to the demand for training medical doctors, there has been a dramatic increase in the number of students enrolled in African medical programs. Consequently, the number of students per microscope in histology laboratories has increased and diminished students' learning opportunities. The weak economic condition in many African countries and a complex bureaucratic process make it a challenging task to procure new student microscopes. Most developed countries have moved away from the TM and employ VM for histology and pathology education (Hortsch et al. 2023). When VM is used for histology instruction in Ghana and other African countries, it often relies on free histology websites that are offered by non-African institutions (Takizawa 2022; Sorensen and Brelje 2023; Hortsch 2023b). The costs of setting up a local VM histology resource are prohibitively expensive for most African universities, a problem that is further compounded by student learners' limited access to stable and affordable Internet connections. The reduction of instruction time has also impacted histology education in Ghana and other African countries. Due to the modular approach, some histology content is taught in different modules in order to provide the material in a better context. However, if this integration is not well executed, it can also result in confusion for some students.

The COVID-19 pandemic forced many universities on the African continent to switch to online teaching (Dhawan 2020; Evans et al. 2020; Darici et al. 2021; Zalat et al. 2021). However, many African schools lacked the necessary infrastructure and resources, and were unprepared for an exclusive online delivery of educational material. Both teachers and students struggled when adapting to this new way of teaching and learning (Keiller et al. 2023). With many free online resources available, histology educators had an advantage when converting their classes to an online-only format. With the COVID-19 pandemic receding, it remains to be seen whether African universities will retain at least some of the technological advances that were forced upon them or whether they will return to the old and antiquated ways of teaching histology to their students.

North African universities have introduced technologies such as VM for histology and pathology education for a much longer time than sub-Saharan African institutions of higher learning (Ayad and Sicurello 2008; Ayad 2011, 2013; Ayad and Yagi 2012; Bacha et al. 2020). Similarly, many South African universities have adopted VM as a major educational resource for histology and pathology teaching (Banach et al. 2008), with some published reports outlining their use of VM for teaching and learning histology (Mars and McLean 1996; Richards et al. 2000). In South Africa, VM is also augmented by other e-learning approaches, such as histological websites and podcasts (McLean 2001; Beylefeld et al. 2008) and the use of mobile electronic devices such as computer tablets (Lazarus et al. 2017).

Histology education in Africa is currently in a phase of transition. Curricular structures are being continuously revised, and new teaching methods are being implemented. Considering the struggles that are associated with acquiring and maintaining light microscopes and the difficulties of using the traditional ways of teaching, many African universities will be forced to adopt e-learning technologies for teaching histology and other basic science, as well as clinical topics. Therefore, it is possible that within the near future VM will become the standard for histology instruction in all of Africa. Numerous private medical schools have already adopted this practice for histology training and have embraced new e-learning technologies and modern educational approaches.

9.3.5 The State of Histology Education in South Asia

South Asian countries are highly diverse in culture, population, and economic status. This segment will first describe histology education in India. India has the largest number of medical schools in a single country and thus educates the largest number of medical professionals in the world (Supe and Burdick 2006; Sood 2008). Subsequently, this segment will outline how histology is taught in Middle Eastern and Southeast Asian countries.

At most medical schools in India, an independent academic course is devoted to the histology education of medical students (awarding an MBBS degree—Bachelor of Medicine and Bachelor of Surgery) and dental students (BDS—Bachelor of Dental Surgery), while less histology is taught to allied health science students pursuing BSc Nursing, BSc Medical Laboratory Technology (MLT), and BSc Perfusion Technology degrees. These bachelor-level courses are designed for students who have completed their higher secondary schooling (equivalent to a high school diploma in the USA).

In 2019, the curricular structure of MBBS courses was changed when the National Medical Commission (NMC) of India introduced a new Competency Based Medical Education (CBME) curriculum (Basheer 2019). The new curricular structure is more integrated and clinically relevant (Rege 2020). Similar to the older Indian medical curriculum, histology remains a major component of the new CBME curriculum (NMC 2018). Histology is taught during Phase 1 (Preclinical phase), the first year of medical education (NMC 2018). As the new CBME curriculum also provides options for 4-week electives in the preclinical sciences (NMC 2020), histology is revisited by a few more advanced students during later phases of their education.

In an MBBS histology course, a typical histology session begins with a 1-h didactic live lecture followed by a 2 h practical session. At most Indian medical schools, approximately 60 h are scheduled for histology education. The didactic lectures (15 h of general histology and 20 h of systemic histology) teach the theoretical foundation of tissue and organ structure at the cellular level. The practical sessions usually use TM to study histologic glass slides. Students will draw their observations in a histology laboratory record book using pink and purple pencils, representing eosin and hematoxylin-stained structures. These drawings are later evaluated and graded by faculty members. The handling of microscopes and the observation of tissue slides are considered by many Indian medical educators as important for developing professionalism and accountability (Karunakaran et al. 2017).

Histology assessments at Indian schools consist of theoretical and practical examinations. In examinations testing theory knowledge, students are usually asked to write short notes on specific topics (general or systemic histology) to assess their understanding of histology concepts. Practical examinations test the identification of spotter slides (bell ringers) and an exercise of drawing a labelled diagram. The marks for each task are added to form the comprehensive histology assessment.

The simple didactic strategies that are often used by Indian histology educators to improve student learning are repeated slide demonstration, practice of drawing, projection of slide images, and periodic short test assessments (Biswas et al. 2017). However, surveys asking about histology teaching in India have shown some shortcomings of the current methods. Students complain about their difficulties in conceptualizing the 3D histological structures, even after acquiring the necessary theoretical knowledge (Biswas et al. 2017). Though some educators view the practice of drawing as beneficial (Biswas et al. 2017), some surveys indicate that most students think that drawing exercises in histology are of no use (Jayanthi et al. 2014).

The NMC of India requires optical microscopes, microtomes, paraffin embedding baths, and hot plates for flattening sections for histology laboratory instruction at Indian medical schools (NMC 2020). A noticeable feature of laboratory histology instruction in India is the infrequent use of VM for teaching or assessments (Chimmalgi 2018; Chimmalgi and Hortsch 2022; Yohannan et al. 2019; Francis et al. 2022). Though computer-assisted learning is recommended by the NMC for teaching disciplines like pharmacology (NMC 2020), it is not recommended for histology or anatomy education. Nevertheless, sporadic use of VM as depicted in Fig. 9.5 has been reported at some institutes and medical schools in India (Yohannan et al. 2019; Francis et al. 2022; Chimmalgi 2018; Hande et al. 2017; Shastri et al. 2016; Raja 2010; Potaliya et al. 2017). The general hesitancy by

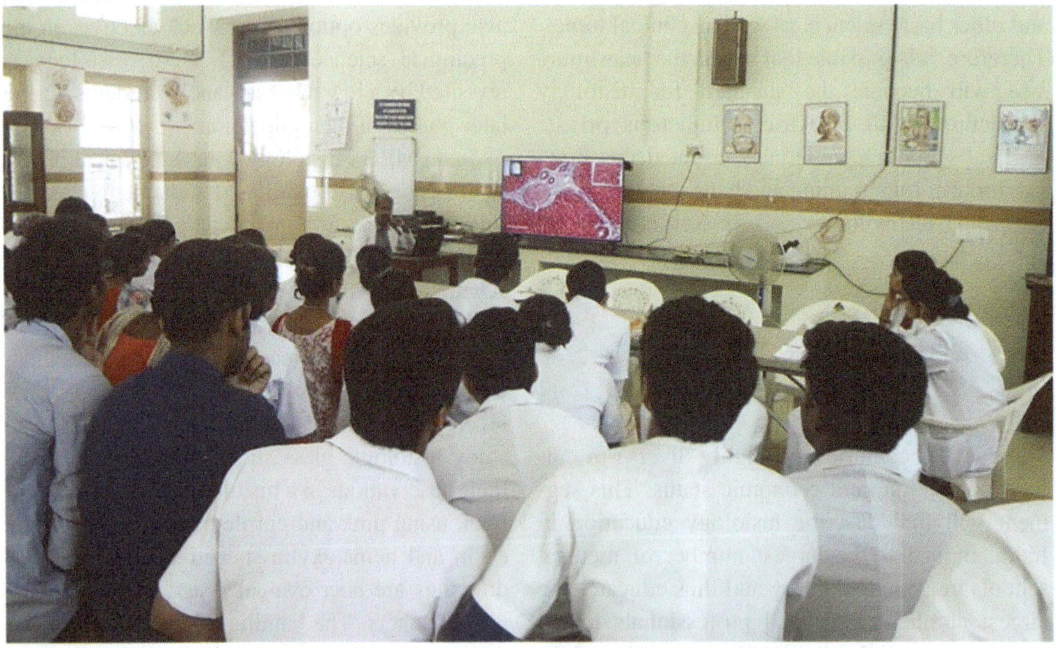

Fig. 9.5 Dr. D. Yohannan using a large television screen to display virtual histology slides to a group of medical students at the Government Medical College in Thiruvananthapuram, Kerala, India (Photo credit: Reproduced with permission from Med Sci Educ 29: 803–17 (2019))

histology educators to adopt VM has multiple reasons, such as an unawareness by medical education policy makers, medical college administrators, and teachers, the current mode of histology assessment, and students' lack of motivation to use new technologies (Yohannan et al. 2019). Instead, glass slides are commonly projected onto a television screen or an overhead projector using a closed-circuit television (CCTV) camera setup. Although didactic lectures are the main modus of teaching, other didactic methods such as flipped classroom (Aristotle et al. 2021) or social media (Maske et al. 2018; Sujatha 2020) have been used in India, usually on an experimental basis.

In general, the e-learning infrastructure at Indian medical colleges is very limited with the exception of a few premiere institutes like the All India Institute of Medical Sciences (Potaliya et al. 2017; Padhi et al. 2021). However, despite often lacking the necessary infrastructure, there is a generally positive perception and awareness of e-learning technologies for medical education in India, especially after the COVID-19 pandemic (Padhi et al. 2021; Singh et al. 2022; Syed et al. 2021). This general trend may lead to further changes and to wider use of VM.

Other countries on the Indian subcontinent generally follow this same pattern for histology education as India. In the late 1990s at an annual research session meeting, several challenges for histology education were identified for developing nations such as Sri Lanka and other South Asian Association for Regional Cooperation (SAARC) countries (Mendis et al. 1999). These include the exponential increase in student numbers, the scarcity of good quality teaching slides, and expenses of maintenance of microscopes and laboratory facilities (Mendis et al. 1999). Similar traditional patterns are seen in Bangladesh (Ahmed et al. 2014) and Nepal (Bhaskhar et al. 2012). A study at a Pakistani university assessed the importance of drawing histology diagrams as a pedagogical exercise and found better academic performance for the drawing student group compared to the non-drawing group (Rafi et al. 2017). This highlights the role of traditional teaching methods in the educational systems of resource-constrained countries. In a survey of Iranian anatomy teachers asking about their familiarity and level of satisfaction with horizontally integrated anatomy courses, 50% of teachers agreed that integration of these courses imparted a better understanding of histology, embryology, and physiology (Dehghan et al. 2012). E-learning methods have also been successfully implemented for histology courses at some Iranian medical schools (Rashidi and Maryam 2012).

The learning environment on the Indian subcontinent differs from countries in the Middle Eastern region. Histology teaching at many Middle East medical schools usually accounts for a fifth of the allotted teaching hours in the anatomy curriculum (Al Mushaiqri 2015). As early as 2010, around 60% of medical colleges in the Gulf Cooperative Council (GCC) countries followed a hybrid Problem-Based Learning (PBL) curriculum, and the majority of the remaining colleges were planning to transition from their traditional curriculum to a hybrid format (Bin Abdulrahman 2008; Hamdy et al. 2010a, b). One PBL teaching strategy that was introduced in 2002 at the College of Medicine and Medical Sciences (CMMS) of the Arabian Gulf University (AGU) in Bahrain is a standardized learning activity called Structured Problem Related Anatomy Demonstrations (SPRAD) (Abu-Hijleh et al. 2005). SPRAD introduces second-, third- and fourth-year medical students to a problem, with diverse anatomical material like skeletal parts, models, prosected wet or plastinated cadaveric specimens, radiographs, and histological slides. This SPRAD model within a PBL program was judged by both students and faculty members to be acceptable, feasible, organized, focused on health problems, and able to provide an improved understanding of anatomy from a basic and applied perspective (Abu-Hijleh et al. 2010). The SPRAD approach also allowed an easy transition to online learning at the AGU during the COVID-19 pandemic (Potu et al. 2022). Many medical schools in Saudi Arabia also successfully introduced a PBL curriculum structure for the preclinical sciences including histology (Telmesani et al. 2011; Al-Drees et al. 2015; Ibrahim et al. 2018).

As a consequence, histology education is more advanced at Middle Eastern colleges and universities when compared with other regions of South Asia. The Qassim University (Alkhamiss 2019) and King Saud University (Alotaibi and ALQahtani 2016) in Saudi Arabia reported that students prefer VM as the instructional technology for histology. Some institutions (e.g., Mohammed Bin Rashid University School of Medicine, United Arab Emirates) successfully use an integrated histopathology curriculum (Lakhtakia 2021) and as a result, mastered the challenges of teaching histology during COVID-19 more easily (Du Plessis et al. 2021). In Israel, a few institutions like the Faculty of Medicine at Haifa (Coleman 2013) and the University of Beer Sheeva (Samueli et al. 2020) have used Whole Slide Imaging (WSI) of their own histological glass slides to create institutional VM slide collections. A recent survey of Universities in Jordan showed that their histology and anatomy teaching remained effective after shifting to an exclusive online platform (Al-Alami et al. 2022).

Histology education in the Southeast Asian Region (SEAR) remains highly diverse. The PBL curriculum at the Faculty of Medicine Universitas Indonesia (FMUI) was modified by adding assignments (Pawitan and Pattiata 2010). Students were asked to draw the cells, structures, and tissues which they observed under the microscope, and these drawings were scored by tutors. These drawing assignments improved the histology test results of participating students (Pawitan and Pattiata 2010). Histology instruction was also integrated with pathological anatomy in an Integrated Practical Class (IPC). This delivered a superior learning outcome when compared with the non-integrated teaching modus (Syaidah et al. 2020). Case-Based Learning was successfully introduced for histology teaching at the Islamic University of Indonesia (Fidianingsih and Wijaya 2015). Schools in Malaysia generally offer conventional histology laboratory instruction to their learners (Aziz et al. 2007). In a 2021 survey about anatomy teaching methods at the Universiti Putra Malaysia, all aspects of their anatomy education environment were generally viewed as positive, except for the practical histology facilities (Jalani et al. 2021). A few years back, a study at the Cyberjaya University College of Medical Sciences, Malaysia, compared a new laboratory manual (outcome-based laboratory manual with photomicrograph images and clinical correlates) with the old one and reported better student scores for the new manual (Ansari et al. 2018). The authors analyzed students' opinions and recommended the use of VM. However, they mentioned that students did not use VM as they considered the traditional method more convenient (Ansari et al. 2018). Learners and teachers at that school were also concerned that VM needs a high degree of technical support and is financially draining and unaffordable (Ansari et al. 2018). However, some schools in the SEAR have reported on the successful installation of VM and WSI. Students at the Penang International Dental School in Malaysia are now using VM (Telang et al. 2016), and this technology is being used in the diagnostic setting at Singapore General Hospital (Cheng et al. 2016). A study at Universiti Sains Malaysia also showed positive results when using VM in the histology laboratory (Simok et al. 2019).

In many countries and regions of Southern Asia, histology education still follows the traditional model with lectures and laboratory sessions using glass slides and light microscopes. However, more schools are now adopting modern didactic strategies and technologies for histology instruction, a process that was accelerated by COVID-19 pandemic restrictions on in-person teaching.

9.3.6 The State of Histology Education in East Asia

The following paragraph surveys histology education in Eastern Asia, with a focus on China, and briefly outlines how histology is taught in South Korea and Japan. In mainland China, students start medical education after graduating from senior high school. This is different from the American and many European medical education systems. In general, Chinese medical schools offer a variety of medical education programs, such as a 5-year, a 5-plus-three integration, and an 8-year program awarding degrees of Bachelor, Master, Doctor of Medicine or PhD in Medicine (Wang 2021; Wu et al. 2021). Medical vocational and veterinary education will not be considered in this overview.

In mainland China, histology is part of the required preclinical education, which is generally completed in the first and/or second and/or third year of a medical curriculum. The timing depends on the specific educational program and school. Traditionally, histology has been taught as a stand-alone course. Recently, the integrated curriculum model has emerged as a new teaching paradigm in China. However, its introduction at schools in China lags behind schools in North America. In 2018, 62% of the surveyed medical schools in China had different types of integrated histology courses (Cheng et al. 2020). Among those courses, 59% of schools integrated histology with other basic medical sciences, 17% with clinical sciences, and 24% with theoretical and practical education (Cheng et al. 2020; Lu et al. 2016; Sherer et al. 2013, 2014). The organ-based integrated curriculum model is providing a deeper level of integration between basic medical courses and/or clinical subjects. In China, some medical schools use this type of integration (Lu et al. 2021; Niu et al. 2015, 2018; Wei et al. 2021; Chen et al. 2005; Huang et al. 2014; Zhu et al. 2020). Such integrated course models are offered to students at Chinese universities for parts or the entire medical program (Lu et al. 2021; Niu et al. 2018; Wei et al. 2021).

Over the past 20 years, the time dedicated to histology education at mainland Chinese medical schools has significantly decreased (Cheng et al. 2020). Nevertheless, it still contains both theoretical and practical sessions. Traditionally, both phases of histology education have been offered in a face-to-face format. However, some experimental online teaching strategies have been tested, specifically flipped classroom and problem-based learning approaches (Cheng et al. 2017; Tian et al. 2014). With the outbreak of COVID-19 pandemic, all schools had to switch to online instruction. A national survey about online teaching formats used in China demonstrated that synchronous live broadcasting was the most popular, and 73% of the medical schools exclusively or partially used this format for theoretical histology sessions (Cheng et al. 2021b). In addition, recorded videos were used by half of the Chinese schools that responded to the survey (Cheng et al. 2021b). These histology lecture videos were offered to students using one of three alternative distribution pathways. The first pathway uses online teaching management platforms, such as Rain Classroom (Tsinghua University, Beijing, China), and Superstar Xuexitong (Superstar Information Technology Development Co., Ltd., Beijing, China) (Zhang et al. 2020). The second pathway uses Massive Open Online Course (MOOC) platforms, such as the Chinese University MOOC (Higher Education Press, Beijing, China), Xuetang Online (Tsinghua University, Beijing, China), and Zhihuishu (Able-Elec Co. Ltd., Shanghai, China). Until July 2022, 43 online courses related to human histology had been uploaded on the above-mentioned MOOC platforms (Cheng et al. 2021b). These courses are freely accessible to students in all of China. The third distribution pathway for lecture videos uses social media platforms, such as QQ and WeChat. A survey by Cheng et al. (2021b) reported that 44% of schools used online teaching management platforms, 24% used Massive Open Online Courses, and 19% used social media platforms as the main distribution technology during the COVID-19 pandemic.

Over the past 20 years, one of the most prominent changes in histology laboratory instruction has been the transition from TM to VM. According to a national survey completed in 2018, 39% of Chinese medical schools exclusively used VM, 12% utilized a combination of TM and VM, and 49% employed only TM (Cheng et al. 2020; Xing et al. 2019). Recently, the COVID-19 pandemic forced many schools that still used TM to switch to VM (Cheng et al. 2021a).

Before 2020, blending online and face-to-face teaching methods were commonly used in China for histology laboratory instruction. During the pandemic, 73% of medical schools implemented online practical sessions for histology. Among these schools, 70% employed synchronous live broadcasting, which is similar to the percentage for online theoretical sessions (Cheng et al. 2021a). However, many histology teachers expressed an interest in switching back to face-to-face laboratory sessions after pandemic restrictions were lifted (Cheng et al. 2021a).

Unlike histology assessment at most North American schools, Chinese medical schools use a variety of assessment formats, including MCQs and short answer questions (Cheng et al. 2017; Sherer et al. 2014; Tian et al. 2014; Li et al. 2021). It should be noted that histology questions relating to clinical aspects have been introduced at some Chinese schools. In the future, more and higher-quality questions of this type are needed to meet the requirements for modern medical education.

In Taiwan, medical education differs from mainland China. Schools of medicine usually offer a 6-year educational program that awards a bachelor's degree (Ministry of Education-Taiwan 2023; Wu et al. 2022). Histology education is generally completed in the second and/or third year of medical school. At National Taiwan University, which is one of the top universities in Taiwan, the histology course has been a stand-alone course with face-to-face teaching as the main instruction format. VM has been used for laboratory teaching, and the assessment includes various question formats (Lee et al. 2020). At National Yung Ming Chiao Tung University, an organ-based integrated curriculum model is used (National Yung Ming Chiao Tung University 2023).

In Japan, medical schools generally offer a 6-year medical education program to high school graduates (Kozu 2006; Suzuki et al. 2008; Teo 2007). In the twenty-first century, the Japanese government has been advocating for the implementation of a national model core curriculum and the common achievement test (CAT) (Kozu 2006; Onishi 2018). In 2005, 89% ($n = 70$) of Japanese medical schools had partially or fully implemented an integrated medical curriculum (Kozu 2006). At the University of Tokyo, histology is currently integrated with other basic sciences (The University of Tokyo 2023); at Kumamoto University, histology is a stand-alone course with VM as the laboratory technology (Kumamoto University 2023). Histology instruction is usually completed in the first or second academic year.

In South Korea, medical schools offer one or both of two medical educational programs; one is a 6-year program for high school graduates, the other is a 4-year program for college graduates. The latter was initiated by the government in 2002 and in 2009 was used by 65% of Korean medical schools (Kim and Kee 2010). In 2017 many Korean medical schools offered a 6-year program (Lee 2017). A graduate-entry program also exists at some schools in China and Japan, however, on a smaller scale (Chou et al. 2012; Kozu 2006). At Seoul National University, histology is completed in the first 2 years of study and VM is used for histology laboratory sessions. During the COVID-19 pandemic, MCQs have been used as the main assessment modality (Kim et al. 2020).

The current state of histology education in Eastern Asia is the result of recent significant changes to the curricular structure, as well as the introduction of student-centered learning and digital technology.

9.3.7 The State of Histology Education in Australia and Aotearoa/New Zealand

Australia and Aotearoa/New Zealand are countries in Oceania in the southern hemisphere with populations of 25.9 million (Australian Bureau of Statistics 2022) and 5.12 million people, respectively (Stats NZ 2022). Australia is ranked 21st and Aotearoa New Zealand is ranked 32nd on the ranking of gross domestic product at purchasing power parity per capita (Worldometer 2022). In Australia, primary and secondary education is from kindergarten to Year 12 (between the UK and the USA), whereas it is from Year 1 to Year 13 in New Zealand. Higher education at universities in Australia and Aotearoa/New Zealand is similar to the UK model with

3–5 year bachelor's degrees, followed by postgraduate studies (master's or doctoral). Students not achieving entry requirements for bachelor's degrees may begin with diploma-level studies, during which they may complete courses that are then credited toward their bachelor's degrees. Additional studies not qualifying for a full degree can also be awarded a certificate. There are 43 universities in Australia and eight universities in Aotearoa/New Zealand, with 22 of these institutions having medical schools.

Histology courses at Australian universities are offered at the undergraduate level (years 1–3) in a variety of different types of studies: chiropractic science (Jonas-Dwyer and Sudweeks 2007), dentistry and oral health (Ariana et al. 2016; Yakin and Linden 2021), exercise science and Chinese medicine (Rich and Guy 2013), laboratory medicine (Caruso 2021), medicine (Loo et al. 1995; Kumar et al. 2006; Kumar et al. 2004; Janssen et al. 2015; McLean 2018), science (Loo et al. 1995), and veterinary science (Mills et al. 2007) programs. While this information has been published by researchers at universities in New South Wales, Queensland, South Australia, Victoria, and Western Australia, there is little or no published information available about histology education at institutions in the Australia Capital Territory, Northern Territory, Tasmania, and in Aotearoa/New Zealand. A search of these institutions' websites indicates that histology is taught in more institutions across Australia and Aotearoa/New Zealand than is reflected in the current literature.

From the available literature, it appeared that almost 60% of histology teaching follows an integrated format (Ariana et al. 2016; Kumar et al. 2004, 2006; McLean 2018), while approximately 40% were offered as stand-alone histology courses (Caruso 2021; Loo et al. 1995; Mills et al. 2007). As histology provides the anatomical basis for the clinical practice of pathology, many institutions in Australia integrate histology with other subjects, specifically pathology. Like many other medical programs around the world, the late 1990s and early 2000s saw many Australian and Aotearoa/New Zealand medical programs switching from individual course-based curricula to integrated problem-based learning curricula (Craig et al. 2010; Trautman et al. 2019). This switch was accompanied by a concurrent reduction in anatomical sciences teaching hours (Craig et al. 2010; Trautman et al. 2019). More recently, driven by the Australian Medical Council (AMC), there has been a shift in medical education in Australia and Aotearoa/New Zealand from bachelor-level training (e.g., Bachelor of Medicine, Bachelor of Surgery; MBBS) to Doctorate level programs (e.g., Doctor of Medicine; MD). Sixteen of the twenty-two medical schools in Australia now offer MD or equivalent degrees (Australian Medical Council 2022). How this has affected histology education in these programs has not been sufficiently investigated.

Histology instruction is delivered in a mixed (traditional in-classroom combined with online instruction) format at 40% of Australian institutions (Ariana et al. 2016; Caruso 2021; Jonas-Dwyer and Sudweeks 2007) and in an online-only mode at 60% of Australian universities (Kumar et al. 2006; Loo et al. 1995; McLean 2018; Mills et al. 2007; Yakin and Linden 2021). Mixed delivery of histology education in Australia usually combines students using TM and/or VM for laboratory work. E-learning resources that are used for histology education at Australian schools include Zoom tutorials (Caruso 2021), Instagram posts (Caruso 2021), the "They Know Anatomy" online gaming platform (Janssen et al. 2015), the Iowa Virtual Slide Box accessed using virtual slide server software (Kumar et al. 2006), the Fabric of Life program (Loo et al. 1995), TurningPoint technology (McLean 2018), a virtual veterinary slide box (Mills et al. 2007), the EnactEd e-learning tool (Rich and Guy 2013), and the Smart Sparrow adaptive learning platform (Yakin and Linden 2021). Assessments are often in the form of online quizzes that are offered using a learning management system (Caruso 2021; Ariana et al. 2016). Based on the published literature, the COVID-19 pandemic has accelerated the adoption of digital technologies for histology education at Australian universities (Pather et al. 2020).

Given the accessibility of VM, it is not surprising that many Australian students have more positive views about VM than about TM for histology laboratory instruction (Yakin and Linden 2021; Caruso 2021; Kumar et al. 2004; McLean 2018; Mills et al. 2007; Ariana et al. 2016; Janssen et al. 2015). The effect of VM on academic performance is less clear with two studies demonstrating moderate to large positive effect sizes (Caruso 2021; Ariana et al. 2016) and two studies demonstrating no significant impact on students' histology learning outcomes (Kumar et al. 2004; McLean 2018). It should be noted that all studies examining educational interventions were single-institution cohort studies (Yakin and Linden 2021; Caruso 2021; Janssen et al. 2015; Kumar et al. 2004; 2006; McLean 2018; Mills et al. 2007; Rich and Guy 2013; Ariana et al. 2016; Jonas-Dwyer and Sudweeks 2007), which limits the generalizability of these results.

Histology education at Australian universities appears to be closely parallel to that in North America, Europe, and many East Asia countries with regard to the adoption and use of technology-enhanced teaching and learning.

9.4 The Global Status of Professional Histology Education and Outlook on Future Developments

The way histology has been taught for a long time to biomedical students is based on microscope technologies and is rooted in its past. For about 150 years the two components of histology education remained fairly static, using lectures for the transfer of histology facts and laboratory sessions with light microscopes and glass slides for the development and training of recognition skills. At many schools worldwide this arrangement is still in place today. However, starting in the most affluent countries and regions of the globe, the last 30 years have seen significant changes in histology education.

These recent transformations in how histology is taught in different geographic regions usually follow a common theme. Most importantly they involve VM displacing the use of TM and often include the curricular integration of histology teaching into a general basic science context (Sherer et al. 2014; Eisenstein et al. 2014; Brauer and Ferguson 2015; Jonassen et al. 2016; McBride and Drake 2018; McLean 2018; Smith et al. 2021; Hortsch 2023a; Schoenherr et al. 2022). However, these changes are happening at different speeds in different countries, and their implementation often appears to be governed by economic factors with countries in North America, Europe, and East Asia, as well as Australia leading in the adoption of novel educational technologies and strategies. Resistance to change by more senior faculty members, who are reluctant to give up traditional teaching modalities, is often an additional impediment for updating histology education (Eng-Tat et al. 2022; Yohannan et al. 2019). Local barriers for adapting new technologies and teaching approaches also include antiquated university and curricular structures.

Overall, there is a common trajectory in how histology education is changing globally. Primary is the adaptation of VM as the main technology for histology laboratory instruction, often in parallel with the use of VM for the teaching and clinical application of pathology (Hortsch et al. 2023). Other educational technologies like histology podcasts, social media, mobile apps, websites, lecture recordings, and more, are also increasingly used, but more at the local level (Hortsch 2016, 2023b; Essig et al. 2020; Maske et al. 2018; Beylefeld et al. 2008; Rosenberg et al. 2006; Ostrin and Duschenkov 2016; Chimmalgi and Hortsch 2022). This supports a more general development toward increased online learning, a movement that was greatly accelerated during the COVID-19 pandemic (Darici et al. 2021; Nikas et al. 2022; Saverino et al. 2022; Hortsch 2023b). The rising use of technology in histology classrooms also helps teachers who are facing increased numbers of students entering professional schools, especially in developing countries.

As mentioned, another development in histology education that is happening at the global scale is the integration of histology into general basic and clinical curricular structures. Fewer universities teach histology as a stand-alone course, but rather coordinate it with other basic science topics and sometimes with clinical instruction (McBride and Prayson 2008; Drake et al. 2009; Eng-Tat et al. 2022). This allows for the structural description of tissues and organs as provided by histology supporting the understanding of their functional and pathological aspects. It can be expected that an integrated curricular structure will remain the predominant form of histology education for medical, dental, and veterinary students and that a stand-alone course structure may only be used for teaching the subject to undergraduate and PhD students.

The major threat to histology education worldwide is the reduction of instructional time as reported in several of the above sections (Hortsch 2023a). Professional schools are challenged to accelerate the transition of their students from the preclinical to the clinical phase of their education resulting in less time being appropriated for basic science instruction. As outlined in this chapter, this development is happening for histology instruction at schools on all continents. From 1967 to 2017, the average scheduled time for histology instruction at North American medical schools decreased by 61.9% (Hortsch 2023a). Interestingly, the two components of histology teaching were unequally affected with time for lectures exhibiting only a 35.6% decrease, whereas laboratory histology instruction time experienced a 75.3% reduction (Hortsch 2023a). More recently, some medical schools have either shifted histology laboratory sessions completely online or have abolished this modus of histology instruction altogether (Barbeau et al. 2013; Gadbury-Amyot et al. 2013; Thompson and Lowrie 2017; Yen et al. 2014; Daniel et al. 2020, 2021). Whereas some authors report no drop in students' learning outcomes (Barbeau et al. 2013; Gadbury-Amyot et al. 2013; Thompson and Lowrie 2017), in at least one example, this has resulted in a significant decrease in students' histology knowledge (Gribbin et al. 2022). With increasing pressure to shorten the time for the basic science education of biomedical students, this development will likely continue at the global level. However, such an approach has several significant consequences. The abolishment of the histology laboratory component would reduce the discipline to a purely descriptive treatment of cell and organ structure and the learning of biological facts. Missing would be the interpretation of micrographic images which encourages a deeper understanding of tissue structure and its relationship to organ function (Hortsch 2023a). In addition, histology education would no longer teach higher levels of analytical thinking and learning (Zaidi et al. 2017). The ongoing reduction of instruction time for histology laboratory sessions would also remove an opportunity for evidence- and team-based learning, two very popular educational approaches for active learning (Lallier 2014; Ettarh 2016; Van Sligtenhorst and Bick 2011).

Another open question about the future of histology education is its relevance and importance for learners who are preparing to work in different biomedical professions. The published literature indicates that basic biomedical knowledge is fundamental for practicing clinicians and also supports clinical reasoning (Woods 2007; Woods et al. 2007). However, except for a few specialists, like pathologists, most physicians, veterinarians, or dentists will not require histological image recognition skills in their daily clinical work. However, as many diseases and pathological conditions are cellular in nature and as more cell-based therapeutic interventions, like stem cell therapy, have an important place in modern medicine, a foundational knowledge of normal cell structure should be highly desirable for most biomedical professions, an opinion shared by many of today's biomedical students (Shaw and Friedman 2012; Zaletel et al. 2016; Carneiro et al. 2023). As a consequence, histology should retain a central role in most professional biomedical science curricula. However, histology educators will have to consider their educational offering to various groups of students, specifically what is taught to which type of learner and which teaching and learning resources and strategies should be offered to specific student groups. With the constant expansion of scientific knowledge about cells, their functional aspects, and their interactions, a differential list of histology learning objectives might be a good first step in delineating what facts and knowledge about cells are desirable for various biomedical careers. Several such lists have recently been published for medical and dental students (Cui and Moxham 2021; Das et al. 2019; Moxham et al. 2018).

Histology education will need to adapt to the needs of future learners, as well as to tomorrow's health care providers. However, the professional demands that these professionals will be facing are uncertain and remain a topic of intense discussion (Woolliscroft 2019). Whatever the future holds, it can safely be assumed that changes to histology education will continue, and very likely, histology educators will be asked to teach more while using less time. This will put increasing demands on educators, the use of modern educational technologies, and novel didactic strategies for histology instruction. Therefore, a continuous careful reevaluation of the role of histology in the education of biomedical professionals remains an urgent mandate.

References

Abali EE, Osheroff N, Buxbaum E, Niederhoffer EC, Symes K, Sanders M (2014) Evolving role of the basic science course director in an integrated curriculum. Med Sci Educ 24:349–351

Abu-Hijleh MF, Chakravarty M, Al-Shboul Q, Latif NA, Osman M, Bandaranayake R, Ganguly PK (2005) Structured problem-related anatomy demonstrations: making order of random teaching events. Teach Learn Med 17:69–73

Abu-Hijleh M, Chakravarty M, Latif N, Osman M, Salem A, Fadel R, Dharap A (2010) The place of anatomy in medical education: guide supplement 41.2-practical application. Med Teach 32:604–606

Agyei-Nkansah A, Adjei P, Torpey K (2020) COVID-19 and medical education: an opportunity to build back better. Ghana Med J 54:113–116

Ahmed R, Shamim KM, Talukder MHK, Alam K, Nargis T (2014) Current practice in the histology practical classes of anatomy in the undergraduate medical education in 15 Bangladesh. Bangladesh J Med Educ 5: 15–19

Al Mushaiqri M (2015) The status of anatomy teaching in the medical schools of the Gulf cooperation council countries: an exploratory study. Ph.D., The University of Western Australia

Al-Alami ZM, Adwan SW, Alsous M (2022) Remote learning during Covid-19 lockdown: a study on anatomy and histology education for pharmacy students in Jordan. Anat Sci Educ 15:249–260

Al-Drees AA, Khalil MS, Irshad M, Abdulghani HM (2015) Students' perception towards the problem based learning tutorial session in a system-based hybrid curriculum. Saudi Med J 36:341–348

Alkhamiss AS (2019) Impact of virtual pathology teaching method on the education of medical students at Qassim University - Saudi Arabia. Jouf Univ Med J 6:35–44

Alotaibi O, Alqahtani D (2016) Measuring dental students' preference: a comparison of light microscopy and virtual microscopy as teaching tools in oral histology and pathology. Saudi Dent J 28:169–173

Amoako-Sakyi D, Amonoo-Kuofi H (2015) Problem-based learning in resource-poor settings: lessons from a medical school in Ghana. BMC Med Educ 15:221

Anonymus (1875) The teaching of histology. Br Foreign Medico-Chirurgical Rev 56:327–339

Ansari RM, Latiff AA, Abmanan N (2018) Impact of outcome-based histology laboratory manual on the performance of year I MBBS students. Educ Med J 10:23–29

Araya MS, Baños JE, Pérez J, Moyano E (2021) Generic competencies: a review of dental curriculum programs in Chile [in Spanish]. Int J Odontostomat 15:928–937

Ariana A, Amin M, Pakneshan S, Dolan-Evans E, Lam AK (2016) Integration of traditional and E-learning methods to improve learning outcomes for dental students in histopathology. J Dent Educ 80:1140–1148

Aristotle S, Subramanian S, Jayakumar S (2021) Effectiveness of flipped classroom model in teaching histology for first-year MBBS students based on competency-based blended learning: an interventional study. J Educ Health Promot 10:152

Australian Bureau OF Statistics (2022) Population: Census [Online]. https://www.abs.gov.au/statistics/people/population/population-census/2021. Accessed 8 Nov 2022

Australian Medical Council (2022) "Accredited Medical Schools" [Online]. Australia: AMC.org.au. https://www.amc.org.au/accreditation-and-recognition/assessment-accreditation-primary-medical-programs/accredited-medical-schools/. Accessed 7 Nov 2022

Avila RE, Samar ME (2011) Virtual histology project: ODONTOWEB [in Spanish]. Int J Odontostomat 5: 13–22

Ayad E (2011) Virtual telepathology in Egypt, applications of WSI in Cairo University. Diagn Pathol 6(Suppl 1):S1

Ayad E (2013) E-education for medical students using WSI in Egypt. Diagn Pathol 8:S39

Ayad E, Sicurello F (2008) Telepathology in emerging countries pilot project between Italy and Egypt. Diagn Pathol 3(Suppl 1):S2

Ayad E, Yagi Y (2012) Virtual microscopy beyond the pyramids, applications of WSI in Cairo University for E-education & telepathology. Anal Cell Pathol 35:93–95

Aziz N, Asari MA, Aziz RN (2007) Teaching of anatomy in an integrated teaching based curriculum at the university science Malaysia. Pak J Med Health Sci 1:131–133

Bacha D, Ferjaoui W, Charfi L, Rejaibi S, Gharbi L, Ben Slama S, Nijim L, Lahmar A (2020) The interest of virtual microscopy as a means of simulation learning in pathological anatomy and cytology. Oncol Radiother 14:23–29

Balemans MC, Kooloos JG, Donders AR, Van Der Zee CE (2016) Actual drawing of histological images improves knowledge retention. Anat Sci Educ 9:60–70

Banach L, Stepien A, Schneider J, Wichrzycka-Lancaster E (2008) Dynamic active telepathology over National Health Laboratory service network, South Africa: feasibility study using Nikon Coolscope. Diagn Pathol 3 (Suppl 1):S3

Barbeau ML, Johnson M, Gibson C, Rogers KA (2013) The development and assessment of an online microscopic anatomy laboratory course. Anat Sci Educ 6: 246–256

Basheer A (2019) Competency-based medical education in India: are we ready? J Curr Res Sci Med 5:1–3

Bechtel W (2006) Discovering cell mechanisms: the creation of modern cell biology. Cambridge University Press, New York

Bergan S, Matei L (2020) The future of the Bologna process and the European higher education area: new perspectives on a recurring topic. In: Curaj A, Deca L, Pricopie R (eds) European higher education area: challenges for a new decade. Cham, Springer Nature

Beylefeld AA, Hugo AP, Geyer HJ (2008) More learning and less teaching? Students' perceptions of a histology podcast. South Afr J High Educ 22:948–956

Bhaskhar V, Malhotra V, Shankar R, Adhlakha A, Hirak JG (2012) Students perception to anatomy education in medical curriculum. Int J Curr Res Rev 4:99–105

Bichat X (1800) Traité des membranes en général et de diverses membranes en particulier. Richard, Caille, Ravier, Paris

Bin Abdulrahman KA (2008) The current status of medical education in the Gulf cooperation council countries. Ann Saudi Med 28:83–88

Biondi AM, Cortese SG (2020) Assessment of the job characteristics of the final year students of the Faculty of Dentistry, University of Buenos Aires [in Spanish]. Rev FAC Odontol Univ Buenos Aires 35:67–71

Biswas S, Sharma S, Chakraborty S (2017) Students' perception of present teaching method of histology—a study from Eastern Part of India. Natl J Integr Res Med 8:61–65

Bloodgood RA (2012) Active learning: a small group histology laboratory exercise in a whole class setting utilizing virtual slides and peer education. Anat Sci Educ 5:367–373

Bohm AA, Von Davidoff M, Huber GC (1904) A textbook of histology including microscopic technic. Saunders & Company, Philadelphia, W.B

Bolender DL, Ettarh R, Jerrett DP, Laherty RF (2013) Curriculum integration = course disintegration: what does this mean for anatomy? Anat Sci Educ 6:205–208

Boutonnat J, Paulin C, Faure C, Colle PE, Ronot X, Seigneurin D (2006) A pilot study in two French medical schools for teaching histology using virtual microscopy. Morphologie 90:21–25

Brauer DG, Ferguson KJ (2015) The integrated curriculum in medical education: AMEE guide no. 96. Med Teach 37:312–322

Burk DT, Lee LM, Lambert HW (2013) Embryology and histology education in north American dental schools: the basic science survey series. J Dent Educ 77:744–756

Calado AM (2019) História do Ensino de Histologia [in Portuguese]. História da Ciência e Ensino 20:455–466

Camacho RP, Sandoval-Monzón R, Yupton LV, Ruiz-García L (2022) Sobreduración y situación académica de los ingresantes a la Escuela Profesional de Medicina Veterinaria de la Universidad Nacional Mayor de San Marcos en el periodo 2004-2013 [in Spanish]. Rev Inv Vet Perú 33:e21300

Campbell G, Demetriou LA, Arnett TR (2010) Virtual histology in the classroom and beyond. Med Educ 44:1124–1125

Campos-Sanchez A, Martin-Piedra MA, Carriel V, Gonzalez-Andrades M, Garzon I, Sanchez-Quevedo MC, Alaminos M (2012) Reception learning and self-discovery learning in histology: students' perceptions and their implications for assessing the effectiveness of different learning modalities. Anat Sci Educ 5:273–280

Cardall S, Krupat E, Ulrich M (2008) Live lecture versus video-recorded lecture: are students voting with their feet? Acad Med 83:1174–1178

Carneiro BD, Pozza DH, TAVARES, I. (2023) Perceptions of medical students towards the role of histology and embryology during curricular review. BMC Med Educ 23:74

Caruso MC (2021) Virtual microscopy and other technologies for teaching histology during Covid-19. Anat Sci Educ 14:19–21

Carvalho VO, Conceicao LSR, Gois MB, JR. (2020) COVID-19 pandemic: beyond medical education in Brazil. J Card Surg 35:1170–1171

Chapman JA (2023) Chapman histology [online]. YouTube. https://www.youtube.com/channel/UCvSvCkHjCbHn8aGd6OPno_A. Accessed 5 Feb 2023

Chapman JA, Lee LMJ, Swailes NT (2020) From scope to screen: the evolution of histology education. Adv Exp Med Biol 1260:75–107

Chen JQ, Xia Q, Zhou R, Yang YJ, Wang JM, Cai ZN, Zhu X, Lu Y, Fu Q, Luo JH (2005) Integrated teaching practice of basic medical sciences in Zhejiang University School of Medicine in China. FASEB J 19: A1354–A1355

Cheng CL, Azhar R, Sng SH, Chua YQ, Hwang JS, Chin JP, Seah WK, Loke JC, Ang RH, Tan PH (2016) Enabling digital pathology in the diagnostic setting: navigating through the implementation journey in an academic medical centre. J Clin Pathol 69:784–792

Cheng X, Ka Ho Lee K, Chang EY, Yang X (2017) The "flipped classroom" approach: stimulating positive learning attitudes and improving mastery of histology among medical students. Anat Sci Educ 10:317–327

Cheng X, Chan LK, Li H, Yang X (2020) Histology and embryology education in China: the current situation and changes over the past 20 years. Anat Sci Educ 13: 759–768

Cheng X, Chan LK, Cai H, Zhou D, Yang X (2021a) Adaptions and perceptions on histology and embryology teaching practice in China during the Covid-19 pandemic. Transl Res Anat 24:1–10

Cheng X, Chan LK, Pan SQ, Cai H, Li YQ, Yang X (2021b) Gross anatomy education in China during the Covid-19 pandemic: a national survey. Anat Sci Educ 14:8–18

Chimmalgi M (2018) Off-line virtual microscopy in teaching histology to the undergraduate medical students: do the benefits correlate with the learning style preferences? J Anat Soc India 67:186–192

Chimmalgi M, Hortsch M (2022) Teaching histology using self-directed learning modules (SDLMs) in a blended approach. Med Sci Educ 32:1455–1464

Chóez OM, González JCP, Rey MMLRD (2018) Knowledge of histology with relevance in the practice of the medicine doctor in Ecuador [in Spanish]. Revista pedagógica de la Universidad de Cienfuegos 14

Chou JY, Chiu CH, Lai E, Tsai D, Tzeng CR (2012) Medical education in Taiwan. Med Teach 34:187–191

Cogdell B, Torsney B, Stewart K, Smith RA (2012) Technological and traditional drawing approaches encourage active engagement in histology classes for science undergraduates. Biosci Educ 19:1–15

Coleman R (2013) The advantages of virtual microscopy for teaching histology. Ital J Anat Embryol 118:58

Cotter JR (2001) Laboratory instruction in histology at the University at Buffalo: recent replacement of microscope exercises with computer applications. Anat Rec 265:212–221

Cracolici V, Judd R, Golden D, Cipriani NA (2019) Art as a learning tool: medical student perspectives on implementing visual art into histology education. Cureus 11:e5207

Craig S, Tait N, Boers D, Mcandrew D (2010) Review of anatomy education in Australian and New Zealand medical schools. ANZ J Surg 80:212–216

Cui D, Moxham BJ (2021) A core syllabus for histology within the medical curriculum-the cell and basic tissues. Clin Anat 34:483–495

Cuschieri S, Calleja Agius J (2020) Spotlight on the shift to remote anatomical teaching during Covid-19 pandemic: perspectives and experiences from the University of Malta. Anat Sci Educ 13:671–679

Daniel M, Monrad SU, Weir S, Kolars JC, Mangrulkar RS (2020) University of Michigan Medical School. Acad Med 95:S249–S253

Daniel M, Morrison G, Hauer KE, Pock A, Seibert C, Amiel J, Poag M, Ismail N, Dalrymple JL, Esposito K, Pettepher C, Santen SA (2021) Strategies from 11 U.S. medical schools for integrating basic science into core clerkships. Acad Med 96:1125–1130

Darici D, Reissner C, Brockhaus J, MISSLER, M. (2021) Implementation of a fully digital histology course in the anatomical teaching curriculum during COVID-19 pandemic. Ann Anat 236:151718

Das M, Ettarh R, Lowrie DJ Jr, Rengasamy P, Lee LMJ, Williams JM, Guttmann GD (2019) A guide to competencies, educational goals, and learning objectives for teaching medical histology in an undergraduate medical education setting. Med Sci Educ 29: 523–534

De Neys W (2006) Dual processing in reasoning - two systems but one reasoner. Psychol Sci 17:428–433

Dehghan M, Anvari M, Abbasi A (2012) Attitudes of anatomy faculties in medical schools toward horizontal integration of courses on anatomical sciences [in Persian]. Iran J Med Educ 12:647–655

Dhawan S (2020) Online learning: A panacea in the time of COVID-19 crisis. J Educ Technol Syst 49:5–22

Dickerson J, Kubasko D (2007) Digital microscopes: enhancing collaboration and engagement in science classrooms with information technologies. Contemp Issues Technol Teach Educ 7:279–292

Dos Santos FS, Osako MK, Perdona GDC, Alves MG, Sales KU (2021) Virtual microscopy as a learning tool in Brazilian medical education. Anat Sci Educ 14:408–416

Drake RL, Mcbride JM, Lachman N, Pawlina W (2009) Medical education in the anatomical sciences: the winds of change continue to blow. Anat Sci Educ 2: 253–259

Drake RL, Mcbride JM, Pawlina W (2014) An update on the status of anatomical sciences education in United States medical schools. Anat Sci Educ 7:321–325

Drees C, Ghebremedhin E, Hansen M (2020) Development of an interactive e-learning software "Histologie für Mediziner" for medical histology courses and its overall impact on learning outcomes and motivation. GMS J Med Educ 37:Doc35

Du Plessis SS, Otaki F, Zaher S, Zary N, Inuwa I, Lakhtakia R (2021) Taking a leap of faith: A study of abruptly transitioning an undergraduate medical education program to distance-learning owing to the COVID-19 pandemic. JMIR Med Educ 7:e27010

Eisenstein A, Vaisman L, Johnston-Cox H, Gallan A, Shaffer K, Vaughan D, O'hara C, Joseph L (2014) Integration of basic science and clinical medicine: the innovative approach of the cadaver biopsy project at the Boston University School of Medicine. Acad Med 89:50–53

Eng-Tat A, Liu P, R J, M C, R LS, Lian WQD, Huppertz B, Khamuani M, Minarcik JR (2022) The challenges of histology education. Clin Anat 36:308

Essig J, Watts M, Beck Dallaghan GL, Gilliland KO (2020) InstaHisto: utilizing Instagram as a medium for disseminating visual educational resources. Med Sci Edu 30:1035–1042

Ettarh R (2016) A practical hybrid model of application, integration, and competencies at interactive table conferences in histology (ITCH). Anat Sci Educ 9: 286–294

Evans JS (2003) In two minds: dual-process accounts of reasoning. Trends Cogn Sci 7:454–459

Evans JS, Stanovich KE (2013) Dual-process theories of higher cognition: advancing the debate. Perspect Psychol Sci 8:223–241

Evans DJR, Bay BH, Wilson TD, Smith CF, Lachman N, Pawlina W (2020) Going virtual to support anatomy education: a STOPGAP in the midst of the Covid-19 pandemic. Anat Sci Educ 13:279–283

Fales-Williams A (2016) From history to histology: the case correlation assignment. Med Sci Educ 26:21–23

Felszeghy S, Pasonen-Seppanen S, Koskela A, Nieminen P, Harkonen K, Paldanius KMA, Gabbouj S, Ketola K, Hiltunen M, Lundin M, Haapaniemi T, Sointu E, Bauman EB, Gilbert GE, Morton D, Mahonen A (2019) Using online game-based platforms to improve student performance and engagement in histology teaching. BMC Med Educ 19: 273

Fernández JE, Lucero R, Plavnik L, Fontana S, Méndez E, Gómez RMA, Hernández LF, Samar ME (2022) La virtualización de la enseñanza de la histología y embriología durante la pandemia COVID-19 [in Spanish]. Rev Fac Odont 32:32–38

Fidianingsih I, Wijaya DP (2015) Case-based learning and simple virtual microscopic slides in teaching histology to first year medical students. 12th Asia Pacific Medical Education Conference (APMEC), 2015 Singapore. National University of Singapore

Filipiak K, Malinska A, Krupa D, Zabel M (2011) Innovative methods of archiving, presentation and providing access to histological sections. Adv Cell Biol 3:41–54

Fonseca FP, Santos-Silva AR, Lopes MA, Almeida OP, Vargas PA (2015) Transition from glass to digital slide microscopy in the teaching of oral pathology in a Brazilian dental school. Med Oral Patol Oral Cir Bucal 20:e17–e22

Francis DV, Charles AS, Jacob TM, Ruban A, Premkumar PS, Rabi S (2022) Virtual microscopy as a teaching–learning tool for histology in a competency-based medical curriculum. Med J Armed Forces India (in press)

Gadbury-Amyot CC, Singh AH, Overman PR (2013) Teaching with technology: learning outcomes for a combined dental and dental hygiene online hybrid oral histology course. J Dent Educ 77:732–743

Garcia M, Victory N, Navarro-Sempere A, Segovia Y (2019) Students' views on difficulties in learning histology. Anat Sci Educ 12:541–549

Garrido F, Barriga A, Leiva-Ordóñez AM, Inzunza O (2021) Repensando el proceso de enseñanza-aprendizaje de la morfología en el mundo post-pandemia [in Spanish]. Revista de Ciencias Médicas 46:60–65

Gartner LP (2003) Anatomical sciences in the allopathic medical school curriculum in the United States between 1967-2001. Clin Anat 16:434–439

Gartner LP, Lee LMJ (2022) Gartner & Hiatt's atlas and text of histology. LWW

Gatumu MK, Macmillan FM, Langton PD, Headley PM, Harris JR (2014) Evaluation of usage of virtual microscopy for the study of histology in the medical, dental, and veterinary undergraduate programs of a UK University. Anat Sci Educ 7:389–398

Gilliland KO (2017) The flipped classroom and learning analytics in histology. Med Sci Educ 27:9–13

Glatz-Krieger K, Spornitz U, Spatz A, Mihatsch MJ, Glatz D (2006) Factors to keep in mind when introducing virtual microscopy. Virchows Arch 448:248–255

Goldberg HR, Dintzis R (2007) The positive impact of team-based virtual microscopy on student learning in physiology and histology. Adv Physiol Educ 31:261–265

Gorelick N, Hancher M, Dixon M, Ilyushchenko S, Thau D, Moore R (2017) Google earth engine: planetary-scale geospatial analysis for everyone. Remote Sens Environ 202:18–27

Gribbin W, Wilson EA, Mctaggart S, Hortsch M (2022) Histology education in an integrated, time-restricted medical curriculum: academic outcomes and students' study adaptations. Anat Sci Educ 15:671–684

Griffith JW (1864) An elementary text-book of the microscope; including a description of the methods of preparing and mounting objects, etc. John Van Voorst, Paternoster Row, London

Gu J, Ogilvie RW (2005) Virtual microscopy and virtual slides in teaching, diagnosis, and research. Taylor & Francis, Boca Raton

Haguenau F, Hawkes PW, Hutchison JL, Satiat-Jeunemaitre B, Simon GT, Williams DB (2003) Key events in the history of electron microscopy. Microsc Microanal 9:96–138

Hamdy H, Telmesani AW, Al Wardy N, Abdel-Khalek N, Carruthers G, Hassan F, Kassab S, Abu-Hijleh M, Al-Roomi K, O'malley K, El Din Ahmed MG, Raj GA, Rao GM, Sheikh K (2010a) Undergraduate medical education in the Gulf Cooperation Council: a multi-countries study (part 1). Med Teach 32:219–224

Hamdy H, Telmesani AW, Al Wardy NA, Abdel-Khalek-N, Carruthers G, Hassan F, Kassab S, Abu-Hijleh M, Al-Roomi K, O'malley K, El Din Ahmed MG, Raj GA, Rao GM, Sheikh J (2010b) Undergraduate medical education in the Gulf Cooperation Council: a multi-countries study (part 2). Med Teach 32:290–295

Hande AH, Lohe VK, Chaudhary MS, Gawande MN, Patil SK, Zade PR (2017) Impact of virtual microscopy with conventional microscopy on student learning in dental histology. Dent Res J (Isfahan) 14:111–116

Hannafin MJ, Land SM (2000) Technology and student-centered learning in higher education: issues and practices. J Comput High Educ 12:3–30

Hightower JA, Boockfor FR, Blake CA, Millette CF (1999) The standard medical microscopic anatomy course: histology circa 1998. Anat Rec 257:96–101

His W (1870) Beschreibung eines Mikrotoms. Arch Mikrosk Anat 6:229–232

Holaday L, Selvig D, Purkiss J, Hortsch M (2013) Preference of interactive electronic versus traditional learning resources by University of Michigan medical students during the first year histology component. Med Sci Educ 23:607–619

Hooke R (1665) Micrographia: or some physiological description of minute bodies made by magnifying glasses with observations and inquiries thereupon. The Royal Society, London

Hortsch M (2016) Taking a SecondLook™ at a time-efficient self-review resource. Med Sci Educ 26:3–4

Hortsch M (2023a) Histology as a paradigm for a science-based learning experience: visits by histology education spirits of past, present, and future. Anat Sci Educ 16(3):372–383. https://doi.org/10.1002/ase.2235

Hortsch M (2023b) The Michigan histology website as an example of a free anatomical resource serving learners and educators worldwide. Anat Sci Educ 16(3):363–371. https://doi.org/10.1002/ase.2239

Hortsch M, Mangrulkar RS (2015) When students struggle with gross anatomy and histology: a strategy for monitoring, reviewing, and promoting student academic success in an integrated preclinical medical curriculum. Anat Sci Educ 8:478–483

Hortsch M, Koney NK-K, Oommen AM, Yohannan DG, Li Y, De Melo R, Leite AC, Carneiro Girão-Carmona VCC (2023) Virtual microscopy goes global: the images are virtual and the problems are real. In: Keenan I, Border S (eds) Biomedical visualization vol. 16 Springer

Huang L, Cai Q, Cheng L, Kosik R, Mandell G, Wang SJ, Xu GT, Fan AP (2014) Analysis of curricular reform practices at Chinese medical schools. Teach Learn Med 26:412–419

Huber GC (1892) Directions for work in histological laboratory. George Wahr (republished by the University of Michigan Library), Ann Arbor

Huber GC (1900) Laboratory work in histology. Ann Arbor, George Wahr (republished by the University of Michigan Library)

Ibrahim ME, Al-Shahrani AM, Abdalla ME, Abubaker IM, Mohamed ME (2018) The effectiveness of problem-based learning in acquisition of knowledge, soft skills during basic and preclinical sciences: medical students' points of view. Acta Inform Med 26:119–124

Ishak A, Alrawashdeh MM, Meletiou-Mavrotheris M, Nikas IP (2022) Virtual pathology education in medical schools worldwide during the COVID-19 pandemic: advantages, challenges faced, and perspectives. Diagnostics (Basel) 12

Jalani SAM, Rushlan MAA, Yen SGS, Hanim SN, Hadie HSM, Abas R (2021) Medical students' perception of anatomy education environment in Universiti Putra Malaysia. Malays J Med Health Sci 17:37–43

Janssen A, Shaw T, Goodyear P, Kerfoot BP, Bryce D (2015) A little healthy competition: using mixed methods to pilot a team-based digital game for boosting medical student engagement with anatomy and histology content. BMC Med Educ 15:173

Jayanthi A, Sarjna MV, Benjamin B (2014) Students' perception of teaching learning method in dissection and histology lab. IOSR J Dental Med Scie 13:24–28

Jiménez JY, Merino LJP, Curay ERR (2020) Flipped classroom: a proposal in the teaching of veterinary histology [in Spanish]. Revista Sinapsis 2

Joaquim DC, Hortsch M, da Silva ASR, David PB, De Melo Leite ACR, Girao-Carmona VCC (2022) Digital information and communication technologies on histology learning: what to expect?-an integrative review. Anat Histol Embryol 51:180–188

Johnson S, Purkiss J, Holaday L, Selvig D, Hortsch M (2015) Learning histology - dental and medical students' study strategies. Eur J Dent Educ 19:65–73

Jonas-Dwyer D, Sudweeks F (2007) Informing students using virtual microscopes and their impact on students' approach to learning. Informing Sci Jl 10:61–70

Jonassen JA, Gilroy AM, Makris J, Florman HM, Allain W, Fischer MA (2016) A first-year medical school course integrating development, structure, and function. Med Sci Educ 26:31–33

Jurjus RA, Butera G, Krum JM, Davis MM, Mills A, Latham PS (2018) Design of an Online Histology and Pathology Atlas for medical students: an instructional aid to self-directed learning. Med Sci Educ 28:101–110

Karunakaran I, Thirumalaikolundusubramanian P, Nalinakumari SD (2017) A preliminary survey of professionalism teaching practices in anatomy education among Indian Medical Colleges. Anat Sci Educ 10: 433–443

Keiller L, Nyoni CN, Van Wyk C (2023) A model for design of online health professions education faculty development courses in sub-Saharan Africa. BMC Med Educ 23:60

Khalil MK, Kirkley DL, Kibble JD (2013) Development and evaluation of an interactive electronic laboratory manual for cooperative learning of medical histology. Anat Sci Educ 6:342–350

Kim KJ, Kee C (2010) Reform of medical education in Korea. Med Teach 32:113–117

Kim JW, Myung SJ, Yoon HB, Moon SH, Ryu H, Yim JJ (2020) How medical education survives and evolves during COVID-19: our experience and future direction. PLoS One 15:e0243958

King TS, Sharma R, Jackson J, Fiebelkorn KR (2019) Clinical case-based image portfolios in medical histopathology. Anat Sci Educ 12:200–209

Kölliker A (1867) Handbuch der Gewebelehre des Menschen für Aerzte und Studirende. Wilhelm Engelmann, Leipzig

Kozu T (2006) Medical education in Japan. Acad Med 81: 1069–1075

Kumamoto University (2023) Kumamoto University Official Website [in Japanese] [Online]. http://syllabus.kumamoto-u.ac.jp/pub/syllabus.html?locale=ja&sylocale=&nendo=2022&jikanwari_shozokucd=42&jikanwaricd=32031&nendoS=2022&jikanwari_shozokucdS=&kaikoS=&kamokuS=%E7%B5%84%E7%B9%94%E5%AD%A6&jikanwaricdS=&tanto_kyoinS=&keywordS=&jitumuS=&limitS=100. Accessed 6 Feb 2023

Kumar RK, Velan GM, Korell SO, Kandara M, Dee FR, Wakefield D (2004) Virtual microscopy for learning and assessment in pathology. J Pathol 204:613–618

Kumar RK, Freeman B, Velan GM, De Permentier PJ (2006) Integrating histology and histopathology teaching in practical classes using virtual slides. Anat Rec B New Anat 289:128–133

Lakhtakia R (2021) Virtual microscopy in undergraduate pathology education: an early transformative experience in clinical reasoning. Sultan Qaboos Univ Med J 21:428–435

Lallier TE (2014) Introducing evidence-based dentistry to dental students using histology. J Dent Educ 78:380–388

Lazarus L, Sookrajh R, Satyapal KS (2017) Tablet technology in medical education in South Africa: a mixed methods study. BMJ Open 7:e013871

Lee HS (2017) Medical education: addressing questions that require attention. Korean J Med Educ 29:199–201

Lee LMJ, Goldman HM, Hortsch M (2018) The virtual microscopy database-sharing digital microscope images for research and education. Anat Sci Educ 11: 510–515

Lee BC, Hsieh ST, Chang YL, Tseng FY, Lin YJ, Chen YL, Wang SH, Chang YF, Ho YL, Ni YH, Chang SC (2020) A web-based virtual microscopy platform for improving academic performance in histology and pathology laboratory courses: a pilot study. Anat Sci Educ 13:743–758

Li E, Guo Y (2014) Collaborative quiz-based teaching strategy in histology. J Vet Med Educ 41:318–320

Li Y, Zheng H, Zhou G, Ma S (2021) Application of online pre-class questions and double feedback improves student performance in the teaching of medical histology. Chin J Anat 44:57–62

Lister JJ (1830) XIII. On some properties in achromatic object-glasses applicable to the improvement of the microscope. Philos Trans R Soc 120:187–200

Longhurst GJ, Stone DM, Dulohery K, Scully D, Campbell T, Smith CF (2020) Strength, weakness, opportunity, threat (SWOT) analysis of the adaptations to anatomical education in the United Kingdom and Republic of Ireland in response to the Covid-19 pandemic. Anat Sci Educ 13:301–311

Loo SK, Freeman B, Moses D, Kofod M (1995) Fabric of life - the design of a system for computer-assisted-instruction in histology. Med Teacher 17:269–276

Lowrie DJ (2020) Histology: an essential textbook. Thieme Publisher, New York

Lu X, Cheng X, Li K, Lee KKH, Yang X (2016) Integration of histology lectures and practical teaching in China. Int J High Educ 5:157–164

Lu M, Wang J, Liu H (2021) The introduction to integrated curriculum for clinical medicine program in Peking University health science center [in Chinese]. Chin J Med Educ 41:780–784

Macdonald A, Cooper M, Cooper C, Salcedo E, Lee L (2021) No need to reinvent - augmenting a flipped class histology curriculum with a commercial online course. FASEB J 35

Mantovani ALS, Lima ARA, Brienze SLA, Dos Santos ER, Da Silva Fucuta P, André JC (2019) Cell biology and histology in medicine: perception on education and student performance. Inter J Health Educ 3:8–16

Mars M, Mclean M (1996) Students' perceptions of a multimedia computer-aided instruction resource in histology. S Afr Med J 86:1098–1102

Marsden JS (2006) An insider's view of the American and UK medical systems. Br J Gen Pract 56:60–62

Maske SS, Kamble PH, Kataria SK, Raichandani L, Dhankar R (2018) Feasibility, effectiveness, and students' attitude toward using WhatsApp in histology teaching and learning. J Educ Health Promot 7:158

Mayer C (1819) Ueber Histologie und eine neue Eintheilung der Gewebe des menschlichen Körpers, Bonn, Adolph Marcus

Mcbride JM, Drake RL (2018) National survey on anatomical sciences in medical education. Anat Sci Educ 11:7–14

Mcbride JM, Prayson RA (2008) Development of a synergistic case-based microanatomy curriculum. Anat Sci Educ 1:102–105

Mclean M (2001) Web pages: an effective method of providing CAI resource material in histology. Med Teach 23:263–269

Mclean M (2018) Flipping histology in an undergraduate problem-based learning medical curriculum: a blended learning approach. Med Sci Educ 28:429–437

Mendis B, Abayaratne R, Ranasinghe AW, Premraj S (1999) The possible use of the internet for the teaching of histology to University of Peradeniya Students. Proceedings & Abstracts of the Annual Research Sessions, 1999 Peradeniya, Sri Lanka. Unviersity of Peradeniya, p 124

Mills PC, Bradley AP, Woodall PF, Wildermoth M (2007) Teaching histology to first-year veterinary science students using virtual microscopy and traditional microscopy: a comparison of student responses. J Vet Med Educ 34:177–182

Ministry of Education-Taiwan (2023) Ministry of Education of Republic of China (Taiwan) Official Website [in Chinese] [Online]. Taipei, Taiwan. Available: https://ulist.moe.gov.tw/Query/AjaxQuery/Discipline/0912#. Accessed 6 Feb 2023

Mione S, Bacher K, Thierens H, Cornelissen M (2011) Virtual electron microscopy in cell biology. Microsc Res Tech 74:251–256
Mione S, Valcke M, Cornelissen M (2013) Evaluation of virtual microscopy in medical histology teaching. Anat Sci Educ 6:307–315
Mione S, Valcke M, Cornelissen M (2016) Remote histology learning from static versus dynamic microscopic images. Anat Sci Educ 9:222–230
Monekosso GL (2014) A brief history of medical education in sub-Saharan Africa. Acad Med 89:S11–S15
Moxham BJ, Emmanouil-Nikoloussi E, Brenner E, Plaisant O, Brichova H, Kucera T, Pais D, Stabile I, Borg J, Scholz M, Paulsen F, Luis Bueno-Lopez J, Alfonso Arraez Aybar L, De Caro R, Arsic S, Lignier B, Chirculescu A (2017) The attitudes of medical students in Europe toward the clinical importance of histology. Clin Anat 30:635–643
Moxham BJ, Mchanwell S, Berkovitz B (2018) The development of a core syllabus for the teaching of oral anatomy, histology, and embryology to dental students via an international 'Delphi Panel'. Clin Anat 31:231–249
Multon S, Pesesse L, Weatherspoon A, Florquin S, Van De Poel J-F, Martin P, Vincke G, Hoyoux R, Maree R, Verpoorten D, Bonnet P, Quatresooz P, Defaweux V (2018) Un "Massive Open Online Course" (MOOC) sur des travaux pratiques en histologie : un objectif, un outil, un public varié ! Retour sur une première expérience [in French]. Ann Pathol 38:76–84
Muñoz MS (2018) Uso de laboratorios virtuales para la enseñanza de la histoembriología humana en la carrera de Enfermería de la Universidad de las Américas [in Spanish]. Revista Educacion Las Americas 7:56–72
Murtomaa H (2009) Dental education in Europe. Eur J Dent 3:1–2
National Yung Ming Chiao Tung University (2023) National Yung Ming Chiao Tung University Official Website [in Chinese] [online]. Taipei . https://med.nycu.edu.tw/%e4%b8%89%e5%b9%b4%e7%b4%9a%e8%aa%b2%e7%a8%8b/. Accessed 6 Feb 2023
Nikas IP, Lamnisos D, Meletiou-Mavrotheris M, Themistocleous SC, Pieridi C, Mytilinaios DG, Michaelides C, Johnson EO (2022) Shift to emergency remote preclinical medical education amidst the Covid-19 pandemic: a single-institution study. Anat Sci Educ 15:27–41
Niu X, Yu S, Guo X (2015) Review and Prospect of the integrated curriculum of basic medical sciences in Shanghai Jiao Tong University. In: 2nd international conference on advanced education technology and management science (AETMS)
Niu X, Su Y, Guo X (2018) Practice of medical integration course at Shanghai Jiaotong university [in Chinese]. China Higher Med Educ:1–3
Nivala M, Lehtinen E, Helle L, Kronqvist P, Paranko J, Saljo R (2013) Histological knowledge as a predictor of medical students' performance in diagnostic pathology. Anat Sci Educ 6:361–367
NMC (2018) Competency Based Undergraduate Curriculum for the Indian Medical Graduate [Online]. Medical Council of India. Available: https://www.nmc.org.in/wp-content/uploads/2020/01/UG-Curriculum-Vol-I.pdf. Accessed 28 Nov 2021
NMC (2020) Electives for the Undergraduate Medical Education Training Program [Online]. Medical Council of India. Available: https://www.nmc.org.in/wp-content/uploads/2020/08/Electives-Module-20-05-2020.pdf. Accessed 14 Sept 2022
Nóbrega TE, Lopes RT, Pereira AC, Silva MAD (2014) As TICs como ferramenta auxiliar no ensino da histologia nos cursos de Odontologia das regiões Nordeste e CentroOeste do Brasil [in Portuguese]. Revista Iberoamericana de Educación en Tecnología y Tecnología en Educación 14:47–54
Norcini JJ, Mckinley DW, Boulet JR, Anderson MB (2006) Educational commission for foreign medical graduates certification and specialty board certification among graduates of the Caribbean medical schools. Acad Med 81:S112–S115
Onishi H (2018) History of Japanese medical education. Korean J Med Educ 30:283–294
Orth J (1878) Cursus der Normalen Histologie zur Einführung in den Gebrauch des Mikroskopes sowie in das Practische Stadium der Gewebelehre. August Hischwald, Berlin
Ostrin Z, Duschenkov V (2016) The pedagogical value of mobile devices and content-specific application software in the A&P laboratory. HAPS Educator 20:97–103
Padhi KS, Balmuchu G, Acharya PS, Singh SR, Joseph T (2021) The perspectives of educators and learners on E-learning: a cross-sectional descriptive study in a medical school. Adv Med Educ Pract 12:1059–1066
Painter SD (1994) Microanatomy courses in U.S. and Canadian medical schools, 1991-92. Acad Med 69: 143–147
Panteli D, Maier CB (2021) Regulating the health workforce in Europe: implications of the COVID-19 pandemic. Hum Resour Health 19:80
Parra ACV (2022) Histología como asignatura en Colombia:percepciones a partir de un sondeo de opinión adocentes, estudiantes y egresados de la carrerade Medicina [in Spanish]. Morfolia 14
Pather N, Blyth P, Chapman JA, Dayal MR, Flack N, Fogg QA, Green RA, Hulme AK, Johnson IP, Meyer AJ, Morley JW, Shortland PJ, Strkalj G, Strkalj M, Valter K, Webb AL, Woodley SJ, Lazarus MD (2020) Forced disruption of anatomy education in Australia and New Zealand: an acute response to the Covid-19 pandemic. Anat Sci Educ 13:284–300
Patrício M, Den Engelsen C, Tseng D, Ten Cate O (2008) Implementation of the Bologna two-cycle system in medical education: where do we stand in 2007? -

results of an AMEE-MEDINE survey. Med Teach 30: 597–605

Pavanelo E, Lima R (2017) Flipped classroom: the analysis of an experience in the discipline of calculus I [in Portuguese]. Bolema 31:739–759

Pawitan JA, Pattiata R (2010) The role of assignments in lab activity test results in histology. South-East Asian J Med Educ 4:41–43

Pawlina W, Ross MH (2018) Histology - a text and atlas. LWW, Wolters Kluwer Health

Potaliya P, Dixit SG, Ghatak S (2017) Impact of team-based virtual microscopy on student learning in histology. Experimental Biology 2017 Meeting. FASEB J, Chicago

Potu BK, Atwa H, Nasr El-Din WA, Othman MA, Sarwani NA, Fatima A, Deifalla A, Fadel RA (2022) Learning anatomy before and during COVID-19 pandemic: students' perceptions and exam performance. Morphologie 106:188–194

Purkyně JE (1834) Der microtomische Quetscher, ein bei mikroscopischen Untersuchungen unentbehrliches Instrument. Archiv für Anatomie, Physiologie und Wissenschaftliche Medicin 1:385–390

Quirk ME, Harden RM (2017) Curriculum planning and development. A practical guide for medical teachers. Elsevier

Rafi A, Rauf A, Anwar MI (2017) Significance of actually drawing microscopic images and its impact on students' understanding of histology. J Dow Univ Health Sci 11:77–81

Raja S (2010) Virtual microscopy as a teaching tool adjuvant to traditional microscopy. Med Educ 44:1126

Rashidi B, Maryam A (2012) Design, implementation and evaluation of electronic teaching of practical and theoretical histology courses: a new experience at Isfahan university of medical science [in Persian]. Iran J Med Educ 11:1214–1222

Rege N (2020) Towards competency-based learning in medical education: building evidence in India. J Postgrad Med 66:9–10

Rich P, Guy R (2013) A "do-it-yourself" interactive bone structure module: development and evaluation of an online teaching resource. Anat Sci Educ 6:107–113

Richards PA, Richards PD, Coetzee HL, Soley JT (2000) The optical microscope--en route to extinction? J Audiov Media Med 23:113–118

Rojas M, Montiel E, Montiel J, Oondarza A, Rodríguez H (1999) Estudio comparado entre metodos de enseñanza tradicional y computacional en histologia humana [in Spanish]. Ver Chil Anat 17

Rosenberg H, Kermalli J, Freeman E, Tenenbaum H, Locker D, COHEN, H. (2006) Effectiveness of an electronic histology tutorial for first-year dental students and improvement in "normalized" test scores. J Dent Educ 70:1339–1345

Sakthi-Velavan S, Zahl S (2023) Integration of virtual microscopy podcasts in the histology discipline in osteopathic medical school: learning outcomes. Anat Sci Educ 16:157–170

Samueli B, Sror N, Jotkowitz A, Taragin B (2020) Remote pathology education during the COVID-19 era: crisis converted to opportunity. Ann Diagn Pathol 49: 151612

Sander B, Golas MM (2013) HistoViewer: an interactive e-learning platform facilitating group and peer group learning. Anat Sci Educ 6:182–190

Sant'Anna CS, Albuquerque CAC, Baraúna SC, Oliveira-Filho GR (2022) Prática deliberada no ensino de histologia na graduação em Medicina: estudo prospectivo randomizado e controlado [in Portuguese]. Revista Brasileira De Educação Médica 46:e082

Saverino D, Zarcone D (2022) Keeping histology students under the microscope during Covid-19 lockdown: lessons learned from the University of Genoa. Anat Sci Educ 15:86–88

Saverino D, Marcenaro E, Zarcone D (2022) Teaching histology and anatomy online during the COVID-19 pandemic. Clin Anat 35:129–134

Schaffer J (1920) Vorlesungen über Histologie und Histogenese nebst Bemerkungen über Histotechnik und das Mikroskop. Wilhelm Engelmann, Leipzig

Schleiden MJ (1838) Beiträge zur Phytogenesis. Müller's Archiv für Anatomie, Physiologie und wissenschaftliche Medicin 5:137–176

Schmidt P (2013) Digital learning programs - competition for the classical microscope? GMS Z Med Ausbild 30: Doc8

Schmidt C, Reinehr M, Leucht O, Behrendt N, Geiler S, Britsch S (2011) MyMiCROscope: intelligent virtual microscopy in a blended learning model at Ulm University. Ann Anat 193:395–402

Schnackenberg HL (2013) Tablet technologies and education. Int J Educ Pract 1:44–50

Schoenherr DT, Dereski MO, Bernacki KD, Khayyata S, Attardi SM (2022) Development and evaluation of an online integrative histology module: simple design, low-cost, and improves pathology self-efficacy. Med Educ Online 27:2011692

Schwann T (1838a) Ueber die Analogie in der Structur und dem Wachsthum der Thiere und Pflanzen. Neue Notizen aus dem Gebiete der Natur- und Heilkunde 5: 33–36

Schwann T (1838b) Ueber die Analogie in der Structur und dem Wachsthum der Thiere und Pflanzen. Neue Notizen aus dem Gebiete der Natur- und Heilkunde 8: 225–229

Schwann T (1838c) Ueber die Analogie in der Structur und dem Wachsthum der Thiere und Pflanzen. Neue Notizen aus dem Gebiete der Natur- und Heilkunde 8: 21–23

Scott I, Logan DC (2004) The birth of cell biology. New Phytol 163:7–9

Selvig D, Holaday LW, Purkiss J, Hortsch M (2015) Correlating students' educational background, study habits, and resource usage with learning success in medical histology. Anat Sci Educ 8:1–11

Shankar PR, Balasubramanium R, Dwivedi NR, Nuguri V (2014) Student feedback about the integrated

curriculum in a Caribbean medical school. J Educ Eval Health Prof 11:23
Shastri D, Anand A, Ravichandran D, Samraj T (2016) Virtual microscopy: here to stay? Natl J Basic Med Sci 6:35–38
Shaw PA, Friedman ES (2012) Clinico-histologic conferences: histology and disease. Anat Sci Educ 5: 55–61
Sherer R, Dong H, Yunfeng Z, Stern S, Jiong Y, Matlin K, Baoping Y, Husain AN, Morgan I, Cooper B, Juan F, Mehta S (2013) Medical education reform in Wuhan University, China: a preliminary report of an international collaboration. Teach Learn Med 25:148–154
Sherer R, Wan Y, Dong H, Cooper B, Morgan I, Peng B, Liu J, Wang L, Xu D (2014) Positive impact of integrating histology and physiology teaching at a medical school in China. Adv Physiol Educ 38:330–338
Shoja MM, Tubbs RS, Loukas M, Shokouhi G, Ardalan MR (2008) Marie-Francois Xavier Bichat (1771-1802) and his contributions to the foundations of pathological anatomy and modern medicine. Ann Anat 190:413–420
Silva RN, Belo MLM (2012) Experiências e reflexões de monitoria: contribuição ao ensino-aprendizagem [in Portuguese]. Scientia Plena 8:1–6
Silva DBL, Liqui LL, Pacheco PM, Guimarães MCC, Pacheco MS (2020) Novas Tecnologias Educacionais: a Elaboração e Apresentação de um Livro Digital de Histologia [in Portuguese]. Informática na Educação: teoria & prática 23:81–94
Simok AA, Hadie@Haji SNH, Abdul Manan@Sulong H, Yusoff MSB, Mohd Noor NF, Asari MA, Kasim F (2019) The impact of virtual microscopy on medical students' intrinsic motivation. Educ Med J 11:47–59
Singh A, Jadon RS, Baitha U, Sethi P, Kaur H, Kumar A, Ranjan P, Wig N (2022) Indian medical student perspectives on online mode of education. J Family Med Prim Care 11:3915–3922
Smith CF, Pawlina W (2021) A journey like no other: anatomy 2020! Anat Sci Educ 14:5–7
Smith CF, Freeman SK, Heylings D, Finn GM, Davies DC (2021) Anatomy education for medical students in the United Kingdom and Republic of Ireland in 2019: a 20-year follow-up. Anat Sci Educ 15:993
Sood R (2008) Medical education in India. Med Teach 30: 585–591
Sorensen RL, Brelje TC (2023) Histology guide [Online]. https://histologyguide.com. Accessed 5 Feb 2023
STATS NZ (2022) National population estimates: at 31 March 2022 [online]. New Zealand government. Available: https://www.stats.govt.nz/topics/population. Accessed 8 Nov 2022
Sujatha K (2020) Use of labeled histology images with key identification for histology teaching learning. Natl J Clin Anat 9:21–24
Supe A, Burdick WP (2006) Challenges and issues in medical education in India. Acad Med 81:1076–1080
Suzuki Y, Gibbs T, Fujisaki K (2008) Medical education in Japan: a challenge to the healthcare system. Med Teach 30:846–850
Sweetman DS (2021) Making virtual learning engaging and interactive. FASEB Bioadv 3:11–19
Syaidah R, Sukmawati D, Edison RE, Jusuf AA (2020) Medical student's perceptions toward integrated histology practical class: a quantitative and qualitative study. Int J Human Health Sci 4:200–205
Syed S, Rastogi A, Bansal A, Kumar A, Jindal A, Prakash A, Agarwal G, Varshney M (2021) Future of e-learning in medical education—perception, readiness, and challenges in a developing country. Front Educ 6:1–8
Takizawa P (2022) Histology@Yale [Online]. Yale University, Department of Cell Biology, New Haven. http://medcell.org/histology/histology.php. Accessed 23 Jan 2023
Tauber Z, Cizkova K, Lichnovska R, Lacey H, Erdosova B, Zizka R, Kamarad V (2019) Evaluation of the effectiveness of the presentation of virtual histology slides by students during classes. Are there any differences in approach between dentistry and general medicine students? Eur J Dent Educ 23:119–126
Tauber Z, Lacey H, Lichnovska R, Erdosova B, Zizka R, Sedy J, Cizkova K (2021a) Students preparedness, learning habits and the greatest difficulties in studying histology in the digital era: a comparison between students of general and dental schools. Eur J Dent Educ 25:371–376
Tauber Z, Lichnovska R, Erdosova B, Zizka R, Cizkova K (2021b) Teaching histology in the age of virtual microscopy and E-resources: is a tailored approach to domestic and international students warranted? Interdisciplinary J Vir Learn Med Sci 12:97–105
Telang A, De Jong N, Van Dalen JV (2016) Media matter: the effect of medium of presentation on student's recognition of histopathology. J Clin Diagn Res 10:JC01–JC05
Telmesani A, Zaini RG, Ghazi HO (2011) Medical education in Saudi Arabia: a review of recent developments and future challenges. East Mediterr Health J 17:703–707
Teo A (2007) The current state of medical education in Japan: a system under reform. Med Educ 41:302–308
Teruel M, Felipe AE, Herrera MF (2017) Teaching strategies to improve the quality of teaching-learning process of student in the histology, embryology and teratology course [in Spanish]. Revista de Educación 27:153–176
The University of Tokyo (2023) The University of Tokyo Official Website [in Japanese] [Online]. https://catalog.he.u-tokyo.ac.jp/detail?code=0543068&year=2022. Accessed 6 Feb 2023
Thompson AR, Lowrie DJ Jr (2017) An evaluation of outcomes following the replacement of traditional histology laboratories with self-study modules. Anat Sci Educ 10:276–285

Thompson AR, Lowrie DJ Jr, Ubani M (2023) The effect of histology examination format on medical student preparation and performance: stand-alone versus integrated examinations. Med Sci Educ 33:1–8

Tian Y, Xiao W, Li C, Liu Y, Qin M, Wu Y, Xiao L, Li H (2014) Virtual microscopy system at Chinese medical university: an assisted teaching platform for promoting active learning and problem-solving skills. BMC Med Educ 14:74

Titford M (2005) The long history of hematoxylin. Biotech Histochem 80:73–78

Titford M (2006) A short history of histopathology technique. J Histotechnol 29:99–110

Torda A (2020) How COVID-19 has pushed us into a medical education revolution. Intern Med J 50:1150

Torres DV, Herranz F, Rocha SZ (2010) 200 years of teaching medicine: the Chilean experience as an example of the changes in medical education during the processes of Independence in Hispanic America [in Spanish]. Bordón Revista De Pedagogía 62:81–91

Trautman J, Mcandrew D, Craig SJ (2019) Anatomy teaching stuck in time? A 10-year follow-up of anatomy education in Australian and New Zealand medical schools. Aust J Educ 63:340–350

Triola MM, Holloway WJ (2011) Enhanced virtual microscopy for collaborative education. BMC Med Educ 11:4

Tuchman AM (1993) Science, medicine and the state in Germany, the case of Baden, 1815–1871. Oxford University Press, New York

Vainer B, Mortensen NW, Poulsen SS, Sorensen AH, Olsen J, Saxild HH, Johansen FF (2017) Turning microscopy in the medical curriculum digital: experiences from the Faculty of Health and Medical Sciences at University of Copenhagen. J Pathol Inform 8:11

Van Sligtenhorst I, Bick RJ (2011) Inclusion of an integrated team based learning session in first year histology is well received, but who fares well and who benefits? Open Med Educ J 4:18–23

Vijayan R (2021) Teaching and learning during the COVID-19 pandemic: a topic modeling study. Educ Sci 11:347

Volkmann H (1966) Ernst Abbe and his work. Appl Opt 5: 1720–1731

Vulcani VAS, Araújo DP, Mascarenhas LJS, Franzo VS, Rocha C (2020) The students' perception on the use of social media in the subjects of anatomy and histology veterinaria [in Portuguese]. Enciclopedia Biosfera, Centro Científico Conhecer – Goiânia 17:16–28

Wanderer C, Oliveira CC, Piemonte MR, Farias ÉLP (2020) Formative assessment: the preparation of digital histology atlas by veterinary medicine academics – experience report [in Portuguese]. Revista Gestão & Saude 22:1–15

Wang W (2021) Medical education in China: progress in the past 70 years and a vision for the future. BMC Med Educ 21:453

Wei L, Li ZC, Xu M, Li S, Wang Y, Fan R, Li M, Cheng H, Wang Q (2021) Evaluation of basic medical curriculum integration based on the training of Chinese excellent category doctors. Biosc Biotech Res Comm 14:8–11

Woods NN (2007) Science is fundamental: the role of biomedical knowledge in clinical reasoning. Med Educ 41:1173–1177

Woods NN, Brooks LR, Norman GR (2007) The role of biomedical knowledge in diagnosis of difficult clinical cases. Adv Health Sci Educ Theory Pract 12:417–426

Woolliscroft JO (2019) Implementing biomedical innovations into health, education, and practice: preparing tomorrow's physicians. Academic Press

Worldometer (2022) GDP per Capita [Online]. Available: https://www.worldometers.info/gdp/gdp-per-capita/. Accessed 8 Nov 2022

Wu H, Xie A, Wang W (2021) Eight-year medical education program: retrospect and prospect of the high-level medical talent training in China. ECNU Rev Educ 4: 190–209

Wu JW, Cheng HM, Huang SS, Liang JF, Huang CC, Yang LY, Shulruf B, Yang YY, Chen CH, Hou MC, Sheu WH (2022) Comparison of OSCE performance between 6- and 7-year medical school curricula in Taiwan. BMC Med Educ 22:15

Xing W, Mei C, Feng R (2019) Construction and application of digital slide network teaching system in histology courses. In: Long S, Dhillon BS (eds) 18th international conference on man-machine-environment system engineering, 2019. Springer Nature, Singapore, pp 195–200

Yakin M, Linden K (2021) Adaptive e-learning platforms can improve student performance and engagement in dental education. J Dent Educ 85:1309–1315

Yen P-Y, Hollar MR, Griffy H, Lee LMJ (2014) Students' expectations of an online histology course: a qualitative study. Med Sci Educ 24:75–82

Yohannan DG, Oommen AM, Umesan KG, Raveendran VL, Sreedhar LSL, Anish TSN, Hortsch M, Krishnapillai R (2019) Overcoming barriers in a traditional medical education system by the stepwise, evidence-based introduction of a modern learning technology. Med Sci Edu 29:803–817

Yohannan DG, Oommen AM, Amogh BJ, Raju NK, Suresh RO, Nair SJ (2022) "Air anatomy" - teaching complex spatial anatomy using simple hand gestures. Anat Sci Educ 15:552–565

Young B, O'dowd G, Woodford P (2013) Wheater's functional histology. Elsevier Churchill Livingstone

Zaidi NB, Hwang C, Scott S, Stallard S, Purkiss J, Hortsch M (2017) Climbing Bloom's taxonomy pyramid: lessons from a graduate histology course. Anat Sci Educ 10:456–464

Zalat MM, Hamed MS, Bolbol SA (2021) The experiences, challenges, and acceptance of e-learning as a tool for teaching during the COVID-19 pandemic among university medical staff. PLoS One 16: e0248758

Zaletel I, Maric G, Gazibara T, Rakocevic J, Labudovic Borovic M, Puskas N, Bajcetic M (2016) Relevance and attitudes toward histology and embryology course through the eyes of freshmen and senior medical students: experience from Serbia. Ann Anat 208:217–221

Zhang H, Liu D, Lang W, Zhang M, Yao H, Wang Y, Lian J, Sun L (2018) The application of flipped class based on MOOC in the teaching of histology and embryology [in Chinese]. J Med Educ Res 17:1126–1129

Zhang W, Li S, Song X, Liang Y (2020) The practice and evaluation of the introduction of flipped teaching in histology laboratory course [in Chinese]. Basic Med Educ 22:729–733

Zhong J, Li Z, Hu X, Wang L, Chen Y (2022) Effectiveness comparison between blended learning of histology practical in flipped physical classrooms and flipped virtual classrooms for MBBS students. BMC Med Educ 22:795

Zhu L, Sun G-C, Guan Y-F (2020) Role of systematic integration teaching reform at the basic medicine teaching stage in Chinese medical education system [in Chinese]. Acta Physiologica Sinica 72:699–706

Zureick AH, Burk-Rafel J, Purkiss JA, Hortsch M (2018) The interrupted learner: how distractions during live and video lectures influence learning outcomes. Anat Sci Educ 11:366–376

The manufacturer's authorised representative in the EU is Springer Nature Customer Service Centre GmbH, Europaplatz 3, 69115 Heidelberg, Germany. If you have any concerns regarding our products, please contact ProductSafety@springernature.com

Printed and bound by CPI Group (UK) Ltd, Croydon, CR0 4YY
28/09/2026
02224075-0020